Smart fibres, fabrics and clothing

Other titles in the Woodhead Publishing Limited series on fibres, published in association with The Textile Institute

Series Editor: Professor J E McIntyre

Regenerated cellulose fibres
Wool science and technology
Silk, mohair, cashmere and other luxury fibres
High-performance fibres
Synthetic fibres
Cotton science and technology
Bast and other leaf fibres

Smart fibres, fabrics and clothing

Edited by
Xiaoming Tao

The Textile Institute

CRC Press
Boca Raton Boston New York Washington, DC

WOODHEAD PUBLISHING LIMITED
Cambridge England

Published by Woodhead Publishing Limited in association with The Textile Institute
Woodhead Publishing Ltd
Abington Hall, Abington
Cambridge CB1 6AH, England
www.woodhead-publishing.com

Published in North and South America by CRC Press LLC
2000 Corporate Blvd, NW
Boca Raton FL 33431, USA

First published 2001, Woodhead Publishing Ltd and CRC Press LLC
© 2001, Woodhead Publishing Ltd
The authors have asserted their moral rights.

This book contains information obtained from authentic and highly regarded sources. Reprinted material is quoted with permission, and sources are indicated. Reasonable efforts have been made to publish reliable data and information, but the authors and the publishers cannot assume responsibility for the validity of all materials. Neither the authors nor the publishers, nor anyone else associated with this publication, shall be liable for any loss, damage or liability directly or indirectly caused or alleged to be caused by this book.
  Neither this book nor any part may be reproduced or transmitted in any form or by any means, electronic or mechanical, including photocopying, microfilming and recording, or by any information storage or retrieval system, without permission in writing from the publishers.
  The consent of Woodhead Publishing and CRC Press does not extend to copying for general distribution, for promotion, for creating new works, or for resale. Specific permission must be obtained in writing from Woodhead Publishing or CRC Press for such copying.

Trademark notice: Product or corporate names may be trademarks or registered trademarks, and are used only for identification and explanation, without intent to infringe.

British Library Cataloguing in Publication Data
A catalogue record for this book is available from the British Library.

Library of Congress Cataloging in Publication Data
A catalog record for this book is available from the Library of Congress.

Woodhead Publishing ISBN 1 85573 546 6
CRC Press ISBN 0-8493-1172-1
CRC Press order number: WP1172

Cover design by The ColourStudio
Typeset by Vision Typesetting, Manchester
Printed by TJ International Ltd, Cornwall, England

Contents

	Foreword	xi
	List of contributors	xiii
	Acknowledgements	xvii
1	**Smart technology for textiles and clothing – introduction and overview** XIAOMING TAO	**1**
1.1	Introduction	1
1.2	Development of smart technology for textiles and clothing	3
1.3	Outline of the book	5
2	**Electrically active polymer materials – application of non-ionic polymer gel and elastomers for artificial muscles** TOSHIHIRO HIRAI, JIANMING ZHENG, MASASHI WATANABE AND HIROFUSA SHIRAI	**7**
2.1	Introduction	7
2.2	Polymer materials as actuators or artificial muscle	9
2.3	Peculiarity of polymer gel actuator	10
2.4	Triggers for actuating polymer gels	10
2.5	Electro-active polymer gels as artificial muscles	15
2.6	From electro-active polymer gel to electro-active elastomer with large deformation	28
2.7	Conclusions	30
	Acknowledgements	30
	References	30

3	**Heat-storage and thermo-regulated textiles and clothing** XINGXIANG ZHANG	34
3.1	Development introduction	34
3.2	Basics of heat-storage materials	35
3.3	Manufacture of heat-storage and thermo-regulated textiles and clothing	41
3.4	Properties of heat-storage and thermo-regulated textiles and clothing	47
3.5	Application	52
3.6	Development trends	54
	References	55
4	**Thermally sensitive materials** PUSHPA BAJAJ	58
4.1	Introduction	58
4.2	Thermal storage and thermal insulating fibres	60
4.3	Thermal insulation through polymeric coatings	68
4.4	Designing of fabric assemblies	75
	References	79
5	**Cross-linked polyol fibrous substrates as multifunctional and multi-use intelligent materials** TYRONE L. VIGO AND DEVRON P. THIBODEAUX	83
5.1	Introduction	83
5.2	Fibrous intelligent materials	83
5.3	Experimental	85
5.4	Results and discussion	86
5.5	Conclusions	91
	References	92
6	**Stimuli-responsive interpenetrating polymer network hydrogels composed of poly(vinyl alcohol) and poly(acrylic acid)** YOUNG MOO LEE AND SO YEON KIM	93
6.1	Introduction	93
6.2	Experimental	95
6.3	Results and discussion	97
6.4	Conclusions	106
	References	107

7	**Permeation control through stimuli-responsive polymer membrane prepared by plasma and radiation grafting techniques** YOUNG MOO LEE AND JIN KIE SHIM	**109**
7.1	Introduction	109
7.2	Experimental	110
7.3	Results and discussion	112
7.4	Conclusions	121
	Acknowledgement	122
	References	122
8	**Mechanical properties of fibre Bragg gratings** XIAOGENG TIAN AND XIAOMING TAO	**124**
8.1	Introduction	124
8.2	Fabrication techniques	125
8.3	Mechanisms of FBG sensor fabrication	127
8.4	Mechanical properties	130
8.5	Influence of the UV irradiation on mechanical properties	133
8.6	Polymeric fibre	141
8.7	Conclusions	145
	Acknowledgements	145
	References	145
9	**Optical responses of FBG sensors under deformations** DONGXIAO YANG, XIAOMING TAO AND APING ZHANG	**150**
9.1	Introduction	150
9.2	Optical methodology for FBG sensors	151
9.3	Optical responses under tension	156
9.4	Optical responses under torsion	158
9.5	Optical responses under lateral compression	161
9.6	Optical responses under bending	165
9.7	Conclusions	166
	Acknowledgements	167
	References	167
10	**Smart textile composites integrated with fibre optic sensors** XIAOMING TAO	**174**
10.1	Introduction	174

10.2	Optical fibres and fibre optic sensors	175
10.3	Principal analysis of embedded fibre Bragg grating sensors	177
10.4	Simultaneous measurements of strain and temperature	181
10.5	Measurement effectiveness	187
10.6	Reliability of FBGs	191
10.7	Error of strain measurement due to deviation of position and direction	192
10.8	Distributed measurement systems	195
10.9	Conclusions	195
	Acknowledgements	197
	References	197
11	**Hollow fibre membranes for gas separation**	**200**
	PHILIP J. BROWN	
11.1	Historical overview of membranes for gas separation	200
11.2	Development of membranes for industrial gas separation	202
11.3	Theories of permeation processes	211
11.4	Phase inversion and hollow fibre membrane formation	211
11.5	Future hollow fibre membranes and industrial gas separation	214
	References	215
12	**Embroidery and smart textiles**	**218**
	BÄRBEL SELM, BERNHARD BISCHOFF AND ROLAND SEIDL	
12.1	Introduction	218
12.2	Basics of embroidery technology	218
12.3	Embroidery for technical applications – tailored fibre placement	220
12.4	Embroidery technology used for medical textiles	221
12.5	Embroidered stamp – gag or innovation?	224
12.6	Summary	225
	References	225
13	**Adaptive and responsive textile structures (ARTS)**	**226**
	SUNGMEE PARK AND SUNDARESAN JAYARAMAN	
13.1	Introduction	226
13.2	Textiles in computing: the symbiotic relationship	226
13.3	The Georgia Tech Wearable Motherboard™	228
13.4	GTWM: contributions and potential applications	236
13.5	Emergence of a new paradigm: harnessing the opportunity	240
13.6	Conclusion	244
	Acknowledgements	245
	References	245

Contents

14	Wearable technology for snow clothing	246
	HEIKKI MATTILA	
14.1	Introduction	246
14.2	Key issues and performance requirements	247
14.3	The prototype	248
14.4	Conclusions	252

15	Bioprocessing for smart textiles and clothing	254
	ELISABETH HEINE AND HARTWIG HOECKER	
15.1	Introduction	254
15.2	Treatment of wool with enzymes	256
15.3	Treatment of cotton with enzymes	263
15.4	Enzymatic modification of synthetic fibres	270
15.5	Spider silk	270
15.6	'Intelligent' fibres	271
15.7	Conclusions	271
	Acknowledgements	272
	References	272

16	Tailor-made intelligent polymers for biomedical applications	278
	ANDREAS LENDLEIN	
16.1	Introduction	278
16.2	Fundamental aspects of shape memory materials	280
16.3	Concept of biodegradable shape memory polymers	281
16.4	Degradable thermoplastic elastomers having shape memory properties	284
16.5	Degradable polymer networks having shape memory properties	287
16.6	Conclusion and outlook	288
	Acknowledgements	288
	References	288

17	Textile scaffolds in tissue engineering	291
	SEERAM RAMAKRISHNA	
17.1	Introduction	291
17.2	Ideal scaffold system	295
17.3	Scaffold materials	296
17.4	Textile scaffolds	298
17.5	Conclusions	306
	Acknowledgements	306
	References	306

	Index	315

Foreword

The history of textiles and fibres spans thousands of years, beginning with the style change from animal skins to the first fabric used to clothe humanity. But during the relatively short period of the past 50 years, the fibre and textile industries have undergone the most revolutionary changes and seen the most remarkable innovations in their history. Chapter One discusses the most important innovations together with the advent of the information industry. In fact, it is the merger of these industries that has led to this book.

We are not talking merely of fabrics and textiles imparting information; indeed, that has been occurring for many, many generations and numerous examples exist from fabrics and tapestries that have told intricate tales of warfare and family life and history, to those imparting information about the wealth and social status of the owners of the fabrics. We are talking about much more. Nor are we referring to fabrics that may have multifunctional purposes, such as fashion and environmental protection, or rainwear, or those fabrics providing resistance to a plethora of threats, such as ballistic, chemical and flame protection. These systems are all passive systems. No, we are talking here about materials or structures that sense and react to environmental stimuli, such as those from mechanical, thermal, chemical, magnetic or others. We are talking 'smart' and 'active' systems. We are talking about the true merger of the textile and information industries.

'Smart textiles' are made possible due to advances in many technologies coupled with the advances in textile materials and structures. A partial list includes biotechnology, information technology, microelectronics, wearable computers, nanotechnology and microelectromechanical machines.

Many of the innovations in textile applications in the past 50 years have started with military applications – from fibreglass structures for radomes, to fragment and bullet resistant body armour, to chemical agent protective clothing, to fibre-reinforced composites – indeed, many of our current defence systems and advanced aircraft would not be possible without these materials. So perhaps it is not surprising that the initial applications for smart textiles have also come either directly from military R&D or from spin-offs. Some of the capabilities for smart textile systems for military applications are: sensing and responding, for example to a biological or chemical sensor; power and data transmission from wearable computers and polymeric batteries; transmitting and receiving RF signals; automatic voice warning systems as to 'dangers ahead'; 'on-call' latent reactants such as biocides or catalytic decontamination *in-situ* for chemical and biological agents; and self-repairing materials.

In many cases the purpose of these systems is to provide both military and civilian personnel engaged in high-risk applications with the most effective survivability technologies. They will thus be able to have superiority in fightability, mobility, cognitive performance, and protection through materials for combat clothing and equipment, which perform with intelligent reaction to threats and situational needs. Thus, we will be providing high-risk personnel with as many executable functions as possible, which require the fewest possible actions on his/her part to initiate a response to a situational need. This can be accomplished by converting traditional passive clothing and equipment materials and systems into active systems that increase situational awareness, communications, information technology, and generally improve performance.

Some examples of these systems are body conformal antennas for integrated radio equipment into clothing; power and data transmission – a personal area network; flexible photovoltaics integrated into textile fabrics; physiological status monitoring to monitor hydration and nutritional status as well as the more conventional heart monitoring; smart footwear to let you know where you are and to convert and conserve energy; and, of course, phase change materials for heating and cooling of the individual. Another application is the weaving of sensors into parachutes to avoid obstacles and steer the parachutist or the cargo load to precise locations.

There are, naturally, many more applications for 'smart' textiles than those applied to military personnel, or civilian police, firemen, and emergency responders. Mountain climbers, sports personnel, businessmen with built-in wearable microcomputers, and medical personnel will all benefit from this revolution in textiles.

You will learn of many more applications for 'smart' textiles in this book. You will find that the applications are limited only by your imagination and the practical applications perhaps limited only by their cost. But we know those costs will come down. So let your imagination soar. The current worldwide textile industry is over 50 million metric tons per year, and if we are able to capture only a measly 1% of that market, it is still worth more than £1 billion.

Dr Robert W. Lewis

Contributors

Pushpa Bajaj,
Department of Textile Technology,
Indian Institute of Technology,
Hauz Khas,
New Delhi,
India
pbajaj@textile.iitd.ernet.in

Bernhard Bischoff,
Bischoff Textile AG,
St. Gallen,
Switzerland
bernhard.bischoff@bischoff-textil.com

Philip J Brown,
School of Materials, Science &
Engineering,
Clemson University,
161 Sirrine Hall,
Clemson,
SC 29634-0971,
USA

Elisabeth Heine,
DWI,
Veltmanplatz 8,
D-52062 Aachen,
Germany
heine@dwi.rwth-aachen.de

Toshihiro Hirai,
Department of Materials Chemistry,
Faculty of Textile Science and
Technology,
Shinshu University,
Tokida 3-15-1,
Ueda-shi 386-8567,
Japan
tohirai@giptc.shinshu-u.ac.jp

Hartwig Hoecker,
German Wool Research Institute at
Aachen University of Technology,
DWI,
Veltmanplatz 8,
D-52062 Aachen,
Germany
hoecker@dwi.rwth-aachen.de

Sundaresan Jayaraman,
Georgia Institute of Technology,
School of Textile and Fiber
Engineering,
Atlanta,
GA 30332-0295,
USA
sundaresan.jayaraman@tfe.gatech.edu

So Yeon Kim,
School of Chemical Engineering,
College of Engineering,
Hanyang University,
Haengdang-dong, Songdong-gu,
Seoul 133-791,
Korea
ymlee@hanyang.ac.kr

Young Moo Lee,
School of Chemical Engineering,
College of Engineering,
Hanyang University,
Haengdang-dong, Songdong-gu,
Seoul 133-791,
Korea
ymlee@hanyang.ac.kr

Andreas Lendlein,
DWI,
Veltmanplatz 8,
D-52062 Aachen,
Germany
lendlein@dwi.rwth-aachen.de

Heikki Mattila,
Fibre Materials Science,
Tampere University of Technology,
PO Box 589,
33101 Tampere,
Finland
heikki.r.mattila@tut.fi

Sungmee Park,
Georgia Institute of Technology,
School of Textile and Fiber
Engineering,
Atlanta,
GA 30332-0295,
USA
sp36@prism.gatech.edu

Seeram Ramakrishna,
Faculty of Engineering,
Department of Mechanical
Engineering,
National University of Singapore,
10 Kent Ridge Crescent,
Singapore 119260
mpesr@nus.edu.sg

Roland Seidl,
Jakob Mueller Institute of Narrow
Fabrics,
Frick,
Switzerland
redaktion@mittex.ch

Bärbel Selm,
Swiss Federal Institute of Materials
Testing,
St. Gallen,
Switzerland
Baerbel.Selm@empa.ch

Jin Kie Shim,
School of Chemical Engineering,
College of Engineering,
Hanyang University,
Haengdang-dong, Songdong-gu,
Seoul 133-791,
Korea
ymlee@hanyang.ac.kr

Hirofusa Shirai,
Faculty of Textile Science and
Technology,
Shinshu University,
Tokida 3-15-1,
Ueda-shi 386-8567,
Japan
tohirai@giptc.shinshu-u.ac.jp

Contributors

Xiaoming Tao,
Institute of Textiles and Clothing,
The Hong Kong Polytechnic
University,
Yuk Choi Road,
Hung Hom,
Hong Kong
tctaoxm@polyu.edu.hk

Devron P. Thibodeaux,
USDA, REE, ARS, MSA,
SRRC-CTCR,
1100 Robert E. Lee Boulevard,
New Orleans,
LS 70124,
USA
tvigo@nola.srrc.usda.gov

Xiaogeng Tian,
Institute of Textiles and Clothing,
The Hong Kong Polytechnic
University,
Yuk Choi Road,
Hung Hom,
Hong Kong
tctaoxm@polyu.edu.hk

Tyrone L. Vigo,
USDA, REE, ARS, MSA,
SRRC-CTCR,
1100 Robert E. Lee Boulevard,
New Orleans,
LS 70124,
USA
tvigo@nola.srrc.usda.gov

Masashi Watanabe,
Faculty of Textile Science and
Technology,
Shinshu University,
Tokida 3-15-1,
Ueda-shi 386-8567,
Japan
tohirai@giptc.shinshu-u.ac.jp

Dongxiao Yang,
Department of Information and
Electronic Engineering,
Zhejiang University,
Hangzhou 310027
China
yangdx@isee.zju.edu.cn

Aping Zhang
Institute of Textiles and Clothing,
The Hong Kong Polytechnic
University,
Yuk Choi Road,
Hung Hom,
Hong Kong
tctaoxm@polyu.edu.hk

Xingxiang Zhang,
Institute of Functional Fibres,
Tianjin Institute of Textile Science
and Technology,
Tianjin, 300160,
China
zhxx@public.tpt.tj.cn

Jianming Zheng,
Faculty of Textile Science and
Technology,
Shinshu University,
Tokida 3-15-1,
Ueda-shi 386-8567,
Japan
tohirai@giptc.shinshu-u.ac.jp

Acknowledgements

The Editor wishes to thank the Hong Kong Polytechnic University for partial support under the Area of Strategic Development Fund and Dr Dongxiao Yang for assistance in compiling this book. The Editor also thanks all contributing authors for their efforts in making this book a reality.

1
Smart technology for textiles and clothing – introduction and overview

XIAOMING TAO

1.1 Introduction

Since the nineteenth century, revolutionary changes have been occurring at an unprecedented rate in many fields of science and technology, which have profound impacts on every human being. Inventions of electronic chips, computers, the Internet, the discovery and complete mapping of the human genome, and many more, have transformed the entire world. The last century also brought tremendous advances in the textile and clothing industry, which has a history of many thousands of years. Solid foundations of scientific understanding have been laid to guide the improved usage and processing technology of natural fibres and the manufacturing of synthetic fibres. We have learnt a lot from nature. Viscose rayon, nylon, polyester and other synthetic fibres were invented initially for the sake of mimicking their natural counterparts. The technology has progressed so that synthetic fibres and their products surpass them in many aspects. Biological routes for synthesizing polymers or textile processing represent an environmentally friendly, sustainable way of utilizing natural resources. Design and processing with the aid of computers, automation with remote centralized or distributed control, and Internet-based integrated supply-chain management systems bring customers closer to the very beginning of the chain than ever before.

Looking ahead, the future promises even more. What new capacities should we expect as results of future developments? They should at least include terascale, nanoscale, complexity, cognition and holism. The new capability of terascale takes us three orders of magnitude beyond the present general-purpose and generally accessible computing capabilities. In a very short time, we will be connecting millions of systems and billions of information appliances to the Internet. Technologies allowing over one trillion operations per second are on the agenda for research. The technology in nanoscales will take us three orders of magnitude below the size of most of today's human-made devices. It will allow us to arrange atoms and molecules inexpensively in most of the ways

1

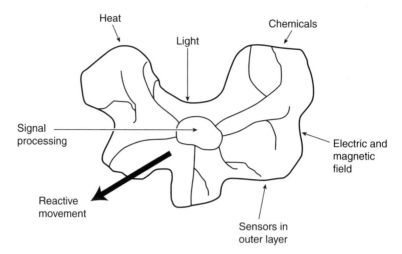

1.1 A single cell living creature is an example of smart structures.

permitted by physical laws. It will let us make supercomputers that fit on the head of a fibre, and fleets of medical nanorobots smaller than a human cell to eliminate cancers, infections, clogged arteries and even old age. Molecular manufacturing will make exactly what it is supposed to make, and no pollutants will be produced.

We are living in this exciting era and feeling the great impacts of technology on the traditional textiles and clothing industry, which has such a long history. Traditionally, many fields of science and engineering have been separate and distinct. Recently, there has been considerable movement and convergence between these fields of endeavour and the results have been astonishing. Smart technology for materials and structures is one of these results.

What are smart materials and structures? Nature provides many examples of smart structures. The simple single-celled living creature may highlight the fundamentals. As shown in Fig. 1.1, various environmental conditions or stimuli act on the outer layer. These conditions or stimuli may be in the form of force, temperature, radiation, chemical reactions, electric and magnetic fields. Sensors in the outer layer detect these effects, and the resulting information is conveyed for signal processing and interpretation, at which point the cell reacts to these environmental conditions or stimuli in a number of ways, such as movement, changing chemical composition and reproductive actions. Nature has had billions of years and a vast laboratory to develop life, whereas humankind has just begun to create smart materials and structures.

Smart materials and structures can be defined as the materials and structures that sense and react to environmental conditions or stimuli, such as those from mechanical, thermal, chemical, electrical, magnetic or other sources. According to the manner of reaction, they can be divided into passive smart, active smart and very smart materials. Passive smart materials can only sense the environmental conditions or stimuli; active smart materials will sense and react to the conditions or stimuli; very smart materials can sense, react and adapt themselves accordingly. An even higher level of intelligence can be achieved from those intelligent materials and structures capable of responding or activated to perform a function in a manual or pre-programmed manner.

Three components may be present in such materials: sensors, actuators and controlling units. The sensors provide a nerve system to detect signals, thus in a passive smart material, the existence of sensors is essential. The actuators act upon the detected signal either directly or from a central control unit; together with the sensors, they are the essential element for active smart materials. At even higher levels, like very smart or intelligent materials, another kind of unit is essential, which works like the brain, with cognition, reasoning and activating capacities. Such textile materials and structures are becoming possible as the result of a successful marriage of traditional textiles/clothing technology with material science, structural mechanics, sensor and actuator technology, advanced processing technology, communication, artificial intelligence, biology, etc.

1.2 Development of smart technology for textiles and clothing

We have always been inspired to mimic nature in order to create our clothing materials with higher levels of functions and smartness. The development of microfibres is a very good example, starting from studying and mimicking silk first, then creating finer and, in many ways, better fibres. However, up to now, most textiles and clothing have been lifeless. It would be wonderful to have clothing like our skin, which is a layer of smart material. The skin has sensors which can detect pressure, pain, temperature, etc. Together with our brain, it can function intelligently with environmental stimuli. It generates large quantities of sweat to cool our body when it is hot, and to stimulate blood circulation when it gets cold. It changes its colour when exposed to a higher level of sunlight, to protect our bodies. It is permeable, allowing moisture to penetrate yet stopping unwanted species from getting in. The skin can shed, repair and regenerate itself. To study then develop a smart material like our skin is itself a very challenging task.

In the last decade, research and development in smart/intelligent materials and structures have led to the birth of a wide range of novel smart products in aerospace, transportation, telecommunications, homes, buildings and infrastructures. Although the technology as a whole is relatively new, some areas have reached the stage where industrial application is both feasible and viable for textiles and clothing.

Many exciting applications have been demonstrated worldwide. Extended from the space programme, heat generating/storing fibres/fabrics have now been used in skiwear, shoes, sports helmets and insulation devices. Textile fabrics and composites integrated with optical fibre sensors have been used to monitor the health of major bridges and buildings. The first generation of wearable motherboards has been developed, which has sensors integrated inside garments and is capable of detecting injury and health information of the wearer and transmitting such information remotely to a hospital. Shape memory polymers have been applied to textiles in fibre, film and foam forms, resulting in a range of high performance fabrics and garments, especially sea-going garments. Fibre sensors, which are capable of measuring temperature, strain/stress, gas, biological species and smell, are typical smart fibres that can be directly applied to textiles. Conductive polymer-based actuators have achieved very high levels of energy density. Clothing with its own senses and brain, like shoes and snow coats which are integrated with Global Positioning System (GPS) and mobile phone technology, can tell the position of the wearer and give him/her directions. Biological tissues and organs, like ears and noses, can be grown from textile scaffolds made from biodegradable fibres. Integrated with nanomaterials, textiles can be imparted with very high energy absorption capacity and other functions like stain proofing, abrasion resistance, light emission, etc.

The challenges lie before us, as the research and development of smart technology and its adoption by industries depend upon successful multidisciplinary teamwork, where the boundary of traditional disciplines becomes blurred and cross-fertilization occurs at a rate much higher than that seen previously. Some of the research areas can be grouped as follows:

For sensors/actuators:

- photo-sensitive materials
- fibre-optics
- conductive polymers
- thermal sensitive materials
- shape memory materials
- intelligent coating/membrane
- chemical responsive polymers
- mechanical responsive materials
- microcapsules
- micro and nanomaterials.

For signal transmission, processing and controls:

- neural network and control systems
- cognition theory and systems.

For integrated processes and products:

- wearable electronics and photonics
- adaptive and responsive structures
- biomimetics
- bioprocessing
- tissue engineering
- chemical/drug releasing.

Research and development activities have been carried out worldwide, both in academic/research institutions and companies. Research teams in North American, European and Asian countries have been actively involved, with noticeable outcomes either in the form of commercial products or research publications.

1.3 Outline of the book

This edited book, being the first on this topic, is intended to provide an overview and review of the latest developments of smart technology for textiles and clothing. Its targeted readers include academics, researchers, designers, engineers in the area of textile and clothing product development, and senior undergraduate and postgraduate students in colleges and universities. Also, it may provide managers of textile and clothing companies with the latest insights into technological developments in the field.

The book has been contributed by a panel of international experts in the field, and covers many aspects of the cutting-edge research and development. It comprises 17 chapters, which can be divided into four parts. The first part (Chapter 1) provides the background information on smart technology for textiles and clothing and a brief overview of the developments and the book structure. The second part involves material or fibre-related topics from Chapters 2 to 9. Chapter 2 is concerned with electrically active polymer materials and the applications of non-ionic polymer gel and elastomers for artificial muscles. Chapters 3 and 4 deal with thermal sensitive fibres and fabrics. Chapter 5 presents cross-linked polyol fibrous substrates as multifunctional and multi-use intelligent materials. Chapter 6 discusses stimuli-responsive interpenetrating polymer network hydrogel. Chapter 7 is concerned with permeation control through stimuli-responsive polymer membranes prepared by plasma and radiation grafting techniques. Chapters 8 and 9 discuss the

Smart fibres, fabrics and clothing

Table 1.1 Outline of the book

Chapter no.	Sensors/actuators	Signal transmission, processing and control	Integrated processes and products	Bio-processes and products
1	✓	✓	✓	✓
2	✓			
3	✓		✓	
4	✓		✓	
5	✓		✓	
6	✓			
7			✓	
8	✓			
9	✓			
10	✓	✓	✓	
11	✓		✓	
12			✓	
13		✓	✓	
14		✓	✓	
15				✓
16	✓			✓
17			✓	✓

principles, manufacturing and properties of optical fibre sensors, with emphasis on fibre Bragg grating sensors.

The third part contains five chapters, with a focus on integrating processes and integrated structures. Chapter 10 provides an overview of the developments and key issues in fibre-optic smart textile composites. Chapter 11 presents hollow fibre membranes for gas separation. Chapter 12 describes embroidery as one way of integrating fibre-formed components into textile structures. Chapters 13 and 14 are on wearable electronic and photonic technologies. Chapter 13 provides insights on adaptive and responsive textile structures (ARTS). Chapter 14 describes the development of an intelligent snowmobile suit.

The fourth part, embracing the last three chapters, is focused on bioapplications. Chapter 15 outlines various bioprocesses for smart textiles and clothing, and Chapter 16 concentrates on tailor-made intelligent polymers for biomedical applications. Chapter 17 describes the applications of scaffolds in tissue engineering, where various textile structures are used for cells to grow.

We have only seen a small portion of the emerging technology through the window of this book. The possibilities offered by this smart technology are tremendous and widespread. Even as the book was being prepared, many new advances were being achieved around the world. It is the hope of the editor and contributors of this book that it will help researchers and designers of future smart fibres, textiles and clothing to make their dreams a reality.

2
Electrically active polymer materials – application of non-ionic polymer gel and elastomers for artificial muscles

TOSHIHIRO HIRAI, JIANMING ZHENG,
MASASHI WATANABE AND HIROFUSA SHIRAI

2.1   Introduction

Many attempts have been made to functionalize polymer materials as so-called 'smart' or 'intelligent' materials (see Fig. 2.1).[1–5] Artificial muscle or intelligent actuators is one of the targets of such attempts. Historically, actuator materials have been investigated mainly in inorganic compounds.[6] Particularly, triggers used for actuation are usually investigated in an electric field application because of the ease of control. Polymer materials investigated from this point of view are very limited and have been known to generate much smaller strain than inorganic materials.[6–8]

On the other hand, polymer materials such as polymer gels have been known to generate huge strain by various triggers such as solvent exchange, pH jump, temperature jump, etc., although the response and durability are rather poor and they have not been used in practical actuators.[1]

In the field of mechanical engineering, the development of micromachining procedure is facing the requirements of the technologies of microfabrication and micro-device assembly, and there are high expectations of the emerging smart materials that can greatly simplify the microfabrication process.[9–12] Under these circumstances, the polymer gel actuator is mentioned as one of the most likely candidates as a soft biological muscle-like material with large deformation in spite of its poor durability.[1,2] Much research has been done on solid hard materials as actuators like poly(vinylidene fluoride) (PVDF), which is a well-known piezoelectrical polymer, and in which crystal structures play critical role for the actuation and the induced strain is very small compared to the gel artificial muscles that will be described in this chapter. Although PVDF needs electrically oriented crystal structure in it, the materials that will be discussed in this chapter do not require such a limitation.

Conventional electrically induced actuation has been carried out mostly on ionic polymer gels. The reason is simply because ionic species are highly

8   Smart fibres, fabrics and clothing

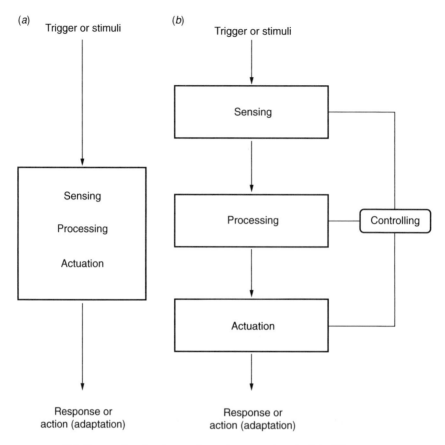

2.1 Concepts of autonomic systems and materials. Three processes (sensing, processing and actuation) are incorporated in materials (in one system): (a) in autonomic materials, while they are separated and must be unified by a controlling system; (b) in conventional autonomic systems.

responsive to the electric field. Ionic gels have proved to be excellent electroactive actuator materials.[1,2]

However, we tried electrical actuation of non-ionic polymer gel or elastomers. Why must non-ionic polymer gel be used for electrical actuation? Because non-ionic polymer gel is superior to ionic polymer gel in several ways, if it can be actuated by an electric field. In ionic gel materials, electrolysis is usually inevitable on the electrodes, and this is accompanied by a large electric current and heat generation. In other words, elecrochemical consumption is inevitable, although this fact has not been mentioned in most papers. In non-ionic polymer gels, no such process is encountered, and this leads to the good durability of the materials. In addition to these advantages, the responding speed and magnitude of the deformation were found to be much faster (10 ms order) and larger (over 100%) than those induced in polyelectrolyte gels. The motion reminds us of real biological muscle.

The concept of the mechanism is simple and can be applied to conventional polymer materials, including materials commonly used in the fibre and textile industries. The concept is also applicable to non-ionic elastomers that do not contain any solvent. The method we present will provide a promising way for developing future artificial muscle. Several concepts developed by other researchers and successfully used for actuating gels are also introduced in comparison with our method.

2.2 Polymer materials as actuators or artificial muscle

Polymer gel is an electroactive polymer material.[13] There are various types of electroactive polymeric materials. As mentioned in the above section, polyelectrolyte is one of them and is most commonly investigated as an electroactive gel. We will come back to discuss this material in more detail in the next section.

Ferroelectric polymer materials like PVDF or its derivatives are mentioned, since they behave as ferroelectric materials (see Fig. 2.2).[14,15] They have crystallinity and the crystals show polymorphism by controlling the preparation method. Much detailed work has been carried out on piezoelectric and/or pyroelectric properties, together with their characteristics as electroactive actuators. These materials have long been mentioned as typical electroactive polymers. Through these materials, it is considered that the strain induced in the polymer materials is not large. The electrostrictive coefficient is known to be small for polymers. These are non-ionic polymers and the induced strain originates from the reorientation or the deformation of polarized crystallites in the solid materials.

There is another type of electrically active polymer that is known as the electroconductive polymer, in which polymer chains contain long conjugated double bonds, and this chemical structure adds electroconductive properties to the polymers. In these cases, the electrically induced deformation is considered to have originated from the electrochemical reactions such as the oxidation and reduction of the polymer chain. For the deformation, some additives such as dopants have been known to be necessary for effective actuation. Therefore, the electrical actuation of these materials has been

$$\left(\begin{array}{c} H \\ | \\ C \\ | \\ H \end{array} - \begin{array}{c} F \\ | \\ C \\ | \\ F \end{array}\right)_n$$

2.2 Chemical structure of poly(vinylidene fluoride) (PDVF).

(a)

[chemical structure of polypyrrole]

(b)

[chemical structure of polyaniline]

2.3 Chemical structures of (a) polypyrrole and (b) polyaniline.

investigated in the presence of water, similar to the case of polyelectrolyte gels. Polypyrrole and polyaniline are typical examples (see Fig. 2.3).[16–19]

2.3 Peculiarity of polymer gel actuator

Polymer gels differ in various ways from hard solid polymer materials.[1,2] The polymer chains in the gel are usually considered to be chemically or physically cross-linked and to form a three-dimensional network structure. For instance, polymer gel is usually a matter swollen with its good solvent, and the characteristics are diversified from a nearly solid polymer almost to a solution with very low polymer content but still maintaining its shape by itself. This extreme diversity in physical properties widens the function of the gel (see Fig. 2.4).

From the standpoint of the actuator, the gel behaves like a conventional solid actuator or biological muscle, or like a shapeless amoeba. The gels also have various actuating modes, symmetric volume change with swelling and de-swelling, asymmetric swelling behaviour, symmetric deformation and asymmetric deformation (see Fig. 2.5). The strain induced in the gel can also be extremely large, depending on the cross-link structure in the gel.[20,21]

2.4 Triggers for actuating polymer gels

As can be expected from the diversified physical characteristics of the gel and the wide variety of the actuating modes, there are various triggers for the actuating polymer gels.

The triggers can be classified into two categories, chemical triggers and physical triggers (see Fig. 2.6). As chemical triggers, solvent exchange includes jumps in solvent polarity (e.g. from good solvent into poor solvent),[22] in pH (e.g. in weak polyelectrolyte gel from a dissociated condition into an associated condition)[23] and in ionic strength (utilizing salting-out or coagulation).[24] These

Highly swollen state   Barely swollen state

High solvent content   Low solvent content
Low polymer content   High polymer content

Solvent behaviour is not   Solvent behaviour is seriously
disturbed by polymer network disturbed by polymer network

2.4 Extreme diversity in physical property widens the function of the gel.

(a) Swelling and de-swelling

(b) Asymmetric swelling or de-swelling

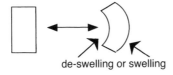

de-swelling or swelling

2.5 Various actuating modes of polymer gels: (a) swelling and de-swelling, (b) asymmetric swelling or de-swelling.

(a) Chemical triggers
- pH change
- oxidation and reduction
- solvent exchange
- ionic strength change

(b) Physical triggers
- light irradiation
- temperature change
- physical deformation
- magnetic field application
- electric field application
- microwave irradiation

2.6 Triggers for polymer and/or gel actuation can be classified into two categories: chemical and physical.

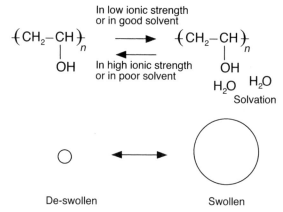

2.7 Chemical triggers including solvent exchange. These types accompany swelling and de-swelling of the solvent, and the deformation is usually symmetric as long as the gel has a homogeneous structure.

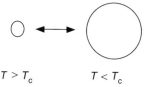

$T > T_c$ $T < T_c$

T_c: phase transition temperature

(a) $+CH_2-CH+_n$
         |
        OCH_3

(b) $+CH_2-CH+_n$
         |
        $C=O$
         |
        $NH-CH-CH_3$
              |
             CH_3

2.8 Temperature jump as a physical trigger: (a) poly(vinyl methyl ether) and (b) poly(N-isopropyl acrylonide).

two types accompany swelling and de-swelling of the solvent, and the deformation is usually symmetric as far as the gel has a homogeneous structure (see Fig. 2.7). Temperature jump, which is a physical trigger, can also induce symmetric deformation in particular polymer gels where the solubility has a critical transition temperature. Typical examples are the gels of poly(vinyl methyl ether) and poly(N-isopropyl acrylamide).[25,26] These gels have high water absorption at low temperatures and de-swell at the characteristic critical temperature around 30–40 °C (see Fig. 2.8). The transition temperature can be controlled by changing chemical structure.[27,28] In the case of urease immobilized gel, the addition of urea, a substrate of

Electrically active polymer materials

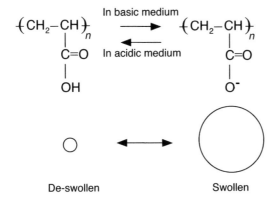

2.9 Chemical trigger can induce swelling and de-swelling of gel, e.g. substrate of urease, urea, is changed into ammonia and the ammonia induces swelling and de-swelling by varying pH.

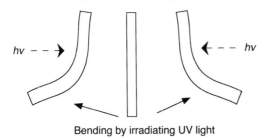

2.10 Light-induced deformation of polymer film. Example shown is the case of PVC film containing spyrobenzopyrane.

urease, induces swelling and deswelling by utilizing the pH change induced by the enzyme reaction (see Fig. 2.9).[29,30]

A physical trigger such as light irradiation is useful for actuating a gel in which the light-induced reversible isomerization occurs and the isomerization accompanies physical strain.[31,32] In this case, the change is usually asymmetric and the gel bends toward or against the direction of the irradiation, depending on the photoinduced reaction (see Fig. 2.10).

In the case of electric field application, the gels usually bend, because the field application induces asymmetric charge distribution and hence the asymmetric strain in the gel.[1,2] Asymmetric charge distribution can easily be induced in polyelectrolyte gels, and this is why polyelectrolyte gel has mainly been investigated as on electroactive polymer material (see Fig. 2.11).

Magnetic field application can also induce a strain in a gel when a structure or species sensitive to the magnetic field is contained in it. We first proposed the idea of applying a super paramagnetic fluid to a gel.[33-35] The gel was found

14 Smart fibres, fabrics and clothing

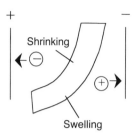

2.11 Electrically induced deformation. In the case of electric field application, the gels usually bend, since field application induces asymmetric charge distribution and hence the asymmetric strain in the gel.

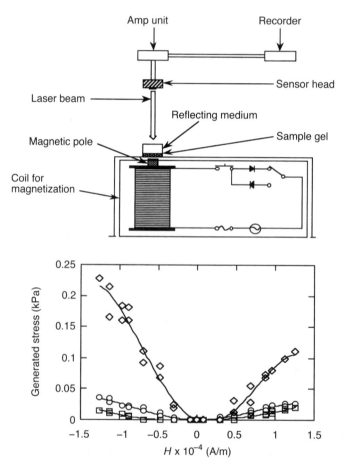

2.12 Magnetic field active gel utilizing super paramagnetic property of a ferro-fluid-immobilized gel. ◇, ferrofluid 75 wt%; ○, ferrofluid 50 wt%; □, ferrofluid 25 wt%.

Electrically active polymer materials 15

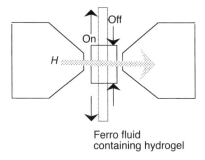

Ferro fluid
containing hydrogel

2.13 Magnetic field induced large deformation. By turning the magnetic field (*H*) on and off, the gel deforms instantly.

to be sensitive to the magnetic field gradient and to induce strain very sensitively, and the structure change in the gel was investigated (see Fig. 2.12). Zryhni and his coworkers investigated the same materials and found discontinuous deformation of the gel by controlling the magnetic field (see Fig. 2.13).[36–38]

2.5 Electro-active polymer gels as artificial muscles

Amongst the polymeric actuator materials mentioned above, polymer gel has an important property as a huge strain generating material. As mentioned in the previous section, the electric field is one of the most attractive triggers for practical actuation. Electroactivity has been mentioned in connection with polyelectrolyte gels, since they contain ionic species. However, ionic species are not only sensitive to an electric field, but also usually electrochemically active, and accompany electrolysis on the electrodes. Electrochemical reactions often result in increased current and heat generation. These processes only dissipate energy, and do not contribute to strain generation. Thus, electrochemical reactions are an undesirable process in most cases. In spite of their many difficulties for practical actuators, polyelectrolyte gels and related materials still remain at the forefront of electroactive polymer materials.

To overcome difficulties in polyelectrolytes, such as electrochemical consumption on the electrodes, we investigated the electroactive properties of the non-ionic polymer gel.

2.5.1 Electroactive polyelectrolyte gels

As pointed out in the previous section, polyelectrolyte gels have been investigated as electroactive actuator materials. The concept originates from the presence of electroactive ionic species in the gels. The ionic species can

16   Smart fibres, fabrics and clothing

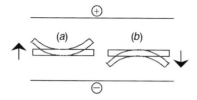

2.14 Bending deformation of a poly(acrylonide-co-sodium acrylate) gel in aqueous solution. Bending direction is changed with sodium acrylate content in the gel. Acrylic acid content was controlled by hydrolysing poly(acrylonide) gel. The mechanism was explained with the results shown in Fig. 2.15.

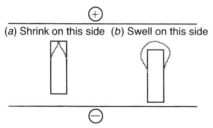

2.15 Electrically induced asymmetric deformation of a poly(acrylonide-co-sodium acrylate) gel. Sodium acrylate content is (a) low, and (b) high. In (a) the gel shrinks on anode side, but swells in (b).

migrate and form localized distribution and/or electrochemical reactions in the gel, which cause its deformation.

2.5.1.1 Poly(acrylic acid) gel

Among polyelectrolyte gels, poly(acrylic acid) (PAA) gel was the first polyelectrolyte investigated as an electroactive polymer gel. Shiga et al. found that PAA gel can be deformed by DC electric field application in the presence of salt.[39–41] A PAA gel rod was immersed in the saline aqueous solution (see Fig. 2.14). The platinum electrodes were apart from the gel surface, and the DC field was applied from both sides of the gel. Shiga et al. found a slow bending motion of the gel, the magnitude of bending depending on the salt and its concentration. They also found an asymmetric deformation of the gel, when the field was applied apart from both ends of the gel rod (see Fig. 2.15). In this case, the gel shrinks at one end and swells at the other end. The motion is explained by asymmetric swelling behaviour under the field. The deformation is explained by the following equation derived by Flory:

$$\pi = \pi_1 + \pi_2 + \pi_3 \qquad [2.1]$$

Electrically active polymer materials

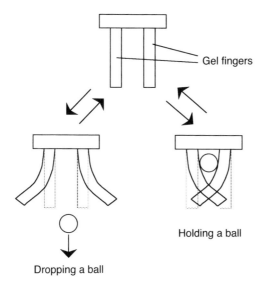

2.16 Gel finger in aqueous solution. Polymer gel contains poly(acrylic) acid and poly(vinyl alchohol). One electrode is occluded in the gel, and the other electrode is exposed in the solution.

$$\pi_1 = - [\ln(1 - v) + v + xv^{1/2}]RT/V_1 \quad [2.2]$$

$$\pi_2 = (v^{1/3} - v/2)RTv_e/V_0 \quad [2.3]$$

$$\pi_3 = (\Sigma c_i - \Sigma c_j)RT \quad [2.4]$$

where π is the overall osmotic pressure which is the summation of the three types of osmotic pressure, π_1, π_2 and π_3, which originate from the interaction between the polymer chain and solvent, from the rubber elasticity of the gel, and from the concentration differences of the ionic species inside and outside the gel, respectively. The parameters v, x, V_0, v_e, V_1, c_i, c_j, R and T are the polymer volume fraction, the interaction parameter between polymer and solvent, the volume of dried polymer, the effective chain number in the network, the molar volume of the solvent, the concentration of species i in the gel, the concentration of species j in outer solution, the gas constant and the absolute temperature, respectively.

The process was considered to originate from an osmotic pressure gap induced by the localization of ionic species of different solvation power. In this movement, electrolysis usually occurred on the electrodes. Shiga et al. optimized the preparation method in order to overcome the difficulty. They put the electrodes on the gel surface, and successfully demonstrated the gel finger in aqueous solution (see Fig. 2.16).

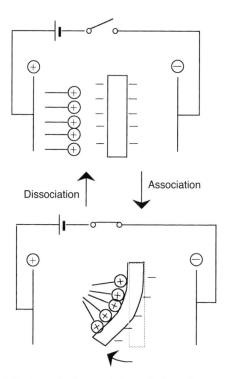

2.17 Association and dissociation of polyelectrolyte gel of poly(2-acrylamido-2-methylpropanesulfonic acid) (PAMPS) with cationic surfactant was found to undergo worm-like motility in aqueous solution.

2.5.1.2 Poly(2-acrylamido-2-methylpropanesulfonic acid) gel

Poly(2-acrylamido-2-methylpropanesulfonic acid) (PAMPS) gel was found to undergo worm-like motility (see Fig. 2.17).[42–44] The principle of this deformation is based on an electrokinetic molecular assembly reaction of surfactant molecules on the hydrogel, caused by both electrostatic and hydrophobic interactions and resulting in anisotropic contraction to give bending towards the anode. When the field is reversed, the surfactant admolecules on the surface of the gel lift off and travel away electrically towards the anode. Instead, new surfactant molecules approach from the opposite side of the gel and form the complex preferentially on that side of the gel, thus stretching the gel. Surfactants such as N-dodecylpyridinium chloride (Cl_2PyCl) were used, which adsorbed within a second and is easily calculated to give a complex formation ratio less than 1×10^{-3}, explaining that the quick and large bending under an electric field is dominated only by the surface complexation and shrinkage of the gel.

Electrically active polymer materials

$$-[(CF_2CF_2)_n-(CFCF_2)_m]-$$
$$|$$
$$OCFCF_2OCF_2SO_3^-Na^+$$
$$|$$
$$CF_3$$
Nafion

$$-[CF_2CF-(CF_2CF_2)_n]-$$
$$|$$
$$(OCF_2CF)_m-O(CF_2)_3COO^-Na^+$$
$$|$$
$$CF_3$$
Flemion

2.18 Ion-exchange polymer–metal composite film of Nafion, of Dupont or Flemion of Asahi Glass Co. Ltd. can bend remarkably by applying low voltage.

2.5.1.3 Perfluorosulfonate ionomer gel

A hydrogel of perfluorosulfonate ionomer (Nafion of Dupont) film, thickness of *ca.* 0.2 mm, was found to be an effective electroactive material (see Fig. 2.18).[45–48] This material can be actuated by a DC field application of low voltage such as 3 volts. Success was attained by the development of the chemical deposition of the electrode on the membrane surface. The principle of the deforming mechanism is somewhat similar to the case of other polyelectrolyte gels. That is, the membrane requires the presence of water and salts, and an encounter of electrochemical consumption is principally inevitable. However, the response time and durability are much higher than with the other gel materials. Moreover, the actuating process is not seriously affected by electrochemical reactions, provided the operating conditions are adequately controlled. Improvement of the efficiency can be considered to originate from the chemical structure of the membrane, and the coexistence of the strong hydrophobicity and strong hydrophilicity in a polymer chain.

2.5.2 Electroactive non-ionic polymer gels

Reviewing the above-mentioned materials, one of the serious defects of polyelectrolyte gels is the electrochemical consumption on the electrode under an electric field application. The electrochemical consumption causes poor durability of the polyelectrolyte gels and limits their application fields.

Therefore, the authors tried to utilize non-ionic polymer gels as actuating materials with large deformation. The results show that the idea works in a far

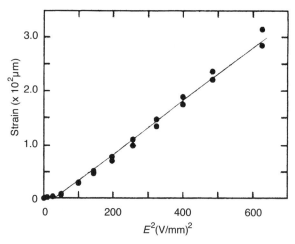

2.19 Dependence of strain in the direction of the field on the electric field.

more efficient manner than expected, but the mechanism turned out to be not the same as they expected initially. The feature will be described below in a little detail.

2.5.2.1 Strain in the direction of the field

Poly(vinyl alcohol)–DMSO gel was prepared by combining physical cross-linking and chemical cross-linking with glutaraldehyde (GA). After the chemical cross-linking, the physical cross-links were eliminated by exchanging solvent into pure DMSO.[49] The chemically cross-linked gel thus obtained has an electronically homogeneous structure. Therefore, the PVA–DMSO gel has no intrinsic polarization in its structure, and electrostrictive strain generation is expected by applying a DC electric field. The results agree with this expectation, and the strain is proportional to the square of the field (see Fig. 2.19).[50] The strain observed reached over 7% in the direction of the field. The response time is very fast, the large strain is attained within 0.1 s, and the shape of the gel is instantly restored by turning off the field. The current observed is around 1 mA at 250 V/mm, which is much smaller than those of polyelectrolyte gels. The current can be depressed by further purification of the polymer and solvent. This performance is much faster than conventional polyelectrolyte gels. We can demonstrate the electro-activated quick strain in the flapping motion by amplifying the strain by 300 times. It is suggested that the flapping motion be accelerated up to 10 Hz, though the demonstration was carried out at 2 Hz (see Fig. 2.20).

Electrically active polymer materials   21

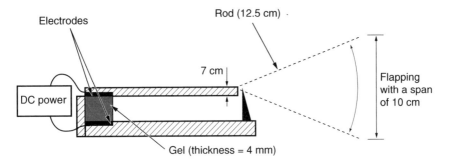

2.20 Flapping motion induced by an electrostrictive deformation of a non-ionic polymer gel swollen with a dielectric solvent.

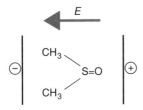

2.21 Structure of dimethylsulfoxide and its orientation by an electric field. Polarized Raman spectroscopy can be employed for investigating the molecular orientation under the field.

2.5.2.2 Electrical orientation of solvent

The strain induced in the direction of the field cannot be explained by the electrostatic attractive force between the electrodes. The effect of the electrostatic field was expected to be less than 25% of the observed strain under our experimental conditions. We therefore have to find another explanation for the strain generation in the gel.

Initially, we expected the orientation of solvent molecule under an electric field to lead the strain generation in the gel, through the changes of interactions between solvent and solute polymer, which forms the gel network.[51]

In order to observe the effect of the electric field on the orientation of the solvent, DMSO, Raman spectroscopy was employed. The molecule has a strong dipole moment, and can be expected to orient along the field direction (see Fig. 2.21).[52] It is oriented very efficiently even in relatively low electric fields, but the orientation decreases over the maximum field intensity (see Fig. 2.22).[53] The deformation of the gel becomes greater in the region of the higher field than that of the maximum orientation, suggesting that the solvent orientation is not directly related to the deformation of the gel.

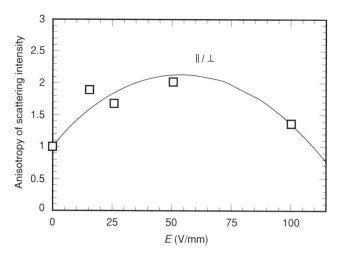

2.22 Electrically induced orientation of DMSO. DMSO orients very efficiently even in relatively low electric field, but the orientation decreases over the maximum field intensity.

2.5.2.3 Bending and crawling motion accompanying huge strain

In observing the contraction along the direction of an electric field, brass plates were used as electrodes. The strain in the perpendicular direction of the field was also observable. In these measurements, the bending deformation of the gels was prevented or completely depressed.

When we carefully observed the gel deformation, solvent flow and some asymmetric deformation was suggested in the gel. But conventional electrodes or a thin metal sheet of 10 μm thickness did not lead to any effective deformation. We used very thin gold electrodes whose thickness was 0.1 μm, and covered both surfaces of the gel with the thin metal sheet. The metal sheet is soft enough and does not disturb even a slight deformation of the gel.

By applying a DC electric field to the gel, the gel bent swiftly and held the deformation as far as the field was on (see Fig. 2.23).[54–56] The bending was completed within 60 ms, and the bending angle reached over 90 degrees. By turning off the field, the strain was released instantly, and the gel resumed its original shape (see Fig. 2.24). The curvature turned out to be proportional to the square of the field (see Fig. 2.25).

Taking the gel size (length 1 cm, width 5 mm and thickness 2 mm) into account, and assuming the gel volume does not change in the deformation, the strain in the gel can be estimated to be over 140% in length.[57] The electric current observed in this motion was less than 30 μA under the field of 500 V/mm.

This response and the huge strain attained in the PVA–DMSO gel is the

Electrically active polymer materials 23

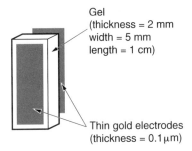

Gel
(thickness = 2 mm
width = 5 mm
length = 1 cm)

Thin gold electrodes
(thickness = 0.1 μm)

2.23 Assembling an electroactive non-ionic gel. The metal sheet was soft enough and does not disturb even a slight deformation of the gel.

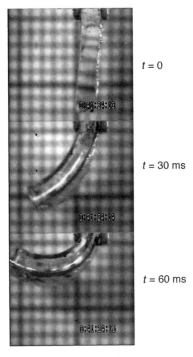

$t = 0$

$t = 30$ ms

$t = 60$ ms

2.24 Swift bending of a non-ionic polymer gel. By applying a DC electric field to the gel, the gel bent swiftly and sustained the deformation while the field was on.

largest value among the electroactive polymer gel materials reported so far. The low current suggests that there is much less energy loss in this motion compared with the conventional polyelectrolyte gels. The energy loss as heat was much less than that of Nafion or Flemion membrane overall, therefore it is far less when the size (thickness and surface area) of the gel is taken into account.

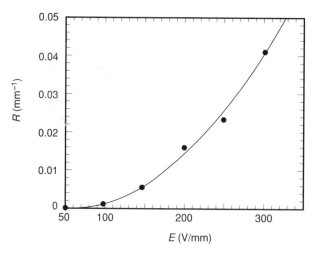

2.25 Dependence of bending curvature on an electric field.

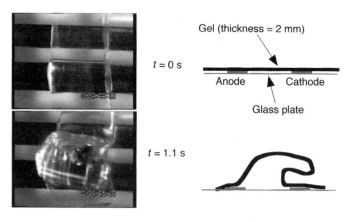

2.26 Crawling motion of a non-ionic polymer gel.

The gel could also show a crawling-type deformation.[54] This is a novel type of motion. The crawling motion was observed when a naked gel was placed on an electrode stripe array. The motion was completed in *ca.* 1 second (see Fig. 2.26).

2.5.2.4 Origin of the asymmetric pressure distribution in the gel

Such a remarkable swift bending or crawling of a non-ionic polymer gel cannot be explained by osmotic pressure gradient, which is usually considered to be the reason for electrically induced bending in polyelectrolyte gel. As pointed out in the previous section, the solvent flow was suggested in the gel. We investigated the effect of an electric field on its flowing property.[54]

Electrically active polymer materials

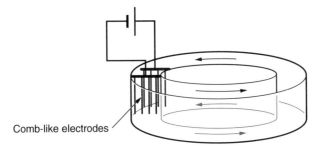

2.27 Electric field-induced flow of a dielectric solvent. A pair of cone-like electrodes were dipped in a circular tray filled with DMSO.

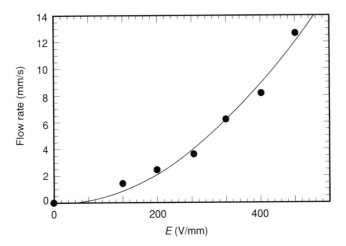

2.28 Dependence of a flow rate on an electric field. The flow rate was proportionally increased with the square of the field.

A pair of comb-like electrodes were dipped in a circular tray filled with DMSO (see Fig. 2.27). When a DC field was applied, the solvent started flowing from anode to cathode. The flowing rate was measured by using polystyrene powder floating on the solvent as a probe. The flow rate increased proportionally with the square of the field (Fig. 2.28). This result suggests that the pressure gradient is generated between the electrodes.

In order to establish a quantitative estimation of the pressure gradient, a theoretical treatment was carried out under some hypotheses shown below:[58]

1. Only one value of ion mobility exhibits for a kind of ion.
2. The turbulence in the gel can be neglected for the calculations of the pressure buildup.
3. The ionizing and accelerating electrodes do not interfere with pressure buildup.

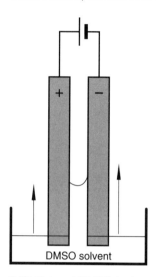

2.29 Solvent DMSO is drawn up between the electrodes.

4 Although only a very small resultant current exists, it is enough to determine the field distribution.
5 Different types of ions do not interfere with each other in the pressure buildup.
6 Surface charges on solvent boundaries have a negligible effect on ion current and field distribution.

The following equation was deduced for the pressure distribution in a gel:

$$p(x) - p(0) = \frac{9}{8} \varepsilon \left(\frac{V - V_0}{d} \right) \frac{x}{d} \qquad [2.5]$$

where $p(x) - p(0)$, ε, V, V_0 and d are pressure gaps between the two positions 0 (on the gel surface) and x (at x in the gel from the gel surface) in the direction of the field, dielectric constant, voltages on the gel surface and at x in the gel, and the thickness of the gel, respectively.

This equation suggests that the pressure gradient generated in the gel is proportional to the dielectric constant of the gel and to the square of an electric field. As the solvent content of the gel is ca. 98% in our experimental system, the dielectric constant of the gel can be assumed to be the same as that of the solvent. By taking the bending elasticity of the gel and the estimated pressure, we could attain excellent agreement between our experimental data and theoretical estimation (see Fig. 2.29).

In order to see the effect of the polymer on the electrically induced deformation, another type of experiment was carried out. A pair of plate electrodes were

Electrically active polymer materials

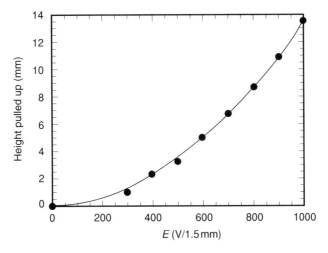

2.30 Dependence of drawn-up height of DMSO on the electric field.

dipped in the solvent, and the DC field was applied between the electrodes. The solvent was pulled up between the electrodes (Fig. 2.30). The height was theoretically estimated by the following equation:

$$h = \tfrac{1}{2}(\varepsilon - \varepsilon_0)(V/d)^2/\rho g \qquad [2.6]$$

where h, ε, ε_0, V, d, ρ and g are liquid surface height, dielectric constant of the gel, dielectric constant of vacuum, voltage applied, distance between the electrodes, density of the gel and gravitational constant.

The curve in Fig. 2.30 was calculated to be one, and is in good agreement with the experimental data. However, when we used a DMSO solution of PVA, the height was much less than that observed in the solvent and, furthermore, was extremely asymmetric on both electrodes (see Fig. 2.31). The solution tends to climb up onto the cathode surface, but not onto the anode, suggesting that the above equation is no longer applicable for the polymer solution.

These phenomena imply that the polymer solution has the tendency to retard the discharging process. The discharge retardation causes the accumulation of the charge on the cathode side in the gel and enhances the pressure gap between the cathode side and the anode side in the gel. Thus, the presence of the polymer network also plays an important role in efficient bending deformation.

For more detailed analysis, further quantitative investigation must be carried out.

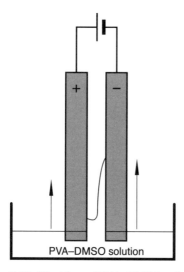

2.31 Climbing of PVA–DMSO solution onto an electrode under the field.

2.5.2.5 Applicability of the 'charge-injected solvent drag' method for conventional polymers

The concept proposed in the previous section can be described as the 'charge-injected solvent drag' method. The advantage of the method is its wide applicability to conventional non-ionic polymeric gel-like materials. We have been working on several non-ionic materials for soft artificial muscles that can be actuated in air.

Here, the case of poly(vinyl chloride) (PVC) will be shown briefly. In the case of PVC, we used plasticizers as solvent. Although tetrahydrofurane is a good solvent for PVC, its boiling point is too low for the preparation of the stable gel at ambient temperature in air. In the example shown, the PVC gel plasticized with dioctylphthalate (DOP) was found to creep reversibly like an amoeba by turning on and off the electric field (see Fig. 2.32).[59] The electrically induced deformation was suggested to be the asymmetric distribution of the injected charge. The mechanism is somewhat similar, but not the same, to that for the non-ionic polymer gel, since the solvent flow has not been confirmed in plasticized PVC.

2.6 From electro-active polymer gel to electro-active elastomer with large deformation

Non-ionic polymer gel swollen with dielectric solvent is shown to be extremely deformed, as is the non-ionic polymer plasticized with non-ionic plasticizer.

Electrically active polymer materials

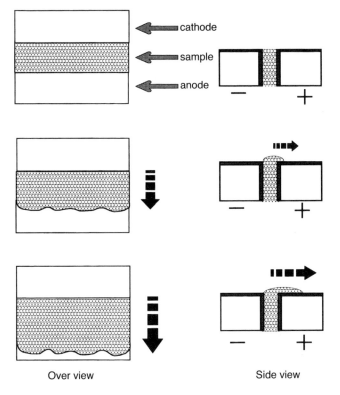

Over view   Side view

2.32 Creeping motion of poly(vinyl chloride) plasticized with dioctylphthalate. PVC gel plasticized with dioctylphthalate (DOP) was found to creep reversibly like an amoeba by turning the electric field on and off.

The mechanism suggested for the gel actuation was 'charge-injected solvent drag', and that for the plasticized polymer was 'asymmetric charge distribution of injected charge'.

The latter mechanism can be applied to the non-ionic elastomers in which the motion of the polymer chain is relatively free and so is the migration of the injected charges. The migration of the injected charge and the balance of the charging and discharging rates must be a critical factor to the deformation provided the electrostatic interaction is a major factor in the actuation.

The experimental results on polyurethane elastomers support the concept described above.[60–63] In addition to our expectations, some novel features of the motion are being clarified in detail, such as memory effect, bending direction control, and so on (Fig. 2.33).

30 Smart fibres, fabrics and clothing

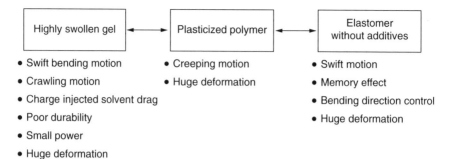

- Swift bending motion
- Crawling motion
- Charge injected solvent drag
- Poor durability
- Small power
- Huge deformation

- Creeping motion
- Huge deformation

- Swift motion
- Memory effect
- Bending direction control
- Huge deformation

2.33 Wide span of large deforming non-ionic electro-active actuator materials from polymer gel, through plasticized polymer, to elastomer with no additives. A prosperous future for the conventional non-ionic polymer materials as autonomic materials can be expected.

2.7 Conclusions

In this chapter, various types of electroactive polymers were introduced. Some of them have a long history as electroactive materials. Recently, however, polymer gels and/or elastomers, which have no intrinsic polarization in their structure and do not contain any ionic species either, have been found to show huge strain by applying electric fields with a low electric current. Energy dissipation occurring as heat is much less than the conventional polyelectrolyte materials. The concepts of 'charge-injected solvent drag' and 'asymmetric charge distribution of injected charge' are proposed as a possible mechanism of the huge deformation. These concepts can be applied to various non-ionic conventional polymers. The author strongly expects that the concepts expand the field of actuator to that of practical artificial muscle, and contribute to the development of the micro-machine or nano-machine in the future.

Acknowledgements

The work described in this article is partly supported by Grants-in-Aid for COE Research (10CE2003) and Scientific Research B(12450382), and by the Regional Science Promoter Program of Nagano Prefecture. The author (TH) expresses his sincere thanks for the support.

References

1 *Polymer Gels. Fundamentals and Biomedical Applications*, ed. DeRossi, Danilo, Kajiwara, Kanji, Osada, Yoshihito and Yamauchi, Aizo, Prenum (1991).
2 Osada, Y and Kajiwara K (eds), *Gel Handbook*, NTS Publishing Co. Ltd. (1997).
3 *Handbook of New Actuators for Accurate Controlling*, Supervised by K Uchino, Fuji

Technosystem Publishing Co. Ltd. (1994).
4 *Organic Polymer Gels*, Kikan Kagaku Sosetsu, No. 8, ed. by the Chemical Society of Japan, Gakkai Shuppan Center (1990).
5 *Recent Trends and Development of Polymer Gels – For Professionals in the Advanced Technology Field*, Supervised by Y Osada and Lin Wang, CMC Co. Ltd. (1995).
6 Uchino K, *Piezo/Electrostrictive Materials*, Morikita Publishing Co. Ltd. (1986).
7 *Jikken Kagaku Koza (Experimental Chemistry Series)*, ed. by the Chemical Society of Japan, **9** (1991).
8 Miyata S and Furukawa T, *Ferroelectric Polymers*, ed. by Polymer Society of Japan, New Polymer Materials, One Point Series 14, Kyoritsu Publishing Co. Ltd. (1989).
9 Micromachine Research Laboratory, The University of Tokyo, *Super Technology – Micromachine*, NTT Publishing Co. (1993).
10 Fujita H, *The World of Micromachine*, K Books Series 85, Kogyochosakai Publishing Co. Ltd. (1992).
11 Actuator Research Group, *New Actuators Targeting Microworld*, K Books Series 103, Kogyochosakai Publishing Co. Ltd. (1994).
12 Kinoshita G, *Soft Machine*, New Corona Series 16, Corona Publishing Co. Ltd. (1992).
13 Bar-Cohen Y, 'Electroactive Polymer Actuators and Devices', *Proceedings of SPIE*, **3669** (1999).
14 Bohannan G W, Schmidt V H, Conant R J, Hallenberg J H, Nelson C, Childs A, Lukes C, Ballensky J, Wehri J, Tikalsky B and McKenzie E, 'Piezoelectric polymer actuators in a vibration isolation application', *Proceedings of SPIE*, **3987**, 331 (2000).
15 Wang H, Zhang Q M, Cross L E and Sykes A O, 'Piezoelectric, dielectric, and elastic properties of poly(vinylidene fluoride/trifluoroethylene), *J. Appl. Phys.*, **74**, 3394 (1993).
16 Mazzoldi A and DeRossi D, 'Conductive polymer based structures for a steerable catheter', *Proceedings of SPIE*, **3987**, 271 (1999).
17 MacDiarmid A G and Epstein A J, 'Polyanilines: A novel class of conducting polymers', *Faraday Discussions*, 1989, n. 88, 317.
18 Baughman R H, 'Conducting polymer artificial muscles', *Synthetic Metals*, **78**, 339 (1996).
19 Gandhi M R, Murray P, Spinks G M and Wallace G G, 'Mechanism of electromechanical actuation in polypyrrole', *Synthetic Metals*, **73**, 247 (1995).
20 Hirai T, Hanaoka K, Suzuki T and Hayashi S, 'Preparation of oriented hydrogel membrane', *Kobunshi Ronbunshu*, **49**, 613 (1989).
21 Hirai T, 'Artificial muscle', *Shin-Sozai (New Materials)*, No. 4 (1992).
22 Suzuki M, 'Nonlinear mechanochemical systems', *Nippon Gomu Kyokaishi*, **60**, 702 (1987).
23 Osada Y, 'Chemical valves and gel actuators', *Adv. Mater.*, **3**, 107 (1991).
24 Hirai T, Asada Y, Suzuki T, Hayashi S and Nambu M, 'Studies on elastic hydrogel membrane. I. Effect of preparation conditions on the membrane performance', *J. Appl. Polym. Sci.*, **38**, 491 (1989).
25 Hirasa O, Morishita Y, Onomura R, Ichijo H and Yamauchi I, 'Preparation and mechanical properties of thermo-responsive fibrous hydrogels made from poly(vinyl methyl ethers)', *Kobunshi Ronbunshu*, **46**, 661 (1989).
26 Okano T, Bae Y H and Kim S W, 'Temperature dependence of swelling of

crosslinked poly(*N*,*N*'-alkyl substituted acrylamides in water', *J. Controlled Release*, **11**, 255 (1990).

27 Tanaka T, Sato E, Hirokawa Y, Hirotsu S and Peeternians J, 'Critical kinetics of volume phase transition of gels', *Phys. Rev. Lett.*, **55**, 2455 (1985).

28 Tokita M and Tanaka T, 'Reversible decrease of gel-solvent friction', *Science*, **253**, 1121 (1991).

29 Kokufuta E, Zhang Y-Q and Tanaka T, 'Saccharide-sensitive phase transition of a lectin-loaded gel', *Nature*, **351**, 302 (1991).

30 Kokufuta E, 'New development of immobilized enzymes', *Hyoumen (Surfaces)*, **29**, 180 (1991).

31 Tanaka T, 'Phase transitions in ionic gels', *Phys. Rev. Lett.*, **45**, 1636–1639 (1980).

32 Irie M, 'Photoresponsive polymers', *Adv. Polymer Sci.*, 28 (1989).

33 Takamizawa T, Hirai T, Hayashi S and Hirai M, 'Magnetostriction of a gel immobilizing superparamagnetic fluid', *Proceedings of International Symposium on Fiber Science and Technology (ISF '94)*, in Yokohama, 293 (1994).

34 Hirai T, 'Electrically and magnetically active gel', *ZairyoKagaku (Materials Science)*, **32**, 59 (1995).

35 Hirai T, Takamizawa T and Hayashi S, 'Structural change induced in the gel immobilizing superparamagnetic fluid', *Polym. Preprints, Jpn.*, **43**, 3126 (1993).

36 Zryhni M, Barsi L and Buki A, 'Deformation of ferrogels induced by nonuniform magnetic fields', *J. Chem. Phys.*, **104**, 20 (1996).

37 Zryhni M, Barsi L, Szabo D and Kilian H-G, 'Direct observation of abrupt shape transition in ferrogels induced by nonuniform magnetic field', *J. Chem. Phys.*, **106**, 5685 (1997).

38 Zryhni M and Horkai F, 'On the thermodynamics of chemomechanical energy conversion realised by neutral gels', *J. Intell. Mater. Syst. and Struct.*, **4**, 190 (1993).

39 Shiga T, Hirose Y, Okada A and Kurauchi T, 'Bending of high-strength polymer gel in an electric field', *Kobunshi Ronbunshu*, **46**, 709 (1989).

40 Shiga T, Hirose Y, Okada A and Kurauchi T, 'Bending of ionic polymer gel caused by swelling under sinusoidally varying electric fields', *J. Appl. Polym. Sci.*, **47**, 113 (1993).

41 Shiga T, Fukumori K, Hirose Y, Okada A and Kurauchi T, 'Plused NMR study of the structure of poly(vinyl alcohol)–poly(sodium acrylate) composite hydrogel', *J. Polym. Sci.: Part B: Polym. Phys.*, **32**, 85 (1994).

42 Osada Y, Okuzaki H and Hori H, 'A polymer gel with electrically driven motility', *Nature*, **355**, 242 (1992).

43 Gong J P, Kawakami I, Sergeyev V G and Osada Y, 'Electroconductive organogel. 3. Preparation and properties of a charge-transfer complex gel in an organic solvent', *Macromolecules*, **24**, 5246 (1991).

44 Gong J P, Kawakami I and Osada Y, 'Electroconductive organogel. 4. Electrodriven chemomechanical behaviors of charge-transfer complex gel in organic solvent', *Macromolecules*, **24**, 6582 (1991).

45 Oguro K, Kawami Y and Takanaka H, 'An actuator element of polymer electrolyte gel membrane-electrode composite', *Bull. Government Industrial Research Institute Osaka*, **43**, 21 (1992).

46 Oguro K, Fujiwara N, Asaka K, Onishi K and Sewa S, 'Polymer electrolyte actuator with gold electrodes', *Proceedings of SPIE*, **3669**, 64 (1999).

47 Mojarrad M and Shahinpoor M, 'Noiseless propulsion for swimming robotic structures using polyelectrolyte ion-exchange membrane', *Proc. SPIE*, 1996 North American Conference on *Smart Structures and Materials*, **2716**, Paper no. 27, 183 (1996).
48 Shahinpoor M, 'Electro-mechanics of ion-elastic beams as electrically-controllable artificial muscles', *Proc. SPIE, Electroactive polymer actuators and devices*, **3669**, 109 (1999).
49 Hirai T, Maruyama H, Suzuki T and Hayashi S, 'Shape memorizing properties of poly(vinyl alcohol) hydrogel', *J. Appl. Polym. Sci.*, **45**, 1849 (1992).
50 Hirai T, Nemoto H, Hirai M and Hayashi S, 'Electrostriction of poly(vinyl alcohol) gel – a possible application for artificial muscle', *J. Appl. Polym. Sci.*, **53**, 79 (1994).
51 Errede L A, Osada Y and Tiers V D, 'Molecular interpretation of chemo-mechanical reponses in polymer liquid systems', *J. Intell. Mater. Systems Struct.*, **4**, 2, 161 (1993).
52 Hirai M, Hirai T, Sukumoda A, Nemoto H, Amemiya Y, Kobayashi K and Ueki T, 'Electrically induced reversible structural change of a highly swollen polymer gel network', *J. Chem. Soc., Faraday Trans.*, **91**, 473 (1995).
53 Hirai T and Hirai M, *Polymer Sensors and Actuators*, ed. Y Osada and D E DeRossi, Chapter 8, p. 245, Springer-Verlag, Berlin, Heidelberg (2000).
54 Hirai T, Jianming Zheng and Watanabe M, *Proc. of SPIE*, SPIE's 6th Annual International Symposium on *Smart Structures and Materials, Electro-active Polymer Actuators and Devices*, **3669**, 209 (1999).
55 Zheng J, Watanabe M, Shirai H and Hirai T, 'Electrically induced rapid deformation of nonionic gel', *Chem. Lett.*, 500–501 (2000).
56 Hirai T, Zheng J, Watanabe M and Shirai H, 'Autonomic gel and its application as artificial muscle', *Zairyo Kagaku* (*Materials Science*), **36**, 186 (2000).
57 Yamaguchi M, Watanabe M and Hirai T, 'Electrical actuation of poly(vinyl chloride) gel', *Polym. Preprints, Jpn.*, **48**, 2124 (1999).
58 Hirai T, Zheng J, Watanabe M and Shirai H, 'Electro-active non-ionic gel and its application', *Proceedings of SPIE*, **3987**, 281 (2000).
59 Hirai T, Zheng J, Watanabe M, Shirai H and Yamaguchi M, 'Electroactive nonionic polymer gel swift bending and crawling motion', *Materials Research Society Symposium Proceedings*, **600**, 267–272 (2000).
60 Hirai T, Sadatoh H, Ueda T, Kasazaki T, Kurita Y, Hirai M and Hayashi S, 'Polyurethane elastomer actuator', *Angew. Makromol. Chem.*, **240**, 221 (1996).
61 Watanabe M, Yokoyama M, Ueda T, Kasazaki T, Hirai M and Hirai T, 'Bending deformation of monolayer polyurethane film induced by an electric field', *Chem. Lett.*, 773 (1997).
62 Watanabe M, Suzuki M, Amaike Y and Hirai T, 'Electric conduction in bending electrostriction of polyurethanes', *Appl. Phys. Lett.*, **74**, 2717 (1999).
63 Watanabe M, Kato T, Suzuki M, Amaike Y and Hirai T, 'Bending electrostriction in polyurethanes containing ions as contaminants or additives', *Jpn. J. Appl. Phys.*, **38**, Part 2, No. 8A, L872 (1999).

3
Heat-storage and thermo-regulated textiles and clothing

XINGXIANG ZHANG

3.1 Development introduction

3.1.1 Introduction

The human body is itself is an automatic thermo-regulated organism. The body constantly generates heat, CO_2 and H_2O by the metabolism of food and muscle activity. The human body controls the release speed of heat by blood vessel dilatation or constriction, muscle and sweat gland activity, etc., and then regulates the body temperature.

It has been shown that the most comfortable skin temperature is 33.4 °C. When the temperature of any part of the skin differs by within 1.5–3.0 °C of this ideal temperature, the human body is unaware of the warmth or coolness. If the difference is more than $\pm$ 4.5 °C, the human body feels discomfort. In addition, a core body temperature of 36.5 °C is required, and a rise or fall of 1.5 °C can be fatal. A balance between the rates of heat loss and heat generated must be maintained. When the thermal resistance of their clothing is insufficient, and a person stays in a relatively low temperature condition for a long time, the body temperature drops. The human body must produce more heat by eating more food or increasing exercise. Also, the body will adjust to the heat loss by constricting the blood capillaries within the skin. Hypothermia may result, if the core temperature is lower than 35 °C. Conversely, in a hot environment, the body must cool itself. The body generates heat and the heat absorbed by the skin from the environment must be dissipated. Under these conditions the body will dilate the blood capillaries, which enables the evaporation of water diffused from the body interior to increase, thereby increasing the cooling effect. This action is known as insensible perspiration. If the environment temperature increases further, the body will activate sweat glands and evaporate liquid water for increased cooling, known as sensible perspiration. In some special environments, due to the high temperature or high thermal resistance of the clothing, hyperthermia can result if the body fails to dissipate sufficient heat.

The traditional heat insulation materials, for example, fabrics such as cotton, hemp, silk, wool and polyester, polyamide and polyacrylic fibre, etc. provide a degree of resistance to body heat loss, which is determined by the number of air pockets present in the fabric. The solar ray selective absorbing textile that has been manufactured since the late 1980s can absorb the near infra-red ray in the sun rays, and convert it to heat, thus enhancing the inner temperature of the clothing. The far infra-red textile that has been produced since the late 1980s can absorb the body's irradiated far infra-red ray and turn it to heat, enhancing thermal resistance. The ultra-violet absorbing and cool-feeling fabric that has been industrialized since the beginning of the 1990s can absorb the ultra-violet and reflect the near infra-red rays from the sun, lowering the inner temperature of the clothing. All of these new functional textiles are one-way thermo-regulated heat insulating materials.

To keep the skin temperature within the 30.4 to 36.4 °C interval, we must put on or take off our clothing according to the external temperature. If clothing could automatically change its thermal resistance according to the temperature, it could control the speed of heat release, and then regulate the inner temperature. We would not need to put it on or take it off very often, and the proper inner temperature would not only increase body comfort, but also reduce fatalities.

3.1.2 Development history

The heat-storage and thermo-regulated textiles and clothing are newer concepts: they are two-way thermo-regulated new products, and have been studied for more than 30 years. The development history of heat-storage and thermo- regulated textiles and clothing is outlined in Table 3.1.

Scientists have increasingly been studying the manufacture process and properties of heat-storage and thermo-regulated textiles and clothing for several decades. There will be more reports about basic, dynamic thermal resistance and its applications in the near future.

3.2 Basics of heat-storage materials

There are three types of heat storage: sensible, latent and chemical reaction heat storage. Latent heat storage is the most important way of storing heat, and is also called 'phase change heat storage'. Latent heat-storage materials have been widely used for about 40 years.

The US National Aeronautics and Space Administration (NASA) accelerated the application research of latent heat-storage materials for the space laboratory in the 1960s. They were used to improve the protection of instruments and astronauts against extreme fluctuations in temperature in

Table 3.1 Development history of heat-storage and thermo-regulated textile and clothing

Time	Researcher	Research introduction	Ref. no.
1965	Mavelous and Desy	Designed a heat-insulating garment containing molten lithium salts pouches. The salts exchanged heat with water in the pouches and the hot water was circulated to tubing throughout the garment.	1
1971	Hansen	Incorporated CO_2 into liquid inside the hollow fibre. When the liquid solidified and the solubility of the CO_2 decreased, the diameter of fibre increased and the heat-insulating ability rose.	2
1981	Vigo and Frost	Incorporation of hydrated inorganic salts into hollow fibres.	3
1983	Vigo and Frost	Incorporation of polyethylene glycol (PEG) and plastic crystal materials into hollow rayon or polypropylene fibre.	4
1985	Vigo and Frost	Coated PEG on the surface of fabrics. The thermal storage and release properties of the fabrics were reproducible after 150 heating and cooling cycles. *Chemical Week* and *Wall Street Journal* reviewed Vigo's research work.	5 6, 7
1987	Bruno and Vigo	The thermal storage and release, anti-static, water absorbence, reliancy, soil release and conditional wrinkle recovery of PEG-coated fabrics were studied.	8
1988	Bryant and Colvin	A fibre with integral microcapsules filled with phase change materials (PCM) or plastic crystals had enhanced thermal properties at predetermined temperatures.	9
	Vigo and Bruno	PEG was cross-linked on the surface of knit fabrics.	10
1989	Vigo et al.	Coated PEG modified the resiliency and resistance to the oily soiling, static charge, pilling, wear life and hydrophilicity of fabric and fibres.	11
1990	Watanabe et al.	Melt spun heat-releasing and heat-absorbing composite fibre using aliphatic polyester as core or island component, fibre-forming polymer as sheath or sea component.	12
1991	Zhang	Designed the heat insulation clothing containing $CaCl_2.6H_2O$ pouches.	13
	Mitsui Corp.	Licensed Vigo's invention and commercialized skiwear and sportswear that contain fabrics with cross-linked polyols.	14
	Neutratherm Corp.	Licensed Vigo's invention and commercialized fabrics incorporating PEG. The fabrics were used for thermal underwear.	14
1992	Mitamura	A composite fibre was melt spun by using polytetramethylene glycol (av. MW 3000) as a core, and poly(ethylene terephthalate) (PET) as a sheath.	15

Heat-storage and thermo-regulated textiles and clothing

Table 3.1 (*cont.*)

Time	Researcher	Research introduction	Ref. no.
1992	Bryant and Colvin	Fabrics coated with a binder containing microcapsules filled with PCM or plastic crystals.	16
1993	Vigo	The antibacterial properties of PEG-coated fabrics were studied.	17
	Umbile	Fabrics were coated with a binder containing microcapsules filled with PEG (av. MW 300–4000).	18
	Momose et al.	A blend of linear chain hydrocarbon, particles of ZrC, thermoplastic elastomer and polyethylene is used as a coating materials for nylon fabrics	19
	Colvin and Bryant	A cooling garment contained pouches containing macroencapsulated PCM.	20
1994	Renita et al.	The wicking effect, antibacterial and liquid barrier properties of non-woven coated with PEG were studied.	21
	Bruno and Vigo	Formaldehyde-free cross-linking PEG-coated fabrics were studied.	22
	Bruno and Vigo	The dyeability of PEG-coated polyester/cotton fabric was investigated.	23
	Tsujito et al.	Carpet coated with a binder containing microcapsules filled with paraffin.	24
	Imanari and Yanatori	Heat insulation clothing with pouches containing paraffin and salts was designed.	25
1995	Pause	The basic and dynamic thermal insulation of fabrics coated with or without microencapsulated PCM was studied.	26
1996	Zhang et al.	The heat storage and crystallizability of PET–PEG block copolymers were studied.	27
	Zhang et al.	Melt spun composite fibre using aliphatic polyester, PET–PEG, paraffin or PEG as core component, fibre-forming polymer as sheath component.	28
	Sayler	PCM with melting point 15–65 °C incorporated throughout the structure of polymer fibres.	29
1997	Zhang et al.	Melt spun heat-storage and thermo-regulated composite fibre using PEG (av. MW 1000–20 000) as core component, polypropylene as sheath component.	30
	Outlast Inc.	Properties of the fibres were studied.	31
	Frisby Inc.	Coated fabrics and wet spinning polyacrylic fibre containing microencapsulated PCM came onto the market.	32
	Vigo	Intelligent fibrous substrates with thermal and dimensional memories were studied.	33
1999	Pause	The basic and dynamic thermal insulation of coated fabrics and non-woven containing microencapsulated linear chain hydrocarbons were studied, and the skin temperature of a simulated skin apparatus in 20 °C and –20 °C conditions were measured.	34

space. Latent heat-storage materials were used in the lunar vehicle and the Sky lab project for the Apollo 15 mission in the 1980s.[31]

3.2.1 Sensible heat-storage materials

There is an obvious temperature change during the heat-absorbing and release process of sensible heat-storage materials. Water, steel and stone are widely used sensible heat-storage materials. Water is the cheapest, most useful sensible heat-storage materials in the temperature interval from 1 °C to 99 °C at 1 standard atmosphere. The absorbing heat content of water for a 1 °C rise in temperature is 4.18 J/g.

3.2.2 Latent heat-storage materials

Latent heat-storage materials are also called phase change materials (PCM). PCM can absorb or release heat with a slight temperature change. PCM may be repeatedly converted between solid and liquid phases to utilize their latent heat of fusion to absorb, store and release heat or cold during such phase conversions. The latent heats of fusion are greater than the sensible heat capacities of the materials.

Material usually has three states. When a material converts from one state to another, this process is called phase change. There are four kinds of phase change, 1 solid to liquid, 2 liquid to gas, 3 solid to gas, 4 solid to solid.

Heat is absorbed or released during the phase change process. The absorbed or released heat content is called latent heat. The PCM, which can convert from solid to liquid or from liquid to solid state, is the most frequently used latent heat-storage material in the manufacture of heat-storage and thermo-regulated textiles and clothing. The melting heat-absorbing temperature interval is from 20 to 40 °C, and the crystallization heat-releasing temperature interval is from 30 to 10 °C. The phase change temperature of hydrated inorganic salts, polyhydric alcohol–water solution, polyethylene glycol (PEG), polytetramethylene glycol (PTMG), aliphatic polyester, linear chain hydrocarbon, hydrocarbon alcohol, hydrocarbon acid, etc., is in this interval.

3.2.2.1 Hydrated inorganic salts

Hydrated inorganic salt is an inorganic salt crystal with n water molecules. The hydrated inorganic salt that can be used in the manufacture of heat-storage and thermo-regulated textiles and clothing usually has a heat-absorbing and -releasing temperature interval of about 20 to 40 °C. Some of the hydrated inorganic salts are listed in Table 3.2. It was observed that incongruent melting and super cooling of most inorganic salt hydrates incorporated into,

Heat-storage and thermo-regulated textiles and clothing

Table 3.2 The hydrated inorganic salts[35]

Hydrated inorganic salts	Melting point (°C)	Melting heat (kJ/kg)	Density kg/m³		Specific heat capacity (J/(kg K))		Heat-storage density (MJ/m³)
			Solid	Liquid	Solid	Liquid	
CaCl$_2$.6H$_2$O	29	190	1800	1560	1460	2130	283
LiNO$_3$.3H$_2$O	30	296					
Na$_2$SO$_4$.10H$_2$O	32	225	1460	1330	1760	3300	300
CaBr$_2$.6H$_2$O	34	138					
Na$_2$HPO$_4$.12H$_2$O	35	205					
Zn$_2$SO$_4$.6H$_2$O	36	147			1340	2260	
Na$_2$SO$_4$.5H$_2$O	43	209	1650		1460	2300	345

or topically applied to, fibres occurred after several heating/cooling cycles. Lithium nitrite trihydrate lasted 25 cycles, calcium chloride hexahydrate lasted only a few cycles, and sodium sulfate decahydrate only one cycle.

3.2.2.2 Polyhydric alcohol

2,2-Dimethyl-1,3-propanediol (DMP), 2-hydroxymethyl-2-methyl-1,3-ropanediol (HMP) and 2-bis-(hydroxyl methyl)-1,3-propanediol are often used as PCM. They produce endothermic and exothermic effects without a change in state at temperatures far below the melting point of the substance; the heat absorption and release of such substances is called the heat of transition. However, they are not suitable for the manufacture of heat-storage and thermo-regulated textiles and clothing due to their phase change temperature being higher than 40 °C. However, if two of the polyhydric alcohols are mixed, the phase change temperature can be in the range 24 to 40 °C.[35,36]

3.2.2.3 PEG and PTMG

Polyethylene glycol is one of the most important PCMs. The melting temperature of PEG is proportional to the molecular weight when its molecular weight is lower than 20 000. The melting point of PEG that has an average molecular weight higher than 20 000 is almost the same. The PEGs with molecular weight lower than 20 000 are listed in Table 3.3. The PEGs with molecular weight from 800 to 1500 have melting points of about 33 °C.

The maximum theory melting point of PTMG is 43 °C. The PTMG is used as PCM in some patents. The melting point of PTMG with molecular weight 3000 is 33 °C, and the heat of melting is 150 J/g.[15] However, the crystallization

Table 3.3 The phase change behaviour of different molecular weight PEG measured with DSC[37]

PEG sample	Average molecular weight	Melting point	Melting heat (kJ/kg)	Crystallization point	Crystallization heat (kJ/kg)
1	400	3.24	91.37	−24.00	85.40
2	600	17.92	121.14	−6.88	116.16
3	1000	35.10	137.31	12.74	134.64
4	2000	53.19	178.82	25.19	161.34
5	4000	59.67	189.69	21.97	166.45
6	6000	64.75	188.98	32.89	160.93
7	10000	66.28	191.90	34.89	167.87
8	20000	68.70	187.81	37.65	160.97

point is supercooling due to the weak interaction between the molecular chain of PTMG. This limits its application.

3.2.2.4 PET–PEG block copolymer

When the average molecular weight of PEG used in the synthesis of PET–PEG block copolymer is higher than 1540, and the PEG weight content in block copolymer is more than 50%, the PEG segment can crystallize alone. The melting point measured by DSC of the PEG segment is 6.86 °C.[27] When the average molecular weight of PEG is 4000, and PEG weight content is 50%, the melting point of the PEG segment is 33.05 °C, and the melting endotherm is 30.56 J/g. The PET–PEG block copolymer can be melt spun into fibre.

3.2.2.5 Linear chain hydrocarbon

Linear chain hydrocarbon is a byproduct of oil refining. The formula is C_nH_{2n+2}. The melting and crystallization points of hydrocarbons with $n = 16 \sim 21$ are in the temperature range 10 to 40 °C, Table 3.4. The commodity linear chain hydrocarbon is usually a mixture of $n \pm 1$ or $n \pm 2$.

Linear chain hydrocarbons are a non-toxic, inexpensive, extensive source of raw materials, suitable for varied usage. They are the most important PCMs in the manufacture of heat-storage and thermo-regulated textiles and clothing.

3.2.2.6 Others

Organic acid, alcohol and ether, for example decanoic acid, 1-tetradecanol, and phenyl ether, at proper phase change temperature, can also be used as PCMs.

Table 3.4 The phase change properties of linear chain hydrocarbons[9,25]

Phase change materials	Number of carbon atoms	Melting point (°C)	Heat of melting (kJ/kg)	Crystallization point (°C)
n-hexadecane	16	16.7	236.58	16.2
n-heptadecane	17	21.7	171.38	21.5
n-octadecane	18	28.2	242.44	25.4
n-eicosane	20	36.6	246.62	30.6
n-heneicosane	21	40.2	200.64	

3.3 Manufacture of heat-storage and thermo-regulated textiles and clothing

3.3.1 Phase change materials or plastic crystal-filled or -impregnated fibres

3.3.1.1 Hydrated inorganic salt-filled or -impregnated fibres

Vigo and Frost filled hollow rayon fibres with $LiNO_3.3H_2O$, $Zn(NO_3)_2.6H_2O$, $CaCl_2.6H_2O/SrCl_2.6H_2O$ and $Na_2SO_4.10H_2O/NaB_4O_7.10H_2O$.[3] When the content of lithium nitrate trihydrate within the rayon is 9.5 g per gram of fibre, the heat endotherm is 302.63, 312.24 and 156.32 J/g, respectively, and the heat exotherm is 221.95, 176.39 and 40.96 J/g, respectively, in the temperature interval − 40–60 °C, after one, ten and 50 heat–cool cycles. It is quite obvious that the decrease of heat capacity of the fibre is greater after more heat–cool cycles.

3.3.1.2 PEG-filled or -impregnated fibres

Although many inorganic salt hydrates within hollow fibres were initially effective in imparting heat-absorbing and -releasing characteristics to hollow fibres, they exhibited unreliable and poor thermal behaviour on repeated thermal–cool cycles. In 1983, Vigo and Frost filled fibres with 57% (wt./wt.) aqueous solution of PEG, with average molecular weight (average MW) of 400, 600, 1000 and 3350.[4] Hollow fibres were filled with PEG by aspirating aqueous solutions of the various different average molecular weights at, or above, room temperature through fibre bundles tightly aligned inside an O-ring until visual observation indicated that the fibres were completely filled. The filled fibres were then placed horizontally and cooled at − 15 °C or lower. Excess moisture was removed from the modified fibres by drying them to constant weight in the presence of anhydrous $CaSO_4$(a) in a desiccator for 24 h. The measured results of DSC show that the heat-absorbing and -releasing capacity of filled hollow polypropylene fibre is 1.2–2.5 times that of untreated fibre, and that of filled hollow rayon is 2.2–4.4 times that of untreated rayon.

3.3.1.3 Polyhydric alcohol filled or -impregnated fibre

Vigo and Frost filled hollow rayon and hollow polypropylene fibre with 2,2-dimethyl-1,3-propanediol (DMP)[36] and impregnated non-hollow rayon with DMP. The heat-absorbing capacity of treated hollow polypropylene fibre was 136.68 J/g in the temperature range 72–102 °C, and the heat-releasing capacity of treated hollow polypropylene fibre was 120.38 J/g in the temperature range 77–47 °C. It is less obvious that there is a decrease of the heat capacity of the fibre after 50 heat–cool cycles. The fibre is not suitable for clothing textiles, due to its high phase change temperature.

3.3.2 Coated fabric

3.3.2.1 Fabrics coated with PEG

Although the PEG and polyhydric alcohols were effective as heat-storage and -release agents in modified fibres/fabrics, they were only suitable for applications that did not require laundering of the fabrics, since they were still water soluble. Vigo and Frost studied fabrics coated with cross-linked PEG in 1985.[5] The average MW of PEG was from 600 to 8000. Fabrics were treated with aqueous solutions containing the following cross-linking agents: 40% solids dimethyloldihydroxy-ethyleneurea (DMDHEU), 40% solids dimethylolethyleneurea (DMEU), 50% solids dimethylolisopropyl carbamate and 80% solids trimethylolmelamine. Acid catalysts used were $MgCl_2.6H_2O$/citric acid in a mole ratio of 20 to 1, $NaHSO_4$ and $Zn(NO_3)_2.6H_2O$. PEG was also present in the pad bath. The cotton, PET and wool fabrics having thermal activity were produced using 50% aqueous solutions of the PEG-600 or PEG-1000 containing 8–12% DMDHEU and 2.4–3.6% mixed catalysts ($MgCl_2.6H_2O$/citric acid). All fabrics were padded to wet pickup of about 100%, dried and cured, then machine washed and dried. The PET treated with 50% aqueous solution of PEG-600, was dried at 60 °C for 7 minutes, and cured at 160 °C for 3 minutes. The weight gains were 42.9%, and the heat-absorbing capacity was 53.92 J/g measured by differential scanning calorimetry (DSC). The heat capacity was 53.08 J/g after ten heat–cool cycles. There was no obvious decrease of the heat capacity of the fibre.[5] The heat-absorbing capacity of untreated polyester fabric was 42.21 J/g and 40.12 J/g.

If the weight gains of the fabric were less than 20%, only the antistatic performance of the fabric was improved, and the thermal activity was not obvious. Fabrics treated with PEG-600/DMDHEU and an appropriate acid catalyst had weight gains of 27–47% and were thermally active compared to the untreated and cross-linked controls. When concentrations of less than 8% DMDHEU were used in the treatments with PEG-600, the weight gains were not sufficient to impart thermal activity to the fabric. Conversely, when higher concentrations of DMDHEU were used (> 12%) or when zinc nitrate catalysts were used at any DMDHEU concentration, extensive cross-linking occurred and the fabric was thermally inactive. When DMDHEU and the mixed acid catalyst were used with PEG of higher molecular weights (> 3350), there was little reaction, resulting in low add-on. Preliminary cross-link density and thermal analysis data indicate that the PEG are insolubilized on the fibres and exhibit thermal activity only when the PEG are of low crystallinity and can react with the polyfunctional cross-linking agent. Further experiments demonstrated that a wide range of one-step curing conditions (with temperatures as low as 80 °C) could be utilized to insolubilize the polymer on fibrous substrates and provide superior thermal storage and release properties than those obtained by the two-step (dry/cure) process that employed high curing temperatures.[14]

However, the DMEHEU or DMEU can be gradually decomposed to release toxic formaldehyde. A formaldehyde-free cross-linking system was discovered using a stoichiometric amount of sulfonic acids and glyoxal to form polyacetals with the same polyols.[22] These polyacetals exhibit the same multifunctional properties and are durable to laundering.

3.3.2.2 Fabrics coated with polyhydric alcohols

Polyhydric alcohols undergo solid-to-solid thermal transitions (SSTs). The often-used plastic crystals are DMP (DSC onset mp 126 °C) and 2-hydroxymethyl-2-methyl-1,3-propanediol (HMP, DSC onset mp 181 °C). The 50% (wt./wt.) aqueous solutions of the plastic crystals were made for subsequent application to fabrics.[5]

Regardless of fibre type, all fabrics immersed in 50% aqueous solutions of the DMP and padded to about 100% wet pickup, then dried at 100 °C, had only slight weight gains after conditioning. Presumably, the high vapour pressure of the DMP precludes the conventional or elevated temperature drying of treated fabrics.[5]

In contrast to the other plastic crystal substance DMP, the HMP did not volatilize from the fabric when it was dried at conventional or elevated temperature to remove excess water. The modified fabrics had heat contents of 87.78–104.5 J/g on heating and 79.42–96.14 J/g on cooling after one to ten thermal cycles. The heat contents of HMP-treated fabrics were 1.7–2.5 times those of untreated fabrics. The fabrics are not suitable for clothing textiles, due to their high phase change temperature.

44 Smart fibres, fabrics and clothing

3.3.2.3 Fabrics coated with a binder containing microcapsules

Bryant adapted a coating to apply to a substrate such as a fabric in order to enhance the thermal characteristics thereof.[16] The coating includes a polymer binder in which are dispersed integral and leak-resistant microcapsules filled with PCM or plastic crystals that have specific thermal properties at predetermined temperatures.

Zuckerman invented a coating composition for fabrics including wetted microcapsules containing PCM dispersed throughout a polymer binder, a surfactant, a dispersant, an antifoam agent and a thickener.[40] The most preferred ratios of components of the coating composition of the invention were: 70 to 300 parts by dry weight of microcapsules for each 100 parts by dry weight of acrylic polymer latex, 0.1% to 1% dry weight each of surfactant and dispersant to dry weight of microcapsules, water totalling 40% to 60% of the final wet coating composition and antifoam agent of from 0.1% to 0.5% dry weight to total weight of the final wet coating composition.

The microcapsules containing $Na_2SO_4.10H_2O$ as PCM were also used for the fabric coating.

Umible had coated a woven polyester with a mixture of microcapsules and polymer binder.[18] The coating layer is composed of 1:1–3 (wt. ratio) mixtures of microcapsules containing PEG with average MW 300–4000 and a polyacrylic binder. The coated fabric evolves heat at 7–11 °C, and absorbs heat at 28–31 °C. Umible thought the fabrics were useful for garments for workers in freezer units and mountaineering.

3.3.2.4 Other coated fabrics

Momose invented a coating agent containing *n*-octadecane 80, *n*-hexadecane 20, thermoplastic elastomer kraton G1650 12, linear polyethylene 20, ZrC 10 and an antioxidant 0.2 part.[19] The mixture was melted and mixed to give a non-bleeding solid with heat content 129.48 J/g and good shape-retaining properties. The mixture was molten at 140 °C and applied to a nylon cloth to give a textile with good heat storage properties, useful for skiwear.

3.3.3 Fibre-spinning

Heat-storage and thermo-regulated textiles can be manufactured by filling hollow fibres or impregnating non-hollow fibres with PCM and plastic crystals or coating fabric surfaces with PEG, plastic crystals or microcapsules. There are still some defects in the wash-resistance, durability and handle of the heat-storage and thermo-regulated textiles produced by these processes. The fibre-spinning process has developed quickly since the 1990s.

3.3.3.1 Composite fibre-spinning

The copolymers of diacid, for example, glutaric acid, hexanedioic acid and decanedioic acid, with diols, for example, 1,3-propylene glycol, 1,4-butanediol and 1,6-hexanediol have heat-absorbing and -releasing properties.[12] The melting points of some of the aliphatic polyesters are in the temperature range 20–40 °C, but their crystallization points are usually beyond the temperature range 30–10 °C.[41] Watanabe used the mixture of two aliphatic polyesters as the core component, PET as the sheath component, and melt span to produce heat-absorbing and heat-releasing synthetic conjugate fibres for heat-insulating garments.[12]

PTMG was used as the core component in the composite fibre-spinning process.[15] A composite fibre that uses PTMG as the core and PET as the sheath was designed.

3.3.3.2 Fibre spun with a mixture

When aliphatic polyester or polyether is used alone as the core or island component in the composite fibre-spinning process, the spinning process is very hard to control, due to its very low melting viscosity. If the aliphatic polyester or polyether is blended with ethylene copolymer, the melting viscosity of the mixture can be high enough for the fibre-spinning process.

Zhang et al. had studied the melt spinnability of PEG alone, and PEG mixed with ethylene–vinyl acetate as the core component, and polypropylene as the sheath.[30] The results showed that the PEG could be spun alone only when the average MW was higher than 20 000. When the average MW was higher than 1000, it could be melt spun well for the PEG and ethylene–vinyl acetate 1:1 (wt./wt.) mixture. Sayler's invention showed that the preferred weight percentages are about 55% PCM, about 15–21% polyolefin, about 7–11% ethylene copolymer, about 7–15% silica particles and about 7.5% microwave-absorbing additives.[29]

The mixture of paraffin and polyethylene was melt spun into the fibre directly. The surface of the fibre was coated with epoxy resin in order to prevent leakage of the paraffin.

3.3.3.3 Fibre with integral microcapsules

Embedding the microcapsules directly within the fibre adds durability as the PCM is protected by a dual wall, the first being the wall of the microcapsule and the second being the surrounding fibre itself. Thus, the PCM is less likely to leak from the fibre during its liquid phase, thus enhancing its life and the repeatability of the thermal response.

Bryant and Colvin produced a fibre with integral microcapsules with PCMs or plastic crystals.[9] The fibre has enhanced thermal properties at predetermined temperatures. The microcapsules can range in size from about 1 to about 10 microns. In fabricating the fibre, the desired microencapsulated PCMs are added to the liquid polymer, polymer solution or base material and the fibre is then expended according to conventional methods such as dry or wet spinning of the polymer solutions and extrusion of the polymer melts. According to the report, the maximum content of PCM in the polyacrylic fibre is about 10%. The minimum filament denier is 2.2dtex.[31]

3.3.4 Heat-storage and thermo-regulated clothing

3.3.4.1 Water circulation clothing

This kind of clothing is usually used as protective clothing for extremely warm or cool environments. It uses ordinary water as a sensible heat-storage material, circulating warm water in winter and circulating cool water in summer through tubing that is incorporated into the body garment or vest. A battery-powdered pump circulates the water through the tubing in the suit load. It has been commercialized for several decades.

3.3.4.2 PCM pouches attached to clothing

These garments were designed by Mavleous and Desy,[1] Zhang,[2] Colvin[3] and Imanari and Yanatori.[25] They use PCMs instead of water as heat-storage materials. They are used in extreme high or low temperature environments for body cooling or warming.

3.3.4.3 Heat-storage and thermo-regulated textile clothing

As the manufacturing technology of heat-storage and thermo-regulated fibres and textiles became more advanced, new types of clothing came onto the market. The PEG-coated fabrics produced by the Mitsui Corporation were used as ski- and sportswear.[14]

Outlast Inc. and Frisby Inc. licensed the thermal regulation technology from Bryant of Triangle Research and Development Corporation in about 1991. Since then, the scientists at Outlast and Frisby have worked diligently to create thermal regulating materials for use in many different products. The microcapsules were either coated onto the surface of a fabric or manufactured directly into polyacrylic fibres. Other heat-storage and thermo-regulated textile products, such as blankets, sleeping bags, underwear, jackets, sports garments, socks, ski boots, helmets, etc., have come onto the market since 1997.[31,32] Output has gradually increased in the last 3 years.

3.4 Properties of heat-storage and thermo-regulated textiles and clothing

3.4.1 Thermal resistance

3.4.1.1 Traditional thermal resistance

The traditional thermal resistance is measured by the standard stationary methods (for example, those involving the use of a guarded hot plate apparatus). But in Pause's opinion, none of these methods is suitable for the determination of materials with PCM, because a long continual thermal stress could activate the phase change which would lead to measurements that deviate significantly from those that should be obtained.[26] Some experiments are still necessary in order to enable us to decide whether this is true.

The thermal resistance of coated fabrics with or without microcapsules was measured by Zuckerman et al.[40] The thermal resistance of coated fabrics with microcapsules is higher than that of untreated fabrics; the results are listed in Tables 3.5 and 3.6. That is mainly due to the action of binder that seals the cloth pore of the fabric.

3.4.1.2 Dynamic thermal resistance

There are phase change materials on the surface of the fibre or in the fibre, so the fabric or fibre is going to absorb heat in the temperature range 20–40 °C, and release it in the range 30–10 °C. Vigo and Frost studied the endotherm and exotherm behaviour of the fabrics and fibres by DSC.[3-5] The plots of DSC measurement are shown in Fig. 3.1 and 3.2.[14] But the DSC measuring results were not translated to the thermal resistance of the textile.

When Neutratherm Corporation studied a wear trial with 50 human subjects

Table 3.5 Thermal resistance of coated fabrics[40]

Materials tested	Acrylic with PCM	Acrylic without PCM	Substrate with PCM	Substrate without PCM
Wt/unit area (g/m²)	270	250	227	207
Stand thickness (mm)	5.40	5.63	0.63	0.61
Compressibility (lb)	12	16	13	20
Raw density (kg/m³)	49	44	360	339
Thermal conductivity (w/m K)	0.0398	0.0342	0.1012	0.1782
Thermal resistance (m² K/w)	0.1281	0.1491	0.0057	0.0029
Specific thermal capacity (kJ/kg K)	3.022	2.391	2.468	1.840

48   Smart fibres, fabrics and clothing

Table 3.6 The basic and dynamic thermal resistance of coated fabrics[26]

Test material	Basic thermal resistance (m^2 K/W)	Dynamic thermal resistance (m^2 K/W)
Membrane material	0.0044	
Membrane material/foam coating	0.0187	
Membrane m./foam/40 g/m^2 PCM	0.0181	0.0863
Membrane m./foam/90 g/m^2 PCM	0.0176	0.1481

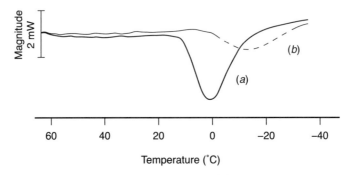

3.1 DSC cooling scans (10 °C/min) of melt-blown polypropylene treated with PEG 1000/DMDHEU and dried/cured: (a) 1.5 min/90 °C and (b) 5 min/100 °C.

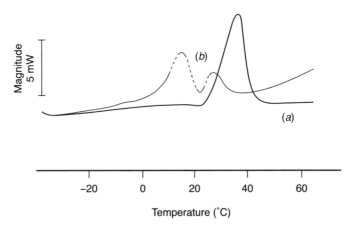

3.2 DSC heating scans (10 °C/min) of woven cotton pritcloth treated with PEG-1450/DMDHEU and dried/cured: (a) 2 min/100 °C and (b) 4 min/80 °C.

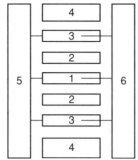

3.3 Testing arrangement. 1: Radiant (panel) heater with temperature sensor, 2: Testing sample, 3: Panel heater/cooler with temperature sensor, 4: Heat insulation, 5: Power supply for panel heater, 6: Computer for recording and processing of measured values.

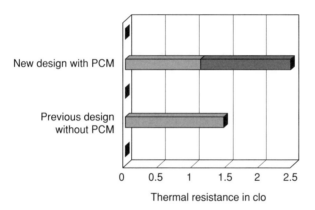

3.4 Comparison of the total thermal insulation of the coated fabrics. ■, Basic thermal resistance; ■, Dynamic thermal resistance.

with cotton thermal underwear containing bound PEG, enhanced wind resistance was the property giving the highest percentage of satisfaction (82%) to the wearers.[14]

In order to measure the basic and dynamic thermal resistance, Pause designed a measurement method;[26] the basic arrangement of the measuring apparatus is presented in Fig. 3.3. The thermal resistance of the coated fabrics with microcapsules is shown in Table 3.6. The total thermal insulation values obtained for the coated fabrics with microcapsules containing linear chain hydrocarbon compared to the thermal insulation of the materials without microcapsules are shown in Fig. 3.4.

Pause compared the thermal insulation effects of a batting (thickness 24 mm) of polyester fibre, outer fabric (0.2 mm) and liner fabric (0.1 mm) with a batting (thickness 12 mm) of acrylic fibre with incorporated PCM, outer fabric (0.2 mm) and liner fabric with PCM-coating.[34] The test results show that the basic thermal insulation of the textile substrate is reduced by approximately 30% when replacing the thick batting made of polyester fibres with a batting of acrylic fibres with incorporated PCM only 12 mm thick. However, the dynamic thermal insulation effect resulting from the heat emitted by the PCM inside the coated layer of the liner materials as well as inside the batting fibres more than doubled the thermal insulation effect of the new garment configuration. The total thermal insulation effect of the new garment exceeds the total thermal insulation effect of the previous garment by approximately 60%.

The dynamic thermal resistance of any one heat-storage and thermo-regulated fabric is not a constant. The dynamic thermal resistance of the fabric changes with time during the measurement process: it should be like a Gaussian distribution, just like the DSC plot. However, it is like a rectangle in Fig. 3.4. Pause's is the only report on dynamic thermal resistance. The main difficulty is in the design of new testing apparatus.

3.4.2 Thermo-regulating properties

There is no standard method for measuring the thermal regulating properties of heat-storage and thermo-regulated textiles and clothing. Much still needs to be done.

Watanabe et al. attached the plain fabric containing aliphatic polyester and the plain fabric containing PET on a metal plate.[12] The temperature of the plate was increased from room temperature to 40 °C, kept constant for a few minutes, and then reduced from 40 °C to 5 °C. The surface temperatures of the fabrics were recorded with an infra-red camera. The difference in the surface temperatures of these two plain fabrics can be calculated from the infra-red picture. The surface temperature of the fabric which is woven by fibre produced by poly(glutaric 1,6-hexanediol) as the core and PET as the sheath is 4.5 °C lower than that of the PET fabric at 40 °C, and 3.2 °C higher than that of the PET fabric at 5 °C.

Vigo and Bruno's preliminary experiments using infra-red thermograph indicated that fabrics containing the cross-linked polyols had a surface temperature difference as much as 15 °C lower than that of a corresponding untreated fabric exposed to a heat source.[14]

Neutratherm TM-treated cotton thermal underwear (based on Vigo's process) was evaluated during skiing and skiing-like conditions, and was found by wearers to be superior to untreated cotton thermal underwear by 75% or more with regard to preventing overheating and chilling due to wind and/or cold weather.[14]

Heat-storage and thermo-regulated textiles and clothing 51

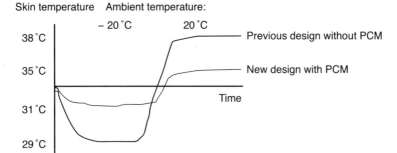

3.5 Comparison of the thermo-regulated effect of the previous and the PCM garment design.

A thermal active fibre was manufactured using PEG (average MW 1000) as the core and PP as the sheath.[30] Zhang et al. manufactured a non-woven using this fibre. The temperature of a drying chamber with fan was kept steady at 50 °C. The temperature of a refrigerator was kept steady at 0 °C. The thermal active non-woven and normal PP non-woven with the same area density were attached on a thin metal plate. The inner temperature of these two non-woven was measured by thermocouple thermometer during the temperature interval up from 0 °C to 50 °C and down from 50 °C to 0 °C. The maximum inner temperature difference of thermal active non-woven and normal PP non-woven is 3.3 °C in a period when the temperature is rising, and 6.1 °C in a period when the temperature is falling.

The garment containing PCM was tested in a climatic chamber at various temperatures to determine the thermo-regulating effect from heat absorption and heat release of the PCM.[34] In this test, the garment samples were attached to a simulated skin apparatus, the temperature of which was measured over time at various ambient temperatures and metabolic heat rates. Based on these tests, time intervals were estimated within which the skin temperature could be stabilized within a desired temperature range. The test results for the two garments under ambient temperature exposures of − 20 °C and + 20 °C are summarized in Fig. 3.5.

The results show that, at both ambient temperatures, the temperature of the simulated skin on the back side of the new garment is stabilized in the comfort temperature range between 31 °C and 35 °C.[34] In Pause's opinion, that means the heat absorption and release of the PCM has created a thermo-regulating effect resulting in a comfortable microclimate temperature during the entire test. On the other hand, in the test of the previous garment, the skin temperature is in the discomfort temperature zones as low as 29 °C and as high as 38 °C.

In other reports, cold weather trials showed a usual drop of 3 °C in the skin temperature for control and only 0.8 °C for the heat-storage and thermo-regulated garment.

3.4.3 Antibacterial properties

The PEG-coated fabrics gain not only absorbed and released heat, but also antibacterial properties.[17,22] The PEG-treated fabric can inhibit the growth of gram-positive *S. aureus* and gram-negative *E. coli* and *P. aeruginosa*.

The mechanism by which PEG-treated fabrics inhibit bacterial growth is being investigated by Vigo.[17] It results from three factors. A slow release of formaldehyde from the DMDHEU cross-linking resin may have an antibacterial effect, as formaldehyde can be used as a disinfecting agent. The PEG may exhibit a form of surfactant behaviour, which also is known to reduce bacterial growth. A third explanation relates to the finish imparting thermal absorption and release properties. The temperature may reach beyond some micro-organisms' growth range, killing those species.

A thermal active non-woven were produced by PEG-treated 100% polypropylene spun bonded-melt blown-spun bonded. The PEG-treated non-woven inhibited bacterial growth.[21] The most probable effects that inhibit microbial growth may be attributable to the surfactant-like properties of the bond PEG, which disrupts cell membranes due to the dual hydrophilic–hydrophobic characteristics of the PEG. This was reported in Vigo and Leonas's recent work.[42]

3.4.4 Other properties

The cross-linked PEG treatment changes a fabric's properties relative to untreated fabrics. Properties imparted include heat-absorbing and -releasing, antibacterial activity, resiliency/antiwrinkling, wear, toughness, absorbency and exsorbency of liquids, improved abrasion and linting resistance, decreased static propensity and increased oily soil release.[8,14,17]

3.5 Application

The possible applications of heat-storage and thermo-regulated textiles and clothing are under development. There are many possible end uses of these textiles.

3.5.1 Casual clothing

Heat-storage and thermo-regulated textiles can be used as casual textiles, for example, face fabric, liner fabric, batting, etc. Thermal underwear, jackets, sports garments and skiwear have now come onto the market.

3.5.2 Professional clothing

Examples are firefighter uniforms, bullet-proof vests, diver's coveralls, space suits, airman suits, field uniforms, sailor suits, etc. Special gloves for extremely high or low temperatures are being studied. Personnel supporting the operations of Air Force One now wear thermo-regulated flight gloves.

3.5.3 Textiles for interior decoration, bedclothes and sleeping bags

Heat-storage and thermo-regulated textiles can be used as curtains, quilts, quilt covers or batting. They are already used as sleeping bags.

3.5.4 Shoe lining

Heat-storage and thermo-regulated textiles can absorb, store, redistribute and release heat to prevent drastic changes in the wearer's head, body, hands and feet. In the case of ski boots, the PCM absorbs the heat when the feet generate excess heat, and send the stored heat back to the cold spots if the feet get chilly. This keeps the feet comfortable. Ski boots, footwear and golf shoes are coming onto the market.

3.5.5 Medical usage

PEG-treated fabric may be useful in medical and hygiene applications where both liquid transport and antibacterial properties are desirable, such as surgical gauze, nappies and incontinence products. Heat-storage and thermo-regulated textiles can keep the skin temperature within the comfort range, so they can be used as a bandage and for burn and heat/cool therapy.

3.5.6 Building materials

Such textiles can be used in architectural structures embedded in concrete, in roofs and in other building materials.

3.5.7 Other uses

They can be used as automotive interiors, and battery warmers. They can also be used in agriculture and geotextiles.

3.6 Development trends

The manufacture, properties and applications of heat-storage and thermo-regulated textiles have been extensively studied since the 1980s but their future development is still not very clear. It is expected that similar importance will be attached to improving comfort to the wearer as was seen in the development of breathable membranes (Gore-Tex, Sympatex).[31] These textiles were selected as one of the '21 inventions that will shape the 21st century' by the editorial staff of *Newsday*.

The traditional heat insulation textile, which works by trapping heat in the dead-air spaces of a material, is meant to retain heat as long and efficiently as possible. The heat-storage and thermo-regulated textile is not a traditional textile or an evolutionary textile with higher heat insulation ability. The heat-storage and thermo-regulated textile keeps you warm when it is cool and cool when it is warm. It is a product that will make you more comfortable during extreme temperature changes. Thus, it is an intelligent textile.[43] When the temperature changes dramatically, the inner temperature of the heat-storage and thermo-regulated garment is stabilized in the comfort temperature range between 31 °C and 35 °C.[32,34]

The thermo-regulated properties of the garment offered against extremes of cold and/or heat are available only for a limited period however, depending on the garment construction and the temperature of the environment. Thus it is more useful in an environment in which the temperature is quickly changing than in a steady temperature environment. However, if the released heat is not absorbed quickly, or the absorbed heat is not released, the textile loses its thermo-regulating function. To improve its durability, there is still much to be done.

3.6.1 Enhancing heat content

The maximum PCM load in the heat-storage and thermo-regulated polyacrylic fibre is 10%.[31] In other words, the maximum heat content of the fibre is 25 J/g. The thermo-regulating function offered against extremes of cold and/or heat is available only for a limited period. The maximum heat content of the fabric is over 40 J/g in patent application and heat content over 50 J/g is possible in technology. The more PCMs there are in the textile, the longer the duration of temperature equalization.

3.6.2 Combining heat-storage and thermo-regulated textiles with trapped-air materials

The experiment results show that a thermo-regulated textile that can regulate inner temperature effectively must have enough high thermal resistance. How to combine the heat-storage and release technology with the trapped-air technology is an area for future research.

3.6.3 Combining with other energy conversion technology

The group VI transition metal carbon compound can convert near infra-red rays into heat.[44] If the particles of ZrC are added to the heat-storage and thermo-regulated textile, they can absorb the near infra-red rays of the sun in daylight and convert them into heat that makes the PCMs in the textile change phase from solid to liquid. The textile releases its heat to keep its temperature relatively steady when the inner temperature is lower than the crystallizing point of the PCMs.[19]

Some of the polyacrylic acid or polyacryl–nitrile copolymers can absorb moisture from the air and release heat. They can be made into fabrics that can regulate inner moisture and temperature but such fabrics are not durable for limited moisture absorbency ability. If the moisture-absorbing fabrics are laminated with heat-storage and thermo-regulated fabrics, the composite fabrics can absorb moisture or heat and prevent drastic changes in the inner temperature. This kind of fabric is useful for ski- and sportswear.

References

1. Mavleous M G and Desy J J, US Pat 3 183 653.
2. Hansen R H, 'Temperature-adaptable fabrics', US Pat 3 607 591.
3. Vigo T L and Frost C M, 'Temperature-sensitive hollow fibers containing phase change salts', *Textile Res. J.*, 1982, **55**(10), 633–7.
4. Vigo T L and Frost C M, 'Temperature-adaptable hollow fibers containing polyethylene glycols', *J. Coated Fabrics*, 1983, **12**(4), 243–54.
5. Vigo T L and Frost C M, 'Temperature-adaptable fabrics', *Textile Res. J.*, 1985, **55**(12), 737–43.
6. 'Fibers with built-in antifreeze', *Chemical Week*, 1986, July 9, 20, 22.
7. 'Fabrics that respond to changes in temperature may emerge from a textile chemist's research', *Wall St. Journal*, 25 (June 13, 1986).
8. Bruno J S and Vigo T L, 'Temperature-adaptable fabrics with multifunctional properties', Book Pap. – Int. Conf. Exhib., AATCC, 1987, 258–64.
9. Bryant Y G and Colvin D P, 'Fiber with reversible enhanced thermal storage properties and fabrics made therefrom', US Pat 4 756 985.
10. Vigo T L and Bruno J S, 'Properties of knits containing crosslinked PEG', Book Pap. – Int. Conf. Exhib., AATCC, 1988, 177–83.

11 Vigo T L and Bruno J S, 'Improvement of various properties of fiber surfaces containing crosslinked polyethylene glycols', *J. Appl. Polym. Sci.*, 1989, **37**, 371–9.
12 Watanabe T, Matsumoto T and Toshikazu A, 'Heat-evolving heat-absorbing synthetic conjugate fibers for insulative garments', JK6-41818.
13 Zhang P R, 'Heat-Storage Garment', CN2107799U.
14 Vigo T L and Bruno J S, 'Fibers with multifunctional properties: a historic approach', *Handbook of Fiber Science and Technology*, Volume III, High Technology Fibers Part C, Menachem Lavin, Jack Preston, Marcel Dekker, Inc., 1993.
15 Mitamura H, 'Synthetic conjugate fibers with latent heat and their manufacture', JK6-200417.
16 Bryant Y G and Colvin D P, 'Fabric with reversible enhanced thermal properties', US Pat 5 366 801.
17 Vigo T L, 'Antimicrobial acting of cotton fabrics containing crosslinked polyols', *Proc. – Beltwide Cotton Conf.*, 1993, (3), 1424–5.
18 Hiroyoshi U, 'Heat-evolving and heat-absorbing fabrics', JK7-70943.
19 Chiaki M, Toshinori F and Wataru S, 'Heat-storing textile products', JK5-214672.
20 Colvin D P and Bryant Y G, 'Micro-climate cooling garment', US Pat 5 415 222.
21 Jinkins R S and Leonas K K, 'Influence of a polyethylene glycol treatment on surface, liquid barrier and antibacterial properties', *Text. Chem. Colorist*, 1994, **26**(12), 25–9.
22 Bruno J S and Vigo T L, 'Thermal properties of insolubilized polyacetals derived from non-formaldehyde crosslinking agents', *Thermochim. Acta*, 1994, **243**(2), 155–9.
23 Bruno J S and Vigo T L, 'Dyeing of cotton/polyester blends in the presence of crosslinked polyols', *Am. Dyest Rep.*, 1994, **83**(2), 34–7.
24 Tsujito Yatake, 'Manufacture of heat-storage materials for bedding and clothing', JK6-235592.
25 Imanari M and Yanatori Midio, 'Heat-storing materials', JK8-29081.
26 Pause B, 'Development of heat and cold insulating membrane structures with phase change material', *J. Coated Fabrics*, 1995, **25**(7), 59.
27 Zhang X X, Zhang H, Hu L et al., 'Crystallizable and heat-storage properties of PET–PEG block copolymers', *Mater. Rev.*, 1996, **10**(6), 63.
28 Zhang X X, Zhang H, Niu J J et al., 'Thermoregulated fiber and its products', China Pat ZL96 105 229 5.
29 Sayler I O, 'Phase change materials incorporated through the structure of polymer fibers', WO98/12366.
30 Zhang X X, Wang X C, Hu L et al., 'Spinning and properties of PP/PEG composite fibers for heat storaging and thermoregulated', *J. Tianjin Inst. Text. Sci. Technol.*, 1999, **18**(1), 1.
31 'This sporting life', *Textile Month*, 1998, (12), 12–13.
32 'Blowing Hot and Cold to Balance Yin and Yang', *Textile Month*, 1999, (12), 16–18.
33 Vigo T L, 'Intelligent fibers substrates with thermal and dimensional memories', *Polymer Adv. Technol.*, 1997, **8**(5), 281–8.
34 Pause B H, 'Development of new cold protective clothing with phase change material', Int Con *Safety and Protective Fabrics*, April 29–May 1, 104, 1998.
35 Hale D V, Hoover M J and O'Neill M J, 'Phase Change Materials Handbook Report', NASA CR61363, Lockheed Missiles and Space Co., Sept. 1971.
36 Zhang Y P, Hu H P, Kong X D et al., *Phase Change Heat Storage Theory and*

Application, Hefei University of Science & Technology of China Publishing House, 1996.
37 Son Change, Moreshouse H L and Jeffrey H, 'An experimental investigation of solid-state phase-change materials for solar thermal storage', *J. Sol. Energy Eng.*, 1991, **113**(4), 244–9.
38 Zhang X X, Zhang H, Wang X C et al., 'Crystallisation and low temperature heat-storage behaviour of PEG', *J Tianjin Inst. Textile Sci. Technol.*, 1997, **16**(2), 11.
39 Vigo T L and Frost C M, 'Temperature-adaptable textile fibers and method of preparing same', US Pat 4 871 615.
40 Zuckerman J L, Perry B T, Pushaw R J et al., 'Energy absorbing fabric coating and manufacturing method', WO 95/34609.
41 Zhang X X, Zhang H, Hu L et al., 'Synthesis and crystallizable properties of aliphatic polyester', *Synthetic Technol. Appl.*, 1997, **12**(3), 10–14.
42 Vigo T L and Leonas K K, 'Antimicrobial activing of fabrics containing crosslinked PEG', *Text. Chem. Colorist Amer. Dyestuff Reporter*, 1999, **1**(1), 42–6.
43 Vigo T L, 'Intelligent fiber substrates with thermal and dimensional memories', *Polym. Adv. Technol.*, 1997, **8**(5), 281–8.
44 Oji Shunsaku, Fujimoto, Masanori Furata, 'Thermally insulating synthetic fibers with selective solar absorpting', JK1-132816.

4
Thermally sensitive materials

PUSHPA BAJAJ

4.1 Introduction

An ideal clothing fabric, in terms of thermal comfort, should have the following attributes: (1) high thermal resistance for protection from the cold; (2) low water vapour resistance for efficient heat transfer under a mild thermal stress conditions; and (3) rapid liquid transport characteristics for transferring heat efficiently and eliminating unpleasant tactile sensations due to perspiration under high thermal stress conditions.

Extremely cold weather protective clothing refers to garments that are used in extreme climates such as on the Trans-Antartica expeditions, in skiwear, mountaineering and for army personnel posted in hilly areas at high altitudes. For this purpose, clothing that provides a high degree of insulation with the least amount of bulk is most desirable. For such applications, breathable membranes and coatings, and high-loft battings of special fibres in conjunction with reflective materials have been suggested.[1-3]

For the International Trans-Antartica Expedition in 1990, the outerwear was made with Gore-Tex fabric constructed of a highly porous membrane bonded to a high performance textile. This outerwear was found to be completely windproof and waterproof, yet 'breathable' to allow the body's perspiration vapour to escape. To encounter the harshest weather in the Antarctic region, where winds can reach 150 miles per hour during storms and temperatures fall to $-46\,°C$, a layered clothing system has been developed to enable the wearer to continually adjust his clothing to conserve and manage his body heat.The protection of an individual in a cold environment would depend on the following main factors: (1) metabolic heat, (2) wind chill, (3) thermal insulation, (4) air permeability and (5) moisture vapour transmission. Survival depends on the balance of heat losses due to (2)–(5) and heat output due to metabolic heat (1).

US Army Research Institute of Environmental Medicine (USARIEM) studies[4] have examined the effects of temperature and wind chill on a human body and indicated that the sensations of warmth and cold at wind velocities between 2 to 32 m/s may be as below:

Thermally sensitive materials

32 °C	Hot
28 °C	Pleasant
22 °C	Cool
16 °C	Very cool
10 °C	Cold
5 °C	Very cold
0 °C	Bitterly cold
− 5 °C	Exposed flesh freezes
− 24 °C	Exposed area of face freezes within one minute
− 32 °C	Exposed area of face freezes within ½ minute

Wind chill depends on the temperature as well as velocity of the wind. In cold climates, consideration of the wind chill effect is very important, because 80% of heat losses are due to wind chill.

Wind chill factor can be derived from the Siple Passel formula (source: Slater K, *J. Text. Inst.* 1986 **77**(3) 157–171):

$$H = (10.45 + \sqrt{V} - V)(33 - T_A) \qquad [4.1]$$

where

H = heat loss from the body in kcal/m^2/h,
V = wind velocity in m/s, and
T_A = air temperature in °C

The skin temperature under calm conditions is taken as 33 °C. The significance of the wind chill factor may be understood by reviewing its effect on the freezing time of exposed flesh. At a wind chill factor of 1000, an individual would feel very cold and at 1200, bitterly cold. When the wind chill factor is 1400, the exposed flesh would freeze in 20 min, at a wind chill factor of 1800, the exposed flesh would freeze in 10 min, and when the wind chill factor is 2400, exposed flesh freezes in 1 min only.

It is evident from Table 4.1 that textile fibres are much better conductors of heat than air (i.e. bad insulators). Therefore, the ideal insulator would be still air. However, there should be some means to entrap the air. Fibrous materials serve this purpose effectively as they have the capacity to entrap huge volumes of air because of their high bulk. For example, a suiting fabric is 25% fibre and 75% air; a blanket is 10% fibre and 90% air; a fur coat is 5% fibre and 95% air. In other words, a textile-based thermal insulator should be capable of entrapping maximum amount of air. The contribution of the fibre itself to thermal insulation comes second. The contribution of convection can be reduced by optimizing the interfibre space. The heat loss due to radiation is not significant. In clothing for extreme cold regions, it is further reduced by incorporating a thin layer of metal-coated film.

Table 4.1 Thermal conductivity of pads of fibres with a packing density of 0.5 m/cc

Fibre	Thermal conductivity (mW/m K)
Cotton	71
Wool	54
Silk	50
PVC	160
Cellulose acetate	230
Nylon	250
Polyester	140
Polyethylene	340
Polypropylene	120
Air (still)	25

Source: *Physical Properties of Textile Fibres*, Morton W E and Hearle J W S, Third edition, Textile Inst., Manchester, UK, 1993, p. 590.

In this chapter, a brief account of polymers, ceramics, phase change materials as melt additives, breathable coatings and design of fabric assembly for thermal insulation is given.

4.2 Thermal storage and thermal insulating fibres

4.2.1 Use of ceramics as melt dope additives

Zirconium, magnesium oxide or iron oxide can be used as particulate fillers in the molten polymer for producing heat generating polyester, polyamide, polyethylene, polypropylene and other functional fibres.

By applying the heat insulation principle, heat-regenerating fibres produced from ceramic composites utilize the far infra-red radiation effect of ceramics, which heat the substrate homogeneously by activating molecular motion. For example, zirconium, magnesium or iron oxide, when blended into manufactured fibres, radiate *ca.* 60 mW far infra-red of wavelengths 8–14 μm at a body temperature of 36 °C. Such fibres have been found suitable for sportswear and winter goods, particularly for those working in extremely cold regions.

There are two possible ways to insulate heat:

1 By using a passive insulating material which encloses the body heat
2 By use of an active material as an additive or coating that absorbs heat from outside.

The far infra-red radiation ceramics activated by body heat belong to the first category of insulating materials.

Thermally sensitive materials

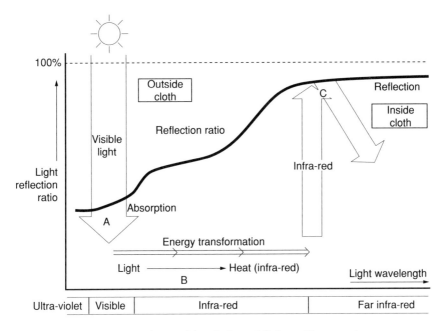

4.1 Heat absorption and insulation of Solar-α (Descente).

The active insulating materials which fall in the second category are those electrically heated materials where the electrical energy of a battery is transformed into heat energy by the heater, with this heat supplied by the heat from the oxidation of iron powder. In another system,[5] the conductive particulate is selected from SnO_2, $BaSO_4$, K titanate and TiO_2, and the far-IR particulate from one or more of ZrO_2, Al_2O_3, TiO_2, Kaolin and MgO for producing thermal storage and thermal insulating fibres.

Zirconium carbide compounds have been used for their excellent characteristics in absorbing and storing heat in a new type of solar system, including domestic water heaters and large-scale generators. Zirconium carbide reflects the light of long wavelengths over 2 μm, but absorbs the light energy of rather short wavelengths (< 2 μm), which make up 95% of sunlight, and converts it into stored heat energy. Descente researchers[6] used this property of zirconium carbide to achieve 'active insulation', by enclosing microparticles of zirconium carbide in polyamide or polyester fibres. They developed the technology to encapsulate zirconium carbide powder within the core of synthetic fibres, in cooperation with Unitika. The clothes made of this fibre, Solar-α (Fig. 4.1) absorb solar visible radiations efficiently and convert them into heat in the form of infra-red radiation which is released in the clothing. The released heat and the heat radiated from the body are reflected by Solar-α and will not escape from the inside to the outside of clothing.

Unitika[7] has also disclosed the manufacture of self-stretchable heat-retentive polyester yarns containing Group IV transition metal carbides. Thus, a blend of 95 parts PET and five parts Zr carbide was melt-spun at 3100 m/min and heat-treated at non contacting-type heater temperature 425 °C to give self-stretchable yarn (B) exhibiting stretch at 185 °C of 4.5% and conventionally spun PET fibres drawn at non contacting heater temperature of 285 °C to give yarn (A) having dry heat shrinkage of 4.5%. Interlacing yarns A and B and subsequent twisting produced woven fabric which exhibited surface temperature 21.1 °C initially, 24.5 °C on exposure of fabric to a source of 100 W for 10 min, and 23.2 °C on storing the fabric for 5 min after exposure to light for 10 min.

Mitsubishi Rayon Co.[8] has reported the development of heat-storage and electrically conductive acrylic fibres with electrical conductivity $> 10^{-6}$ S/cm and clothes for winter clothing and sportswear. These bicomponent fibres comprise a core–sheath structure, a core of P(AN/MA/Sod. methallyl sulphonate) containing 15–70 vol % white electrical conductive ceramic particles (e.g. W-P) and P (acrylonitrile/vinyl acetate) as a sheath.

Toyo Kogyo Co.[9] disclosed the production of insulative antistatic polyester fibres. A blend of PET and 20% pellets comprising 4:1 (wt. ratio) mixture of ceramic powder containing 70 parts Sb_2O_5 and 30 parts SnO_2 (particle size < 1 μm) and one part adipic acid-butylene glycol polymer was melt-spun and drawn. A unit of this fibre showed surface temperature of 53.5 °C after exposure to light (300 W) for 60 s and surface resistivity 3.0×10^9 Ω vs. 46.5 °C and 5.2×10^{12} Ω, respectively for fibres containing no ceramic powder.

Kuraray[10] claimed the formation of PET fibres with heat-storing/heat insulating properties. Such fibres contain heat-storing agents obtained by impregnating porous materials with organic compounds having melting point 10–50 °C, heat of melting ≥ 20 mJ/mg and crystallization temperature 10–45 °C. Thus, a composition containing PET and decanoic acid-impregnated SiO_2 support was spun to give fibres with heat evolution of 40 mJ/mg.

Insulative fibres with improved heat retention properties have been made by Unitika.[11] A melt-kneaded mixture of calcium laurate treated six parts α-alumina (A 50N) with average particle size of 1.0 μm; four parts of Zr carbide were blended with 90 parts of PET as the core, and PET as the sheath were together melt-spun to give fibres showing a surface temperature of 26.9 °C on exposure to a white light source (100 W) at 20 °C and 65% relative humidity and heat conductivity 29 W/cm °C.

In another patent, production of thermally insulative fabrics from fibres[12] comprising 98 parts nylon 6 and two parts Na stearate-treated α-type Al_2O_3 (AKP-30) particles with average particle diameter 0.4 μm has been disclosed. The mixture was melt spun at 250 °C, drawn, heat-treated at 165 °C, and made into a woven fabric with heat conductivity 2.4×10^{-4} W/cm °C.

Heat-ray-radiating fibres[13] with high warming and heat insulating properties have been produced. The fibres, useful for clothes, bed clothes, etc., contain n-type semiconductors showing volume resistivity 30–500 Ω cm. Thus, a non-woven fabric was manufactured from multifilament yarns comprising PET and 3% Al-doped electrical conductive ZnO powder.

Thermally insulative undergarments from blends of metal containing spandex and other fibres have been produced. A blend containing 15% spandex fibres having alumina, silica, titania and Pt at 10:82:3:5 wt. ratio, and 85% cotton showed heat conductivity 4.906×10^{-4} W/cm °C and heat retention 28.6%. A ladies' stocking of a blend[14] comprising 6.4% spandex fibres containing alumina, silica, titania and Pt at 10:82:3:5 wt. ratio, 24.1% acrylic fibres and 13.5% nylon fibres also showed heat conductivity 3.525×10^{-4} W/cm °C and heat retention 42.3%.

Kuraray[15] has also produced heat storing fibres which contain 0.5–40% mixtures of Group VIII transition metal oxide and TiO_2 at 1/(0.3–2.0 ratio). Thus, Septon KL 2002 (hydrogenated isoprene–styrene triblock rubber) containing 15% powder Fe_3O_4 and 15% powder TiO_2 as the core and PET as the sheath were spun at core/sheath = 1/3, stretched, and heat-set to give multifilaments (75 denier/24 filaments), which were plain-woven, scoured, dyed in black and irradiated with a reflector lamp for 40 s at inside temperature 20 °C (inside temperature is the temperature of air between the specimen and a heat insulator in a test apparatus). The fabric showed surface temperature of 54.8 °C and inside temperature 33.0 °C immediately after the irradiation, and 22.0 °C and 25.3 °C, respectively, 5 min after the irradiation.

Shimizu[16] has demonstrated the use of substances containing high amounts of H_2O of crystallization, e.g. powdered borax, $Al(OH)_3$, Zn borate, $Ca(OH)_2$ or powdered alum as fillers for producing self-extinguishing far-IR radiating heat insulative fibres. A blend comprising 100 parts master batch containing PET and 30% $Al(OH)_3$ and 500 parts PET was pelletized and spun to give self-extinguishing fibres containing 5% $Al(OH)_3$.

4.2.2 Hollow fibres

Use of hollow fibres in place of cylindrical fibres affects the physical properties of the product. A reduced heat transfer is of special interest under the changed physical properties for fibre-filled products such as pillows, mattresses, sleeping bags, etc. In such products, synthetic hollow fibres have been given preference for many years.[17]

Kawabata[18] also used an empirical approach to show that the thermal conductivity of single fibres along their axis is about ten times as great as heat conductivity in a direction transverse to the fibre axis. This thermal anisotropy and its implications on the effects of fibre orientation within the web become especially interesting when considering the hollow fibres now often used in insulation batts.

Lightweight hollow fibre thermally insulating fabrics for winter clothing have been produced by Unitika.[19] A composition containing 98 parts nylon 6 and two parts sodium stearate treated-α-Al_2O_3 particles (AKP-30) was melt-spun to form fibres with degree of hollowness 30%, drawn and made into woven fabric exhibiting heat transmission 1.6×10^{-4} W/cm °C.

Nylon fibres[20] having a hollow, square-shaped cross-section (A) with two perpendicular protrusions (B) at each end for insulative garments are produced as follows: A composition containing 96 parts nylon 6 and four parts zirconium carbide as the (A) section-forming component and nylon 6 only as the (B) portion-forming component were melt-spun together at 275 °C to give hollow non-circular fibres with heat retention temperature by a specified test as 26.1 °C.

Hollow insulative lightweight polyester fibres have also been produced by Toray Industries.[21] The fibres have a circular cross-section with a triangular hollow portion having a specified dimension. A polyester was melt-spun through a spinneret with each hole comprising three arc-shaped slits, cooled by blowing air onto the fibres, lubricated, and drawn to give fibres with degree of hollowness 20.2% and lustre 98.5%.

Hollow hygroscopic insulative PET fibres with good pilling resistance[22] have been produced. A composition comprising 100 parts ethylene glycol-sodium 5-sulfoisophthalate-terephthalic acid copolymer (I) and three parts polyethylene glycol as the core, and PET as the sheath were together melt-spun at 50:50 ratio, made into a woven fabric and subsequently treated with an alkali solution to dissolve I and give a lightweight insulative fabric with water absorption 146 mm and good pilling resistance.

Heat-insulative hollow PET fibres with pores extending from the surface of the fibres to the hollow portion of the fibres and with the hollow portion containing slowly releasable hygroscopic polymer gels have been disclosed by Teijin.[23] The fibres are useful for undergarments and gloves. Poly(ethylene terephthalate) was melt-spun through a spinneret and drawn to form hollow fibres with degree of hollowness 40%. They were made into a tricot, then treated with NaOH solution to cause weight loss of 20% to give fibres having ≥ 1 pore with width 0.2–2.0 µm and length 10–15 µm per 100 µm. In another system, the knit was immersed in an aqueous solution containing Na acrylate 20, acrylic acid 5, and polyethylene glycol dimethacrylate 10 parts in a container for 10 min at 0.2 torr, subsequently immersed in H_2O for 3 min at 100 °C, dried and heat-treated for 1 min at 120 °C to give a knit with weight increase 23.5% and exhibiting moisture absorption 15.0% and average body temperature retention rating (5 best, 1 worst) 4.5 by a specified test.

Porous insulative polypropylene fibres[24] have been produced by melt-spinning of polypropylene (I) with melt index (MI) 0.8, heat-treated at 150 °C for 24 h, and drawn to give fibres with porosity 80% and average pore diameter 2.2 µm.

Thermally insulating hollow side-by-side bicomponent polyester fibres have been produced.[25] PET with [η] 0.64 and PET with [η] 0.58 were together melt-spun through a ring-shaped spinneret at 1:1 throughput ratio, drawn, crimped, heat-treated at 120 °C in the relaxed state, and cut to give hollow fibres with the degree of hollowness 5–30% and latent crimp 18/25 mm on heat-treating the fibres at 180 °C.

The author[26] has also documented the developments in acrylic fibres, wherein hollow acrylic fibres find their use in making insulative garments and in preparation of hollow fibre membranes for water treatment in food, pharmaceutical and electronic industries and in nuclear power plants. These fibres are also useful as precursors to carbon fibres, because they may dissipate heat more effectively than the conventional acrylic precursors.

Hollow acrylic fibres can be produced by a wet or dry spinning method or as bicomponents having sheath/core structure wherein the core is disposed or dissolved to get the hollow structure in the form of an internal linear continuous channel. Toray Industries has disclosed a process to manufacture hollow acrylic fibres by spinning a liquid containing acrylonitrile–methacrylic acid copolymer as a sheath and polyvinyl alcohol as a core. The resultant hollow acrylic fibre was used to produce hollow carbon fibres for use in composites.

4.2.3 Insulating structures with phase change materials

Phase change materials (PCM) absorb energy during the heating process as phase change takes place, otherwise this energy can be transferred to the environment in the phase change range during a reverse cooling process. The insulation effect reached by the PCM is dependent on temperature and time. PCMs as such are not new; they exist in various forms in nature. The most common example of a PCM is that of water at 0 °C, which crystallizes as it changes from a liquid to a solid (ice). A phase change also occurs when water is heated to a temperature of 100 °C, at which point it becomes steam. More than 500 different phase change materials have been documented by NASA.[27]

However, use of the special thermophysical properties of PCMs to improve the thermal insulation of textile materials only became possible by entrapping them in microcapsules, each with a diameter of 1 μm to 6 μm.

Pause[28] has developed heat and cold insulating membranes for using on a roof structure. For this purpose, PCMs with a crystallization temperature between − 25 °C and 20 °C (Table 4.2) were selected to minimize the heat emission into the outside environment. Further, to reduce heat absorption from the outside environment, PCMs with melting points of 25 °C to 40 °C were chosen.

Table 4.2 Characteristics of phase change materials (PCM). Temperature range from −25 °C to 20 °C

PCM	Crystallization point (°C)
Dodecane	−15.5
Tridecane	−8.8
Tetradecane	−0.2
Pentadecane	4.8
Hexadecane	−12.2
Heptadecane	−16.5
PCMs for temperature range 25 °C to 40 °C	
	Melting point (°C)
Octadecane	28.2
Nonadecane	32.1
Eicosane	36.8
Heneicosane	40.5

Source: Ref. 28.

Vigo and Frost[29] pioneered the use of phase change materials for developing temperature-adaptable fabrics. Polyester, cotton, nylon 66 and wool fabrics were treated with an aqueous solution of polyethylene glycol (mol. wt. 600 or 3350) or one of the crystal materials (2,2-dimethyl-1,3-propanediol or 2-hydroxymethyl-2-methyl-1,3-propanediol). Such fabrics exhibited up to 250% greater thermal storage and release properties than the untreated ones. The energy storage capacity of hollow polypropylene fibres and viscose rayon filled with carbowax during thermal cycling between 230 and 330 K has also been demonstrated.

Vigo and Bruno[30] developed *NeutraTherm* fabrics, the first ever 'phase change clothing', by coating the fabric with PEG 1000, a polyethylene glycol waxy solid. The phase change of the fabrics lasts only about 20 min; after that, the garments revert to their original insulating properties. The PEG 1000 structure changes from soft to solid in keeping with the temperature, absorbing heat when it softens, and releasing heat when it solidifies. It is now reported that the lack of durability of such topical treatments to laundering and leaching, which remove the water-soluble polyethylene glycols, can be overcome by insolubilizing low-molecular-weight polyethylene glycols (average molecular weights 600–1000) on fabrics by their reaction with dimethylol dihydroxyethylene urea under conventional pad-dry-cure conditions to produce textiles that are thermally adaptable even after laundering.

Acrylic fibres with increased latent heat retention have been prepared[31] by treating wet-spun acrylic fibres in the swollen gelled state with substances having latent heat and having properties for transition from the solid state to the liquid state, drying the fibres and heat-treating the fibre. Acrylonitrile-Me acrylate-sodium-2-(acrylamido)-2-methylpropanesulfonate copolymer was wet-spun, treated with an aqueous solution containing 50 g/l polyethylene glycol (I) to get a weight add-on of 6.2% dried, and heat-treated at 125 °C to give fibres with heat retention 33% as per the JIS L-1096 test.

Gateway Technologies[32] has reported the application of heneicosane, eicosane, nonadecane, octadecane, and heptadecane as phase change materials for insulative fabrics. This material can be 425% as warm as a high bulk wadding. Schoeller Textil AG[33] Switzerland has also explored the application of micro-encapsulated phase change materials like waxes, which offer 'interactive' insulation for skiwear, snowboarding and ski boots. They store surplus heat and the wax liquefies; when temperatures drop, the microcapsules release the stored heat.

There are some hydrogels than can modulate the swelling ratio in response to environmental stimuli such as temperature, pH, chemical, photo-irradiation, electric field, etc. The collapse of a gel in response to environmental changes predicted by Dusek and Patterson,[34] was also investigated extensively by Tanaka et al.[35] Thermosensitive hydrogel collapses at elevated temperature through the lower critical solution temperature (LCST). The volume change occurs within a quite narrow temperature range. The permeability of water through the gel can be changed by an 'on–off' switch according to the environmental temperature. Therefore, such materials can be used as coatings for temperature adaptable fabrics, drug delivery systems and enzyme activity control.

Lee and Hsu[36] reported the swelling behaviour of thermosensitive hydrogel based on *N*-isopropyl acrylamide–trimethyl methacryloyloxy ethyl ammonium iodide (TMMAI) copolymers, a nearly continuous volume transition and associated phase transition from low temperature, and a highly swollen gel network to high temperature, a collapsed phase near its critical point between 31 and 35 °C. The gel with little TMMAI content (Table 4.3) showed a good reversibility and, the higher the TMMAI content (2–6%) in the copolymer, the higher the gel transition temperature.

$$CH_2=C(CH_3)-C(=O)-O-(CH_2)_2-N^+(CH_3)_3 \, I^-$$

TMMAI

Zhang and Peppas[37] have also studied the interpenetrating polymeric network (IPN) composed of the temperature-sensitive poly(*N*-isopropyl

Table 4.3 Characterization of NIPAAm/TMMAI copolymer gels

Sample	Feed composition (%)		Cloud point effect[a]	Cloud point temp. (°C)	Swelling ratio (g H_2O/g dry gel)
	NIPAAm	TMMAI			
TMM0	100	0	st	30–35	14.1
TMM1	99	1	st	30–35	19.3
TMM2	98	2	st	40–50	24.1
TMM3	97	3	vw	40–50	31.9
TMM5	95	5	vw	50–60	44.5
TMM6	94	6	vw	50–60	53.2

[a]st, strong; vw, very weak.
Source: Ref. 36.

acrylamide) (PNIPAAm) and the pH-sensitive polymethacrylic acid (PMAA). The results showed that these hydrogels exhibited a combined pH and temperature sensitivity at a temperature range of 31–32 °C, a pH value of around 5.5, the pk_a of PMAA. The LCSTs of the polymers were significantly influenced by the polymer concentration. The addition of saccharides into aqueous polymer solutions decreased the LCST of all polymers. As the polymer concentration increased, the saccharide effects became more pronounced. A monosaccharide, glucose, was more effective than a disaccharide, maltose, in lowering the LCST, especially in Pluronic solutions.[38]

The phase transitions of N-isopropylacrylamide and the light-sensitive chromophore-like trisodium salt of copper chlorophyllin copolymers induced by visible light has also been reported by Suzuki and Tanaka.[39] The transition mechanism in such gels is due only to the direct heating of the network polymers by light. As this process is extremely fast, such systems might be used as photo-responsive artificial muscles, switches and memory devices. All these phase change materials have great potential as coating materials in making skiwear, clothing for cold regions and high altitudes, etc.

4.3 Thermal insulation through polymeric coatings

Heat transfer can take place by means of conduction, convection or radiation across a barrier from the hot side to the cooler side. In conduction, heat passes from a hotter to a colder region along the static material. Convection is the transfer of heat by a flow of gases or liquids of different temperatures. In radiation, thermal energy is transmitted as electromagnetic waves.

PVC and PU are some of the widely used thermal insulators. They are invariably used in their foamed, expanded forms. The reason for this is the contribution of entrapped air or other gases to the total thermal insulation.[40]

Thermally sensitive materials

Table 4.4 Thermal conductivity of different substances

Materials	Thermal conductivity (mWm^{-1}K^{-1} at 20 °C)
Aluminium	200 000
Glass	1000
PVC tiles	700
Water	600
Brick	200
Fibres	200
Carpet	50
Clothing fabrics	40
Air (still)	25

Source: Ref. 40.

The total heat conduction λ_F of a polymeric material is the sum of components λ_G, λ_R, λ_S and λ_C, where G = gas in the cell, R = radiation across the cell of the foam, S = conduction through the wall and C = convection of the gas.

The highest resistance to heat flow comes from the air entrapped in the cell. Still air has one of the lowest thermal insulations, 25 mW/m K (Table 4.4). Therefore, polymeric materials in the form of foam have more entrapped air and show improved thermal insulation, the λ_S factor, which depends on the type of polymer itself and contributes approximately 30% of the total resistance to heat flow. Therefore, the type of polymer used in thermal insulators is the next major factor to be considered while designing a thermal insulator. The other two contributions, i.e. λ_R due to the radiation across the cell of the foam and λ_C due to the convection of the gas, is kept to optimum levels by proper design of the cell structure of the foam coating.

Requirements for extreme cold weather are:

- breathable membranes and coatings
- down and feather mixtures
- high-loft batting of special fibres
- reflective materials
- combinations.

4.3.1 Why breathable fabrics?[41–44]

To avoid the condensation of perspiration in a garment, breathable fabrics are required. The garment can also become damp in rain or snowfall or in sportswear. The moisture diffuses through the clothing and can reach the human skin. As the moisture evaporates, the wearer feels cold and uncomfortable.

Table 4.5 Water vapour permeability data of polymeric coatings/films

Film	Cup method 20 °C, 95% RH (g/m² 24 h)
Normal compact PUR	286
Microporous fluorocarbon	2140
Hydrophilic polyester	1290
Coated microporous PUR	1500–2000
Coated hydrophilic PUR	1000–1700
Grabotter film directly coated to the textile	1700–2000

Source: Gottwald L, 'Water vapor permeable PUR membranes for weatherproof laminates', *J. Coated Fabrics*, Jan 1996 **25**, 168–175.

A breathable fabric is waterproof and breathable because of the enormous difference between the size of a water droplet and a water vapour molecule. The former is 100 μm, in diameter whereas the latter is 0.0004 μm, i.e. there is a factor of around 250 000 between the two sizes.[41]

Waterproof/moisture permeable fabrics from several synthetic fibres have been designed for skiwear, track suits, rainwear, and mountaineering clothing. The application of a one-piece overall made from a woven nylon fabric (30-den flat nylon warp yarns; weft of 900-den air jet textured yarns) in the 1986 dogsled race in Alaska has been described. A nylon glove with a Gore-Tex liner was also designed for this purpose (Table 4.5).

Breathable fabrics can be classified into three main categories:

- coated fabrics
- laminated
- high-density woven fabrics.

4.3.2 Waterproof breathable coatings

The microporous barrier layer 'breathes' primarily through a permanent air permeable pore structure. Diverse techniques have been used to manufacture microporous coatings and films. The most important methods are listed below.

4.3.2.1 Mechanical fibrillation

For certain polymers, biaxial stretching produces microscopic tears throughout the membrane, which imparts a suitable microporous structure. For example, PTFE membranes are used in the Gore-Tex (two-layer and three-layer laminates). The thin microporous membrane is made from a solid PTFE sheet by a novel drawing and annealing process. In drawn form, the tensile strength is increased threefold. The manufacturers claim that these PTFE membranes contain approximately nine billion pores per square inch, with a pore volume of up to 80% and a maximum pore size of 0.2 µm.

4.3.2.2 Solvent exchange

In this process, a polymer dissolved in a water miscible solvent is coated thinly onto the fabric. The porous structure is developed by passing through a coagulation bath where water displaces the solvent. For example, the textile substrate is coated with polyurethane solution, where DMF is used exclusively as the solvent, and the fabric is passed through a coagulation bath containing water, where water displaces the solvent to give a porous structure.

4.3.2.3 Phase separation

The coating polymer is applied from a mixture of a relatively volatile solvent with a proportion of higher boiling non-solvent. Precipitation of the polymer as a microporous layer occurs as the true solvent evaporates faster during the subsequent drying process. For example, the polyurethane-based Ucecoat 2000 system operates on this principle. Here, a lower boiling solvent (methyl ethyl ketone) evaporates preferentialy as the fabric passes through the oven, thereby increasing the concentration of the non-solvent in the coating. When the concentration of the non-solvent reaches a critical level, the polyurethane precipitates out in a highly porous form and the remainder of the solvent and the non-solvent evaporate from the coating as the fabric passes through the oven.

The fabrics useful for apparel for keeping the body warm in cold weather or for sporting have porous polyurethane-based films. Nylon fabric penetrated with Asahiguard 710 was coated with polyurethane, hardener, DMF and hydrophobic silica (0.016 µm size) and washed with water. It was then printed with polyurethane and sodium stearate-treated α-alumina, dried at 100 °C for 2 min, and later impregnated with F-type water repellent, dried and set at 170 °C for 30 s. This treated fabric showed moisture permeability 7630 g m^2 24 h and thermal conductivity 4.1 W cm °C $\times$ 10^{-4}.

4.3.2.4 Solvent extraction

Finely divided water soluble salts or other compounds can be formulated into the polymer and subsequently extracted from the dried film or coating with water to get a microporous structure.

4.3.2.5 Electron bombardment

A process has been developed for rendering the solid coated fabrics microporous by bombarding the polymer coating with electron beams. The technique

Table 4.6 Water vapour resistance of clothing fabrics for comparison

Fabrics	Water vapour resistance (mm still air)
PVC coated	1000–2000
Waxed cotton	1000+
Leather	7–8
Typical non-wovens	1–3
Woven microfibre (nylon or polyester)	3–5
Closely woven cotton	2–4
Ventile L28	3–5
Two-layer PTFE laminates	2–3
Three-layer laminates (PTFE, PE)	3–6
Microporous PU (various types)	3–14
Open pores	3–5
Sealed pore	6–14
Hydrophilic PU coated	4–16
Witcoflex staycool in nylon, polyester	9–16
On cotton, poly/cotton	5–10

Source: Handbook of technical textiles, Ed Horrocks A R and Anand S C, Woodhead Publishing, UK, 2000, 304.

involves feeding the coated fabric between two electrodes, generating high voltage electrons which can be focused into discreet beams which drill through the coating without damaging the fabric beneath.

4.3.2.6 UV-EB curing

The sunbeam process is used to manufacture microporous films and coatings in-situ by cross-linking suitable monomers with electron beam or UV light. Radiation curing has the following advantages over the conventional methods:

- low energy consumption
- low environmental pollution
- fast curing and thus fast processing
- short start-up times.

4.3.2.7 Crushed foam coating

Mechanically foamed and thickened lattices are coated onto fabric and dried. Large surface pores are formed, which are compacted by calendering through a pressure nip which yields a microporous fabric. The water-vapour permeability values of various polymeric coatings are given in Table 4.6.

Woods has studied the relation between clothing thickness and cooling

Thermally sensitive materials

Table 4.7 Some commercially available insulation materials

Products	Manufacturer	Characteristics	End use
Drylete®	Hind	Combines Hydrofil nylon and hydrophobic polyester to push moisture away from the body and then pull it through fabric for quick drying	Skiwear, running and cycling apparel
Gore-Tex®	W. L. Gore	The original waterproof/breathable laminated fabric with pores large enough for water to escape, small enough to block rain	Outerwear, hat, gloves, footwear, running and cycling apparel
Hydrofil®	Allied	A new super-absorbent hydrophilic nylon that sucks moisture away from the skin	Linings, long underwear, cycling apparel
Silmond®	Teijin	A durable water repellent and windproof fabric made from polyester microfibres	Outerwear
Synera®	Amoco	A strong, lightweight polypropylene fabric that transports moisture away from the skin	Long underwear, jacket lining
Supplex® Tactel	DuPont	Extremely soft, quick drying nylon in smooth or textured version apparel	Outerwear, skiwear, running and cycling apparel
Thermolite®	DuPont	An insulation made of polyester fibres that have been coated	Skiwear, climbing apparel
Thermoloft®	DuPont	A semithick insulation that contains hollow polyester fibres that trap warm air	Skiwear, outdoor apparel
Thinsulate®	3M	Thin insulation made of polyester and polypropylene fibres	Skiwear, gloves, footwear
Thintech®	3M	A waterproof/breathable laminated fabric that is highly resistant to contamination by detergents or perspiration	Skiwear, outerwear

Source: Ref. 40.

during motor-cycling in the range from − 1 to 24 °C. To maintain a normal body temperature in winter on a motor-cycle without a protective shield, much thicker clothing (mean clothing thickness of 20 mm) inside windproof oversuit is needed at 5 °C, but it is only effective if perspiration does not accumulate inside the clothing.

4.3.3 Waterproof breathable membranes

4.3.3.1 Composition

Microporous film fully impregnated with a hydrophilic polymer, Thintech, is sandwiched between the outer and inner fabrics of a garment, followed by lamination of Thintech to their Thinsulate Thermal Insulation.

4.3.3.2 Application

It prevents rain, sleet or snow from penetrating a garment and stops the build-up of moisture within the garment.

4.3.4 Antistatic and heat energy-absorbing coatings

The coating comprises 5–90% ultrafine Sb–Sn oxide particles (from < 0.1 μm Sb_2O_3) and 10–90% binder resin.

Polyester cloth treated with aqueous acrylic emulsion (acrylic melamine resin) containing 20% of 0.005 μm ATO particles and baked to form a cloth showing the half-life of an electric charge of one sec. The front and back temperatures are 27 °C and 26 °C, respectively, after irradiating with 300 W lamp at 20 °C and 60% RH.[45]

Heat-insulating, moisture-permeable, and water-resistant nylon[46] has been produced by coating it with polyurethane containing 2–4% of inorganic fine powder (Sb oxide-doped SnO_2) and Zr carbide having photoabsorption thermal conversion capability. The coated nylon fabric showed water resistance pressure $(x) \geq 7000 \, g/m^2$ 24 h and moisture permeability $(y) \geq 0.6 \, kg/cm^2$.

Bonding of IR-ray radiating ceramics has also been attempted for producing thermal insulative textiles. Insulating fabrics were prepared by immersing fabrics in emulsions containing IR ray-radiating ceramics and polymers and adding trivalent metal salts to the emulsion to cause bonding of the ceramic to the fabrics.[47] Thus, a fabric was immersed in a dispersion containing synthetic rubber latex (solids 15%) and 10% (on fabric) IR ray-radiating ceramic, and stirred as 10% $Al_2(SO_4)_3$ was added dropwise, washed and dried to give an odour-absorbing insulating fabric with ceramic content of 10%.

The thermal insulation capabilities of cold protective clothing materials may be significantly improved by the application of MicroPCM, that is capsules containing small amounts of PCM (phase change materials). The PCM have the ability to changing their state within a certain temperature range, storing energy during the phase change and then releasing it at crystallization temperature. It is shown that, by applying a coating of 50 g/m^2 onto a fibrous batting of 180 g/m^2, the insulation capacity is comparable to a batting of 360 g/m^2. Micro PCM may thus be used to increase the insulation capabilities against the cold, with a resulting material which will be significantly thinner, allowing much larger freedom of movement.[28]

4.4 Designing of fabric assemblies[48-55]

In colder environments, layered fabrics rather than a single fabric are used in most cases (Table 4.7). Under such conditions, the two most important characteristics of fabrics are water vapour and heat transport. In order to gain insight into the transient states of such transport across a set of layers of fabrics, Yasuda et al.[49] developed an MU water vapour transport simulator (MU–WVT simulator). They reported that water transport was not influenced significantly by surface characteristics – the hydrophilic or hydrophobic nature of fabrics. On the other hand, when liquid water contacted a fabric, such as in the case of sweating, the surface wettability of the fabric played a dominant role in determining the water vapour transport rate through the layered fabrics. In such a case, the wicking characteristic, which determines how quickly and how widely liquid water spreads out laterally on the surface of/or within the matrix of the fabric, determines the overall water vapour transport rate through the layered fabrics.

The overall water vapour and heat transport characteristics of a fabric should depend on other factors such as the water vapour absorbability of the fibres, the porosity, density and thickness of the fabric, etc. In this study, the major emphasis was on the influence of the chemical nature of the fibres that constitute a woven fabric. Dynamic water vapour and heat transport in the transient state were investigated for fabrics made of polyester, acrylic, cotton and wool fibres (Table 4.8). The overall dissipation rate of water vapour depends on both the vapour transport rate and the vapour absorption by fibres, which are mutually interrelated. Water vapour transport is governed by the vapour pressure gradient that develops across a fabric layer (Fig. 4.2). When a fabric is subjected to given environmental conditions, the actual water vapour transport rate differs greatly depending on the nature of the fibres, even

Table 4.8 Characteristics of fabrics

Property	Polyester	Acrylic	Cotton	Wool
Fabric density[a] (cm^{-1})				
Ends	36.4	37.0	38.1	34.9
Picks	22.5	22.1	23.8	24.4
Thickness[a] (mm × 10)	2.7	2.9	3.1	3.2
Weight[a] (g/m^2)	121.7	116.4	125.0	113.9
Porosity[a] (%)	67.5	66.4	74.8	73.4
Air permeability[a] (cm^3/cm^2/s)	77.8	81.0	28.2	54.9
Advancing contact angle of water	116	b	c	116
Water vapour absorption rate (g/g/min × 10^4)	3.3	6.1	10.1	15.0

[a]According to JIS (Japanese Industrial Standard) L 1096. [b]Water droplet was absorbed by the fabric gradually. [c]Water droplet was absorbed by the fabric immediately.
Source: Ref. 49.

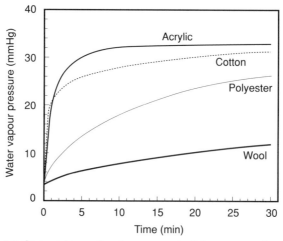

4.2 Comparison of water vapour build-up curves in chamber 1 in the sweat (liquid water) case. *Source*: ref. 49.

when other parameters are nearly identical, such as density, porosity and thickness.

Hong et al.[50] have studied in-depth dynamic moisture vapour transfer through clothing assemblies. Cotton, polyester and a blend in plain woven, pure finish fabrics were studied to determine their influence on dynamic surface wetness and moisture transfer through textiles. A simulated sweating skin was used, over which were placed test fabrics incorporating a clothing hygrometer to continuously measure dynamic surface wetness. Moisture vapour concentration and its rate of change at both inner and outer fabric surfaces were determined. At the inner fabric surface facing the sweating skin, all the cotton fabric exhibited the slowest build-up of moisture vapour concentration, followed by cotton/polyester blend. The all-polyester fabric showed the highest rate of change in moisture vapour concentration. The newly modified clothing hygrometer provides a sensitive method for ascertaining dynamic surface wetness on both fabrics and clothing as worn.

Dynamic moisture vapour transfer through textiles has also been investigated by Kim.[51] The influence of semipermeable functional films on the surface vapour pressure and temperature of fabric–film–fabric assemblies has been reported. The film added between two fabric layers affects vapour pressure and temperature changes on both surfaces of the assembly under simulated body-clothing conditions. The physical characteristics of films greatly influence the level of transient, comfort-related variables such as changes of inner surface vapour pressure and temperature. Mixed cotton and polyester layers of a film reveal that the film can negatively affect the moisture sorption of the assembly, not only by blocking the air spaces for moisture diffusion but, more importantly, by ineffective heat dissipation from the cotton inner fabric for continued moisture sorption. Using film and fabric assemblies, this study reinforces the interdependence of moisture sorption and temperature barrier effects during dynamic moisture transfer through the clothing assemblies. The percentage increases of moisture uptake of C/P (inner/outer) and P/C (inner/outer) without any film (none) are the same throughout the experiment (0.7, 1.6, 2.2 and 2.7%). The presence of films, however, produces different moisture uptake behaviours in the C/P and P/C (Fig. 4.3).

The thermo-insulating properties of lofty non-woven fabrics, which have been reported in several papers, depend on the nature of pore size and distribution in a fabric, which are a function of fibre fineness and material density. Jirsak et al.[52] compared in a more comprehensive way the thermo-insulating properties of both perpendicular-laid and conventional cross-laid nonwoven fabrics. The total thermal resistance R of a textile fabric as a function of the actual thickness of the material:

$$\lambda_{total} = \lambda_{sa} + \lambda_r \qquad [4.2]$$

$$R[m^2 K W^{-1}] = L/\lambda_{total} \qquad [4.3]$$

where λ_{total} is the thermal conductivity (Wm^{-1}K^{-1}) as defined in Eq. (4.2), λ_{sa} is the combined thermal conductivity of solid fibre and air and L is the thickness of the sample (m). The thickness of the material depends on the compression P(Pa):

$$L = L(P) \qquad [4.4]$$

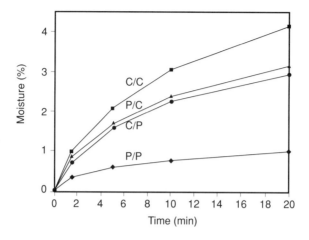

4.3 Dynamic moisture uptake, as a function of time, measured on fabric–film–fabric assemblies. *Source*: ref. 51. C/C = cotton/cotton double-knit assembly, P/P = polyester/polyester assembly.

Thus, materials showing the same λ_{total} value, those with higher compressional resistance, will exhibit better thermal resistance under the same compression. The perpendicular-laid fabrics, due to their unique fibre orientation, demonstrate a higher degree of compressional resistance in the thickness direction than the cross-laid ones.

Farnworth[53] also presented a theoretical model of the combined conductive and radiative heat flows through fibrous insulating materials and compared them with experimental values of the thermal resistances of several synthetic fibre battings and of a down and feather mixture (Table 4.9). No evidence of convective heat transfer is found, even in very low-density battings. The differences in resistance per unit thickness among the various materials may be attributed to their different absorption constants.

Crews et al.[54] evaluated the use of milkweed floss as an insulative fill material and compared its performance to other insulators. Seven identical jackets were constructed using different fill materials matched on a per unit weight basis. The insulation (clo) values for the jackets were measured using a standing, heated manikin as an environmental chamber. Thickness (loft), compression, resiliency and hand were also measured. The results show (Table 4.10) that milkweed floss blended with down has insulative properties similar to down. Down is superior to milkweed floss in loftiness and compressibility, which influence product performance, but the properties of milkweed floss can be enhanced by blending it with down.

Table 4.9 Physical and thermal properties of all samples studied at minimal compression (0.16 kPa)

Sample	Fibre material	Fibre radius (μm)	Thickness (mm)	Mass (kg/m^2)	Fibre volume (%)	Thermal resistance m^{2}K/W
Polarguard	Polyester	11.7	26	0.270	0.76	0.48
Hollofil	Polyester	13.2	28	0.350	0.94	0.59
Thinsulate M400	Polypropylene	1 to 3	7.7	0.405	5.8	0.25
Thinsulate C150	Polypropylene and polyester	1 to 3 20	5.2	0.175	~3.0	0.16
Down and feathers	40% and 60%	—	42	0.545	~1.3	1.22

Source: Ref. 53.

Table 4.10 Insulative value for filling material type before and after cleaning

Fill type	Insulation value I_T (clo)	
	Before cleaning	After cleaning
50% Milkweed floss/50% down	2.36	2.26
100% Down	2.28	2.25
100% Quallofil	2.26	2.19
100% Milkweed	2.24	2.16
50% Milkweed/50% feathers	2.24	2.06
100% Thinsulate	2.15	2.04
60% Milkweed/40% olefin	2.04	1.97

Source: Ref. 54.

References

1. Bajaj P and Sengupta A K, 'Protective clothing', *Textile Progr.*, 1992, **22**(2/3/4), 1–110.
2. Ukponmwan J O, 'The thermal insulation properties of fabrics', *Textile Progr.*, 1993, **24**(4), 1–54.
3. Mathur G N, Hansraj and Kasturia N, 'Protective clothing for extreme cold region', *Indian J. Fibre Text. Res.*, 1997, **22**, 292–6.
4. Godman R F, 'Tolerance limits for military operations in hot and/or cold environments', paper presented in 12th Commonwealth Defence conference on *Operational Clothing and Combat Equipment*, Ghana, 1978.
5. Jian R, 'Thermal storage and thermal insulating fibres and their manufacture and applications', CN 1,229, 153, 22 Sep 1999, *Chem. Abstr.*, 2000, **133**, 31770.

6 Hongu T and Phillips Glen O, *New Fibres*, 2nd edition, Woodhead Publishing Ltd., UK, 1997, 63–5.
7 Unitika Ltd, Japan, 'Self-stretchable heat-retentive polyester yarns containing gr. IV transition metal carbides and mixture yarns containing them and wool like fabrics therefrom: JP 11,21,720,26 Jan 1999, *Chem. Abstr.*, 1999, **130**, 140435y.
8 Mitsubishi Rayon Co. Ltd. Japan, 'Heat-storageable and electrically conductive fibres and threads and clothes' JP 2000 96,346, 4 April 2000, *Chem. Abstr.*, 2000, **132**, 238309b.
9 Toyo Kogyo Co. Ltd., Osaka 'Insulative antistatic fibres and products thereof', JP 03,193,970,23 Aug. 1991, *Chem. Abstr.*, 1991, **115**, 258198s.
10 Kuraray Co. Japan, 'Heat-storing, heat-insulating fibres with lasting thermal properties', JP 08,246,227, 24 Sep 1996, *Chem. Abstr.*, 1996, **125**, 331455m.
11 Unitika Ltd., Japan, 'Insulative fabrics with improved heat retention properties', JP 08,127,961, 21 May 1996, *Chem. Abstr.*, 1996, **125**, 117354j.
12 Unitika Ltd. 'Thermally insulative fabrics for winter clothing and sportswear' JP 07,300,741, 14 Nov 1995, *Chem. Abstr.*, 1996, **124**, 120007h.
13 Kuraray Co. Ltd. Japan, 'Heat-ray radiating fibres with high warming and heat insulating properties', JP 2000 154,419, 6 2000, *Chem. Abstr.*, 2000, **133**, 5794m.
14 Suwanii Kk, 'Thermally insulative stockings from blends of metal containing spandex fibres and other fibres', JP 06,41,801, 15 Feb 1994, *Chem. Abstr.*, 1994, **120**, 325666m.
15 Kuraray Co. 'Heat-storing fibres and their use in composite fibres', JP 06,158,419, 07 June 1994, *Chem. Abstr.*, 1994, **121**, 159320t.
16 Shimizu S, 'Self-extinguishing far-IR radiating heat insulative fibres containing substances having high amount of water of crystallization', JP 11,350,253, 21 Dec 1999, *Chem. Abstr.*, 2000, **132**, 51093j.
17 Beyreuther R, Schauer G and Hofmann H, 'Melt spinning of hollow fibres'. In *Progress in Textiles: Science & Technology Vol. 2 Textile Fibres: Developments and Innovations*, IAFL publications, New Delhi, India, 2000, 299–353.
18 Kawabata S, 'Measurement of anisotropic thermal conductivity of single fibre', *J. Textile Machin. Soc., Jpn*, 1986, **39**(12), 184.
19 Unitika Ltd. Japan, 'Light weight hollow fibre thermally insulating fabrics for winter clothings and sportswear', JP 08,60,486, 5 Mar 1996, *Chem. Abstr.*, 1996, **124**, 319582x.
20 Unitika Ltd. Japan, 'Thermally insulating heat-retentive hollow noncircular synthetic conjugate fibres containing zirconium carbide particles and exhibiting heat retention properties', JP 11,43,820, 16 Feb 1999, *Chem. Abstr.*, 1999, **130**, 183725k.
21 Toray Industries, 'Hollow insulative lightweight polyester fibres with good lustre' JP 06,228,815, 16 Aug 1994, *Chem. Abstr.*, 1994, **121**, 302951g.
22 Kuraray Co., 'Hollow hygroscopic insulative synthetic fibres with good pilling resistance' JP 05, 279, 911, 26 Oct 1993, *Chem. Abstr.*, 1994, **120**, 166703c.
23 Teijin Ltd. Japan, 'Heat-insulative hollow fibres with hot spring bath effect', JP 09,111,664, 28 April 1997, *Chem. Abstr.*, 1997, **127**, 52176x.
24 Mitsubishi Rayon Co., 'Porous insulative polypropylene fibres and their manufacture', JP 04,300,318, 23 Oct 1992, *Chem. Abstr.*, 1993, **118**, 126420k.
25 Toray Industries Inc. Japan, 'Thermally insulating hollow side-by-side bicomponent polyester fibres and their manufacture', JP 09,250,028, 22 Sept. 1997, *Chem. Abstr.*,

1997, **127**, 249330z.
26. Bajaj P, 'New developments in production of acrylic fibres'. In *Progress in Textiles: Science & Technology Vol. 2 Textile Fibres: Developments and Innovations*, IAFL publications, New Delhi, India, 2000, 534–614.
27. Hale D V, Hoover M J and O'Neil M J, *Phase Change Materials Handbook*, Contract NAS 8-25183. Alabama: Marshall Space Flight Center (September 1971).
28. Pause B, 'Development of heat and cold insulating membrane structures with phase change material', *J. Coated Fabrics*, July 1995, **25**, 59–68.
29. Vigo T L and Frost C M, 'Temperature adaptable fabrics', *Textile Res. J.*, 1985, **55**, 737–43.
30. Vigo T L and Bruno J S, 'Temperature adaptable textiles', *Textile Res. J.*, 1987, **57**, 427–9, September 1989, 34.
31. Kanebo Ltd., 'Acrylic fibres with increased latent heat retention and their manufacture', JP 06,220,721, 09 Aug 1994, *Chem. Abstr.*, 1994, **121**, 302948m.
32. Gateway Technologies Inc., *High Performance Textiles*, Aug 1995, 5–6.
33. Rupp J, 'Interactive textiles regulate body temperature', *Intern. Textile Bull.*, 1999 **45**(1), 58–9.
34. Dusek K and Patterson D, 'Transition in swollen polymer networks induced by intramolecular condensation, *J. Polym. Sci.*, Part A-2, 1968, **6**, 1209.
35. Tanaka T, Nishio T, Sun S and Nishio S U, 'Collapse of gels in an electric field', *Science*, 1982, **218**, 467.
36. Lee W F and Hsu C H, 'Thermoreversible hydrogel V: synthesis and swelling behaviour of the N-isopropylacrylamide-co-trimethyl methacryloyloxyethyl ammonium iodide copolymeric hydrogels', *J. Appl. Polym. Sci.*, 1998, **69**, 1793–803.
37. Zhang J and Peppas N A, 'Synthesis and characterization of pH- and temperature-sensitive Poly(methacrylic acid)/Poly(N-iso propyl acryl amide) interpenetrating polymeric networks', *Macromolecules*, 2000, **33**, 102–7.
38. Kim Y H, Kwan I C, Bae Y H and Kim S W, 'Saccharide effect on the lower critical solution temperature of thermo-sensitive polymers', *Macromolecules*, 1995, **28**(4), 939–44.
39. Suzuki A and Tanaka T, 'Phase transition in polymer gels induced by visible light', *Nature*, 26 July 1990, **346**, 345–7.
40. Sarkar R K, Gurudatt K and De P, 'Coated textiles as thermal insulators – Part II', *Man-made Textiles in India*, 2000, **XLIII**, 279–85.
41. Davies S and Owen P, 'Staying dry and keeping your cool', *Textile Month*, Aug 1989, 37–40.
42. Tanner J C, 'Breathability, comfort and Gore-Tex laminates', *J. Coated Fabrics*, 1979, **8**, 312–22.
43. Reischl U and Stransky A, 'Comparative assessment of Gore-tex and Neoprene vapor barriers in a firefighter turn-out coat', *Textile Res. J.*, 1980, **50**, 643–7.
44. Pause B, 'Measuring water vapor permeability of coated fabrics and laminates', *J. Coated Fabrics*, 1996, **25**, 311–20.
45. Sumitomo Osaka Semento Kk Japan, 'Antistatic and heat-energy absorbing coating, fibres and wearing apparels therewith', JP 08,325,478, 10 Dec 1996, *Chem Abstr.*, 1997, **126**, 105406k.
46. Unitika Ltd, Japan, 'Heat insulating, moisture permeable and water resistant

fabrics having polyurethane coatings', JP 08,246,339, 24 Sept 1996, *Chem. Abstr.*, 1997, **126**, 9259u.
47 Yamane, Sachiko, 'Bonding IR-ray radiating ceramics to textiles', JP 03,27,176, 5 Feb 1991, *Chem. Abstr.*, 1991, **114**, 187500a.
48 Fan J and Keighley J H, 'An investigation on the effects of: body motion, clothing design and environmental conditions on the clothing thermal insulation by using a fabric manikin', *Intern. J. Clothing Sci. Technol.*, 1991, **3**(5), 6–13.
49 Yasuda T, Miyama M and Yasuda H, 'Dynamic water vapor and heat transport through layered fabrics Part II: Effect of chemical nature of fibres', *Textile Res. J.*, 1992, **62**(4), 227–35.
50 Hong K, Hollies N R S and Spivak S M, 'Dynamic moisture vapor transfer through textiles, Part I: Clothing hygrometry and the influence of fibre type', *Textile Res. J.*, 1988, **58**, 697–706.
51 Kim J O, 'Dynamic moisture vapor transfer through textiles, Part III: Effect of film characteristics on microclimate moisture and temperature changes', *Textile Res. J.*, 1999, **69**(3), 193–202.
52 Jirsak O, Sadikoglu T G, Ozipek B and Pan N, 'Thermo-insulating properties of perpendicular-laid versus cross-laid lofty nonwoven fabrics', *Textile Res. J.*, 2000, **70**, 121–8.
53 Farnworth B, 'Mechanisms of heat flow through clothing insulation', *Textile Res. J.*, 1983, **53**, 717–24.
54 Crews P C, Sievert S A, Woeppel L T and McCullough E A, 'Evaluation of Milkweed Floss as an insulative fill material', *Textile Res. J.*, 1991, **61**, 203–10.
55 Hansraj, Nishkam A, Subbulakshmi M S, Batra B S and Kasturia N, *Proc. 'Technical Textiles'*, organized by SASMIRA, Mumbai, India, 31 Jan 1998, 24–32.

5
Cross-linked polyol fibrous substrates as multifunctional and multi-use intelligent materials

TYRONE L. VIGO AND
DEVRON P. THIBODEAUX

5.1 Introduction

The concept of 'intelligent' or 'smart' materials was critically reviewed within the context of fibrous materials having these attributes. It is generally agreed that the development of such materials occurred first with shape memory metal alloys (in the 1960s), followed by studies on polymeric intelligent gels (1970s). Also, a few studies were initiated in the late 1970s and early 1980s on intelligent fibrous materials.[1] This study will briefly review fibrous materials that possess this special attribute, which is usually defined as a profound change caused by an external stimulus that may or may not be reversible. Also examined in detail are our recent investigations on fibrous materials containing bound polyols that have at least two intelligent attributes: thermal and dimensional memories. Particular attention is focused on the techniques for measurement and characterization of load and pressure development in modified fabrics wet with polar solvents, and possible applications for these materials.

5.2 Fibrous intelligent materials

Although intelligent fibrous substrates were first described and demonstrated in Japan in 1979 with shape memory silk yarn,[2] it is our opinion that the discovery of crease recovery in cellulosic fabrics in the wet and dry state by Marsh and coworkers in 1929 was actually the first example of intelligent materials that even preceded intelligent shape memory alloys.[3] However, this discovery, which later led to the commercialization of durable press fabrics, was never identified or promoted as an intelligent textile concept.

The shape memory silk yarn was modified by immersion in a hydrolysed protein solution, followed by subsequent drying, crimping, immersion in water and heat setting at high pressure in the wet state. On subsequent heating at 60 °C, the yarn becomes bulky and crimped, but it reverts to the untwisted and uncured state upon drying.² Two other important developments that were patented in the late 1980s can be classified as intelligent fibrous materials. The first was polyester or polyamide fabrics, which absorbed visible solar radiation and converted it into heat as infra-red radiation released inside the garment due to the microencapsulation of zirconium carbide; it was irreversible since it did not reverse at high temperatures to provide cooling.² The second was thermochromic garments (colour change induced by incorporation of compounds that respond to changes in environmental temperature) produced by Toray Industries⁴ called SWAY, which have reversible colour changes at temperature differences greater than 5 °C and are operable from -40 °C to $+80$ °C.

Most of the other recent investigations on intelligent fibrous materials have been primarily concerned with shape memory effects. Some relevant examples are polyester fabrics with durable press properties produced by high energy radiation cross-linking with polycaprolactone and acrylate monomers⁵ and temperature-sensitive polyurethanes prepared by copolymerization of polytetramethylene glycol and methylene diisocyanate, 1,4-butanediol.⁶ More novel intelligent fibre applications and responses to external stimuli were described in a 1995 American Chemical Society symposium. These polymeric smart materials included fibre optic sensors with intelligent dyes and enzymes whose fluorescence changes with changes in glucose, sodium ion and pH, and elastomeric composites embedded with magnetic particles, which undergo changes in their moduli when exposed to external magnetic fields and thus can be used for vibration or noise reduction in transportation and machinery.⁷ Smart fibrous composite sensors made of carbon and glass have also been described for prevention of fatal fracture by changes in their insulating and conductive properties.⁸

Fibrous materials containing bound or cross-linked polyols were patented and initially characterized by multifunctional properties imparted to the modified materials by a single process.⁹,¹⁰ Major property improvements include thermal adaptability, high absorbency and exsorbency, enhanced resistance to static charge, oily soils and flex abrasion, and antimicrobial activity. The most notable of these property improvements was temperature adaptability and reversibility due to the phase change material (PCM) characteristics of the cross-linked polyol bound to the fibres. However, this unique and new fibre property was not initially defined as an intelligent material attribute. Moreover, a related invention, the disclosure of fibres with thermal adaptability due to attachment by microencapsulation of hexadecane and n-eicosane (PCMs with high latent heat and thermal reversibility), was also not characterized as an intelligent fibrous material.¹¹

5.3 Experimental

The modified fabrics containing the bound polyols, derived from using formulations with different molecular weight, acid catalysts and cross-linking resins as previously described,[13] were woven cotton and 65/35 cotton polyester printcloth or sheeting (125–135 g/m^2). The surface temperatures were measured with a Hughes Probeye Thermal Video System 3000 with an image processor, infra-red camera head, argon gas cylinder and colour screen monitor. Untreated and modified fabrics (6 in^2 or 150 mm^2) were mounted on a pin frame surrounded by a wooden enclosure containing a 150 watt light bulb whose heat output was controlled by a voltage regulator device (Variac) with the camera sensitivity set at 2.0 and emissivity value set at 0.95, which was appropriate for fabric surfaces. Surface temperatures (°F) were measured every 10 min and for up to 30 min with the infra-red camera 1 m from the fabric by positioning the cross-hairs of the camera in the centre of the fabric with the untreated control value at 147 °F (64 °C).

For the work and power measurements previously described,[13] a 65/35 cotton/polyester cross-tuck knit (177 g/m^2), woven cotton printcloth (125 g/m^2), and 50/50 hydroentangled pulp/polyester non-woven (80 g/m^2) containing 40–60% bound polyol were evaluated. Each of the fabrics was attached to a metal alligator clip, suspended vertically. The amount of shrinkage, work (W) performed, and power (P) generated were determined with weights attached after wetting with various solvents (water, acetone and ethanol). The amount of shrinkage was also determined without weights attached for each of the different types of fabrics for each solvent.

For the tension and pressure measurements, a desized, scoured and bleached 100% cotton interlock knit fabric (205 g/m^2) was treated with 50% PEG-1450–11% DMDHEU (Fixapret ECO)–5.1% citric acid–5.1% $MgCl_2.6H_2O$ to a wet pickup of 110%, then cured for 3 min at 110 °C, washed in warm water to a pH of 8.5 to neutralize acid catalyst, excess water removed, then oven-dried for 7 min at 85 °C. The resultant fabric had a wt. gain of 65% and had shrinkage in one direction (longitudinal) when wet with various polar solvents (28% with water; 25% with 1% aq. NaCl; 10% with 95% ethanol; and 7% with acetone). Three inches of a 1 × 6 inch strip of the treated fabric was clamped between the jaws of an Instron (Model 4201) having a 10 lb load cell, completely wet with each of the four solvents and tension load/stress measured after initial wetting to a maximum time of 15 min.

5.4 Results and discussion

5.4.1 Thermal memory

The characterization of fibrous materials containing cross-linked polyols as smart materials with a thermal memory was first discussed by Vigo in an American Chemical Symposium which he co-organized in 1995[7] and later described in more detail in conference proceedings and in refereed journals.[12,13] This thermal memory or thermal reversibility is unique and has been determined and documented by differential scanning calorimetry and infra-red thermography. The latter measures changes in surface temperature on exposure to a heat source. As shown in Fig. 5.1, cotton fabrics containing bound polyol had substantial reduction in surface temperature relative to an untreated control fabric (surface temperature 147 °F or 64 °C) after all fabrics were exposed to a heat source for a minimum of 30 minutes. Even more impressive is that the lowest amounts of bound polyol in the fabric gave the greatest decrease in surface temperature. Weight gains were 34% for fabrics containing bound PEG-1000, 37% for fabrics containing bound PEG-2000, 28% for fabrics containing bound PEG-4000 and only 25% for fabrics containing bound PEG-6000.

The decrease in surface temperature in all treated fabrics usually occurred within 30 seconds and was greater with higher molecular weight cross-linked polyols bound to the fabric, due to the increasing crystallinity of the polyol with increasing molecular weight. Surface temperature decrease on the modified fabrics was 17–28%. Thermal memory and reversibility have been documented in commercial skiwear marketed in Japan, with the crosslinked polyol as one of the fabric components; thermocouples in this component demonstrated that the skiers had optimum thermal comfort by an increase in fabric surface temperature when idle and lowering of the surface temperature after skiing, thereby preventing discomfort due to perspiration and body heat.[14]

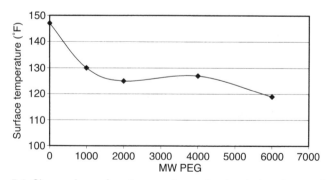

5.1 Change in surface temperature of untreated and cross-linked polyol-treated fabrics after 30 min exposure to a heat source.

Applications for thermally adaptable fabrics containing cross-linked polyols have been previously described, primarily with regard to apparel and insulation.[10] However, there are many other applications that could evolve based on the 'intelligent' thermal memory of these modified fibrous materials. The wear life of many automotive and aerospace components could be extended by the moderation of temperature fluctuations during their use. Camouflage uniforms made of this material would be useful for the military because the wearer would have a different surface temperature on garments normally detected by infra-red sensors. The thermal buffering capacity of the modified fabrics could be used advantageously in various healthcare applications, such as in burn therapy and to improve skin temperature and circulation in patients with maladies such as arthritis and diabetes. Changes in the fabric surface temperature could also be used to fabricate remote sensing devices for detection in industrial, residential, horticultural, agricultural and other environments where temperature change detection would be useful. These modified fabrics could be used to minimize plant stress due to temperature changes and fluctuations as well as protection of livestock and other animals. Many other 'thermal memory' applications for these modified materials as sensors and/or actuators are envisioned with the molecular weight of the bound polyol defining the magnitude and temperature range desired.

5.4.2 Dimensional memory

It was known from our earlier studies that certain types of fabric constructions containing bound polyol were more prone to wet shrinkage. The magnitude of this effect with various solvents such as hexane, ethanol, acetone and water, was investigated with three different fabric constructions: knit, woven and non-woven.[13] An increase in the polarity of the solvent resulted in increasing reversible wet shrinkage (highest with water, no shrinkage with hexane) that was most pronounced with knit fabric constructions. To demonstrate the potential for converting chemical into mechanical energy, different quantities of weights were affixed to modified knit, woven and non-woven fabrics that were subsequently wet with water. The work performed (W) and power generated (P) were determined. As shown in Fig. 5.2, knit fabrics are far superior to woven and non-woven fabrics in this characteristic, due to its optimum wet shrinkage in the presence of an applied force. Work performed by the wet knit fabric was 30–50 times greater than that of the non-woven and woven fabrics when each had the same amount of weight attached (50 g). It was even an order of magnitude greater when the knit fabric had twice the weight attached compared to the other two fabrics. Similar results were obtained for power generated (P). These results were used to present the concept and design of an intelligent medical bandage that would contain a removable and

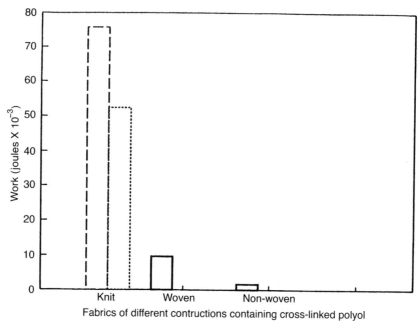

5.2 Work (W) performed by fabrics of different constructions containing cross-linked polyol by lifting attached weights after wet with water for 30 s. —— +50 g, - - - +100 g, ... +200 g (from ref. 13).

disposable component of knit cross-linked polyol fabric affixed to a non-disposable component by metal fasteners or other methods.[15] However, no pressure or tension measurements were made to determine the amount of tension or force developed for these types of intelligent fabrics with dimensional memories activated by wetting with polar solvents.

The tensile load (lb) and stress developed (psi) were measured for untreated and treated fabric strips after wetting with deionized water (Table 5.1). The tension developed for the untreated fabric was negligible, but it was very significant for the treated fabric. Multiple measurements were made and there was some detectable variability (a range of 4.5–5.6 lb and corresponding values of 130–196 psi). In addition to the development of high tensile stress, the modified fabrics wet with water achieved these values rapidly (within 90 seconds of wetting) and maintained them over the measurement period of 5 min (Fig. 5.3). Although there was somewhat more variability for treated fabrics that were wet with 1% aq. NaCl with regard to load and tensile stress developed (3.5–6.2 lb and 120–217 psi), these fabrics also developed such tension rapidly (Fig. 5.4). Thus, tension developed rapidly in treated fabrics wet with water and aqueous salt solution and was generally proportional in

Table 5.1 Maximum load and tension developed in fabrics after wetting with aqueous solvents[a]

Fabric	Solvent/solution	Max. load (lb)	Max. tension (psi)
Untreated control	H_2O	—	—
Treated	H_2O	5.6	196
Treated	1% aq. NaCl	6.2	217

[a]All fabrics were clamped between jaws of Instron (1 × 3 in length) that were wet within 30 s and tension developed measured from initial wetting up to 5 min.

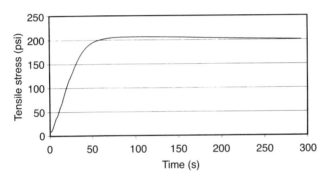

5.3 Development of tension over time with modified fabric after wetting with water.

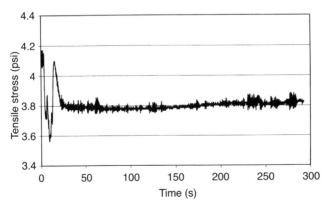

5.4 Development of tension over time in untreated control fabric after wetting with water.

magnitude to the wet shrinkage that these fabrics experienced when they were not constrained.

The behaviour of the modified fabrics when wet with the two organic solvents shown in Table 5.2 and Fig. 5.5 and 5.6 (95% ethanol and acetone, respectively) differed from those wet with aqueous media in three fundamental aspects: (1) maximum tension took longer to achieve with these solvents than those wet with aqueous media; (2) the tension developed was also less than those exposed to aqueous media; and (3) the reversal of load and pressure occurred over relatively short time periods due to the high vapour pressure and accompanying evaporation of solvent. Fabrics wet with ethanol took about 450–550 seconds to reach maximum stress (138 psi generated), then reversed this effect due to slow solvent evaporation (Fig. 5.5). Fabrics wet with acetone reach maximum tension more rapidly (150 seconds with a corresponding load of 1.6 lb and 56 psi), but rapidly reversed back to zero load after 250 seconds due to the complete evaporation of the solvent (Fig. 5.6).

Table 5.2 Maximum load and tension developed in fabrics after wetting with polar organic solvents[a]

Fabric	Solvent	Max. load (lb)	Max. tension (psi)
Treated	95% ethanol	3.9	138
Treated	Acetone	1.6	56

[a]All fabrics were clamped between jaws of Instron (1 × 3 in length) that were wet within 30 s and tension developed measured from initial wetting up to 15 min.

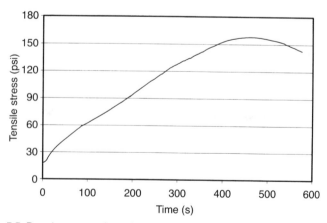

5.5 Development of tension over time in modified fabric after wetting with ethanol.

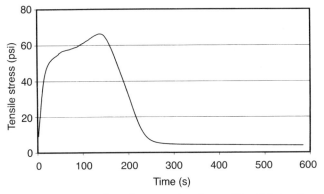

5.6 Development of tension over time in modified fabric after wetting with acetone.

Intelligent fabrics containing cross-linked polyols that have reversible dimensional memories on exposure to polar solvents can serve in a variety of applications. A variety of sensors could be developed that detect different types of polar solvents and solutions related to the amount of wet shrinkage and/or tension developed. Alternatively, fabric sensors could be fabricated to detect varying amounts of water vapour and liquid water by wet shrinkage and/or accompanying tension developed on wetting. Several types of actuator systems based on these modified fabrics are also envisioned. In the biomedical and healthcare area, wetting with solvents such as sterilized water, salt solutions and alcohol could be used to activate medical pressure bandages and possibly tourniquets; alternatively, blood (plasma) could activate pressure development and thus stop undesirable bleeding. These fabrics could be used to produce micromotors and a variety of valves or other mechanical devices that are activated by the reversible wet shrinkage that occurs with a variety of polar solvents.

5.5 Conclusions

Fabrics of woven, non-woven and knit constructions containing cross-linked polyols have been shown to have thermal memories over a wide temperature range and can be adapted to various thermal environments by the molecular weight of the bound polyol and the degree of cross-linking on the fibrous substrate. In contrast, only specific fabric constructions (cross-tuck and interlock knits) containing the cross-linked polyols are able to generate substantial tensile load when wet with polar solvents. The load and stress generated are usually proportional to the degree of wet shrinkage when the fabrics are not constrained. Maximum and rapid tension development occurs with water and aqueous salt solutions, while ethanol and acetone are somewhat less effective with faster reversibility due to rapid evaporation of these low-boiling, high vapour pressure solvents. Numerous applications for the cross-linked polyol fabrics with thermal and dimensional memories are envisioned with two major areas of use as sensors and actuators.

References

1. Vigo T, Intelligent fibrous materials, *J. Textile Inst.* 1999, **90**(3), 1–13.
2. Hongu T and Philips G O, *New Fibers*, New York, Ellis Horwood, 1990.
3. Vigo T, *Textile Processing and Properties: Preparation, Dyeing, Finishing and Performance*, Amsterdam, Elsevier, 1997.
4. Toray Industries, Spiro-oxazine compounds and preparation thereof and photochromic article shaped, US Pat 4 784 474, Nov. 15, 1988.
5. Wantabe I, Fabrics with shape memory, Japan Kokai Tokkyo Koho JP 09,95,866, April 8, 1997.
6. Lin J R and Chen L W, Study on shape-memory behavior of polyether-based polyurethanes, *Appl. Poly. Sci.*, 1998, **69**, 1563–74.
7. Dagani R, Polymeric 'smart' materials respond to changes in their environment, *Chem. Eng. News*, 1995, **73**(38), 30–3.
8. Muto N, Yanagida H, Nakatsju T, Sugita M, Ohtsuka Y and Arai Y, Design of intelligent materials with self-diagnosing function to prevent fatal fracture, *Smart Mater. Struct.*, 1992, **1**(4), 324–9.
9. Vigo T L, Frost C M, Bruno, J S and Danna, G F, Temperature adaptable textile fibers and method of preparing same, US Pat 4 851 291, July 25, 1989.
10. Vigo T L and Bruno J S, *Handbook of Fiber Science and Technology, Vol. III, High Technology Fibers, Part C*, New York, Marcel Dekker, 1993.
11. Triangle Research and Development Corp., Bryant Y G and Colvin, D P, Fibre with reversible enhanced thermal storage properties and fabrics made therefrom, US Pat 4 756 958, July 12, 1988.
12. Vigo T L, Intelligent fibrous flexible composites containing crosslinked polyethylene glycols, 4th Int. Conf. *Composites Engineering*, New Orleans, U. of New Orleans, 1996.
13. Vigo T L, Intelligent fibrous substrates with thermal and dimensional memories, *Poly. Adv. Technol.*, 1997, **8**, 281–8.
14. Private communication, Mitsui Co. and Phenix Co., Tokyo, Japan, 1991.
15. Vigo T L, 6th Int. Conf. *Composites Engineering*, Las Vegas, 1998.

6
Stimuli-responsive interpenetrating polymer network hydrogels composed of poly(vinyl alcohol) and poly(acrylic acid)

YOUNG MOO LEE AND SO YEON KIM

6.1 Introduction

Recently, there have been many investigations of stimuli-responsive hydrogels, which are three-dimensional polymer networks exhibiting sensitive swelling transitions dependent on various stimuli, e.g. electric field,[1,2] pH,[3-7] temperature,[8,9] ionic strength[10] or other chemicals.[11,12] They have been considered to be useful in separation and medical applications, especially in drug delivery systems.

The schematic diagram illustrating the diffusion of bioactive materials through stimuli-sensitive hydrogels is shown in Fig. 6.1. Among these, pH and/or temperature-responsive systems have been potential candidates because these factors could be the most widely available in the environment of the human body. In an aqueous system, the temperature dependence of the swelling of a polymer gel is closely related to the temperature dependence of polymer–water and polymer–polymer interaction. Katono et al.[13] investigated thermal collapse from interpenetrating polymer networks (IPNs) composed of poly(acrylamide-co-butyl-methacrylate) and PAAc, and obtained on–off release profiles as a function of temperature. Poly(N-isopropylacrylamide) (PNIPAAm) hydrogels demonstrated negative temperature sensitivity with lower critical solution temperature (LCST) in aqueous solution, which showed the decrease of swelling with an increase of the temperature above 30–32 °C.[14]

In the case of pH-dependent drug delivery systems, many researchers have focused on the swelling properties of hydrogels. The acidic or basic components in the hydrogels led to reversible swelling/deswelling because they changed from neutral state to ionized state and vice versa in response to the change of pH. Brannon and Peppas[15] reported a pH-sensitive drug release system composed of a copolymer of hydroxyethyl methacrylate (HEMA) and methacrylic acid (MAAc) or maleic anhydride. The poly(NIPAAm-co-vinyl terminated poly(dimethylsiloxane)-co-AAc) hydrogel synthesized by Hoffman and Dong[16] permitted the release of drugs at pH 7.4 and showed release-off at

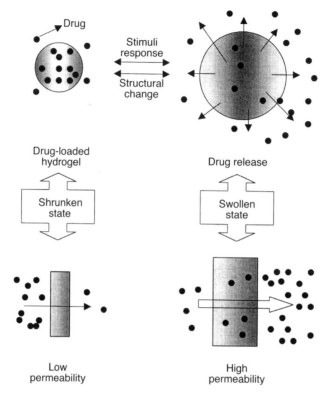

6.1 The diffusion of bioactive materials through stimuli-sensitive IPN hydrogels.

pH 1.4, and was proved to be a suitable carrier for indomethacin through the gastrointestinal (GI) tract.

In addition, electric current-sensitive hydrogels actuated by electric stimulus are particularly interesting in connection with the mechanical energy triggered by an electric signal.[17-20] Electric current-sensitive hydrogels are usually made of polyelectrolytes and an insoluble but swellable polymer network which carries cations or anions. This system can transform chemical free energy directly into mechanical work to give isothermal energy conversion. A typical function of a gel containing ionic groups is to bend reversibly under the influence of an electric field, making it useful in some actuators driven by an electric field. From the viewpoint of mechanical engineering, great hopes are set on these materials as new actuators,[17-20] especially in the fields of robotics and medical welfare instruments.

The present chapter will review the pH/temperature-sensitive properties of novel PVA/PAAc IPN hydrogels synthesized by a unique sequential method through UV irradiation and a freezing–thawing process.

6.2 Experimental

6.2.1 Materials

Acrylic acid monomer (AAc) obtained from Junsei Chemical Co. was purified by inhibitor remove column (Aldrich Chem. Co.) Poly(vinyl alcohol) (PVA; DP = 2500, degree of deacetylation = 99%) was purchased from Shinetsu Co. Methylenbisacrylamide (MBAAm) as a cross-linking agent and 2,2-dimethoxy-2-phenylacetophenone (DMPAP) as a photoinitiator were purchased from Aldrich Chem. Co. Indomethacin (anhydrous, MW = 357.8), used as a model drug, was obtained from Sigma Chem. Co. As five representative solutes, theophylline (99%, Aldrich Chem. Co.), vitamin B_{12} (Sigma Chem. Co.), riboflavin (Junsei Chem. Co.), bovine serum albumin (Young Science Inc.) and cefazolin sodium were used. Cefazolin sodium was kindly donated by Chongkeundang R&D Center in Seoul, Korea.

6.2.2 Synthesis of PVA/PAAc IPN hydrogels

The PVA/PAAc IPNs were prepared by sequential method. PAAc as an initial network was synthesized inside a PVA solution using UV irradiation. PVA as a secondary network was then formed by repetitive freezing–thawing process. The whole reaction scheme of PVA/PAAc IPNs is described in Fig. 6.2. PVA was dissolved in deionized water and heated at 80 °C for 2 h to make 10 wt% aqueous solution. AAc monomers were mixed with 0.2 wt % DMPAP as a photoinitiator and 0.5 mol% MBAAm as a cross-linking agent. The mixture was combined with PVA solution to yield PVA–PAAc. The feed composition and designation of each sample are listed in Table 6.1.

The mixed solutions were poured onto Petri dishes and irradiated using a 450 W UV lamp (Ace Glass Co.) for 1 h under N_2 atmosphere. The irradiated samples were placed at -50 °C for 6 h and at room temperature for 2 h. These freeze–thawing cycles were repeated eight times.

Table 6.1 Designation and feed contents of PVA/PAAc IPN

Sample name	PAAc (mol%)	AAc (g)	MBAAm (g)	DMPAP (g)
IPN37	70	5.70	0.061	0.0084
IPN46	60	4.89	0.052	0.0098
IPN55	50	3.26	0.035	0.0065
IPN64	40	2.16	0.023	0.0043
IPN73	30	1.62	0.017	0.0032

96 Smart fibres, fabrics and clothing

6.2 The preparation of PVA/PAAc IPN hydrogels by UV irradiation and freezing–thawing method.

6.2.3 Characterization

Fourier transform-infra-red (FTIR) spectroscopy (Nicolet Model Magna IR550) and differential scanning calorimetry (DSC, Du Pont Instruments DSC 910) were used to confirm the structure of the IPNs. The equilibrium water content (EWC) was measured in pH 4 and pH 7 buffer solution. The EWC was calculated using the equation in reference.[21] In addition, the swelling ratio of each sample was evaluated as EWC/100.

6.2.4 Loading and releasing of drug

The indomethacin, as a model drug, was loaded into IPN disks by the solvent sorption method. The indomethacin release experiment was conducted in various conditions, changing the pH and/or the temperature of the release medium, as described elsewhere. The permeation studies were carried out using a two-chamber diffusion cell.

6.3 Results and discussion

6.3.1 Synthesis of PVA/PAAc IPNs

The sequential method by UV irradiation, followed by freezing and thawing, would be a novel method of preparing the stimuli-sensitive polymer hydrogel. The structure of the IPN samples prepared was confirmed by FTIR spectra.

6.3.2 Swelling measurement of PVA/PAAc IPNs

The swelling ratios of all PVA/PAAc IPNs were relatively high, and they showed reasonable sensitivity to both pH and temperature. Temperature-dependent swelling behaviours at pH 7 are shown in Fig. 6.3(a) and (b). Measurements were carried out with hydrogels after immersing the samples for 20 h at 25 °C. The IPN hydrogels showed different kinds of thermo-sensitivity. All IPN samples showed positive temperature-sensitive systems before reaching their equilibrated state as illustrated in Fig. 6.3(a). After equilibrium swelling, however, the swelling ratio of IPN46 increased with temperature, showing a positive temperature-sensitive system, while IPN64 and IPN55 exhibited negative temperature-sensitive systems. The swelling behaviours of IPN64 and IPN55 are similar to that of virgin PVA homopolymer.[22] This is ascribable mainly to the swelling tendency of PVA itself, rather than to polymer–polymer interactions between PVA and PAAc, such as hydrogen bonding and ionic repulsion in PAAc. The PVA content, even in IPN55 hydrogels, is supposed to be greater than the PAAc content because the

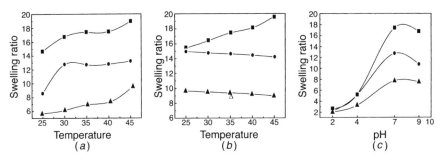

6.3 Temperature/pH-dependent swelling behaviours of PVA–PAAc IPN: (a) temperature-dependent swelling behaviour before equilibrium (6 h) at 25 °C, (b) temperature-dependent swelling behaviour after equilbrium (24 h) at 25 °C, (c) pH-dependent swelling behaviours at 25 °C; (■) IPN 46, (●) IPN 55, (▲) IPN 64.

residual acrylic acid monomer, which did not take part in polymerization during the synthesis of the IPNs, was removed from the IPNs by washing after the whole procedure. Accordingly, the swelling changes in IPN46 must be related to the hydrogen bonding between two polymers and the ionic repulsion of carboxylate ion produced due to the breakage of hydrogen bondings, while those in IPN64 and IPN55 depend mainly on the swelling tendencies of PVA composition.

As shown in Fig. 6.3(c), pH-dependent swelling behaviour was observed at 35 °C with changes in pH. The pK_a value of PAAc is known to be 4.8.[21] Therefore, at pH 2 and pH 4 buffer solutions, carboxyl groups in PAAc associate and form hydrogen bonding with PVA chains; thus, the swelling ratios are small. Carboxylate ions formed at pH greater than 4, however, induce the repulsion between them, and the hydrogen bondings between PVA and PAAc should decrease, resulting in the increase of free volume in the matrix and, therefore, a rapid increase in swelling ratios. The small deswelling that occurred at pH 9 is considered to be due to an increase of ionic strength in the buffer solution at pH 9, which might inhibit the polymer–water interactions inside the hydrogel matrix. The swelling ratio increased with the PAAc content within the PVA/PAAc IPNs.

The relationship between the changes in molecular structure in the IPN hydrogel and electrolytes in the buffer solution is shown in Table 6.2. The ionization of carboxylic acid in the IPN hydrogel was an important factor in the pH-dependent swelling. However, the interaction between ionic groups in the IPN chains and electrolytes in the buffer solution became more intense when the pH of the buffer solution changed from 4 to 7; therefore, when a pH was below the pK_a value of the PAAc (at pH 4) in which the carboxylic acid group in the IPN hydrogel was in a hydrogenated state, the effect of the ionic strength on the swelling was less than a pH above the pK_a value of PAAc (at pH 7) in which the carboxylic group in the IPN hydrogel was in an ionized state. For the buffer solution at pH 7, in which the ionic strength was adjusted to 0.01, the swelling ratio of each IPN sample became high, ranging from 15.2 to 33.4, while the increase of electrolyte concentration to 0.1 in the medium produced a drastic collapse of all the IPN hydrogels. However, the swelling of all IPN hydrogels at pH 4 was not dependent on ionic strength, as can be seen in Table 6.2.

6.3.3 Release experiment of drug from PVA/PAAc IPNs

All the samples exhibited higher release rates at higher temperatures. These results are in correspondence with the temperature-dependent swelling behaviours of PVA/PAAc IPN hydrogels. All IPN samples showed positive temperature-sensitive systems before they reached the equilibrated state. The

Table 6.2 Effect of pH and ionic strength in the swelling medium on the equilibrium swelling ratio of IPN hydrogels

Sample	pH	Ionic strength	Swelling ratio
IPN37	7	0.01	32.3 ± 0.92
IPN46	4	0.1	6.8 ± 0.44
		0.01	6.0 ± 0.91
	7	0.1	16.3 ± 1.80
		0.01	28.0 ± 0.36
IPN55	4	0.1	4.2 ± 0.47
		0.01	4.4 ± 0.30
	7	0.1	11.9 ± 1.80
		0.01	21.5 ± 1.50
IPN64	4	0.1	5.5 ± 1.32
		0.01	4.9 ± 1.20
	7	0.1	10.1 ± 0.67
		0.01	17.4 ± 0.58
IPN73	7	0.01	14.5 ± 1.13

release of the incorporated drugs within IPN hydrogels was delayed for about 10–15 hours, which was not enough time for all the samples to be fully equilibrated. The release profiles indicated positive temperature-sensitive release behaviours, and thus the composition of IPN was negligible in the temperature-dependent release system.

The changes of release rate for IPN46, IPN55, IPN64 hydrogels in response to stepwise temperature changes between 25 and 45 °C are shown in Fig. 6.4. High release rates are obtained at 45 °C, while low release rates are observed during the lower temperature period, as mentioned above. The released amounts from IPN46 and IPN55 are almost the same, while the released amount from IPN64 is small due to the low swelling. In the case of IPN64, the fact that the hydroxyl groups in PVA formed hydrogen bonding and had little chance to experience the ionic repulsion from carboxylate groups was proved.

The pH-sensitive release behaviours of indomethacin were observed at 25 °C with change in pH. Indomethacin is expected to diffuse through the water-swelling region in the polymer gel. At pH 4, PAAc is in the form of carboxylic acid, which produces hydrogen bonding with the hydroxyl group of PVA, resulting in the decrease of drug release. However, at pH 7, carboxylate ions formed in PAAc induced the drastic increase of swelling which exhibited high release rate. The release amounts from IPN46 and IPN55 were almost identical, and IPN64 gave lower values due to significant low swelling compared to the others at pH 4. It is considered that the release rate from IPN hydrogels is nearly independent of the composition of samples at pH 4. In the

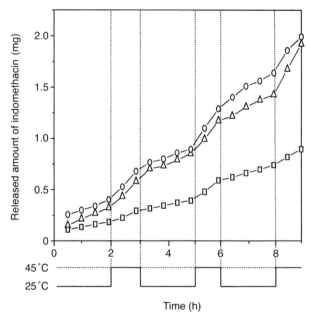

6.4 Release change of indomethacin from IPN hydrogels in response to stepwise temperature change between 25 °C and 45 °C; (○) IPN46, (△) IPN55, (□) IPN64.

case of the drug release experiment at pH 7, more PAAc contents in IPN contributed to the higher release amount.

Oscillatory release profiles at 25 °C with changing pH levels between 4 and 7 are shown in Fig. 6.5. The plot gives the reversible change of release rate with pH fluctuation. IPN46 and IPN55 showed lower release at pH 4 and higher at pH 7, and continued reversible behaviour during the experiment. However, IPN64 did not show sharp reversible change because it did not have enough carboxylic groups.

6.3.4 Solute permeation through PVA/PAAc IPNs

Solute permeation experiments were performed using five representative solutes with different molecular weights and hydrodynamic sizes. The characteristics of selected solutes are listed in Table 6.3. In the experimental buffer solution in which pH was adjusted to 4 or 7, theophylline,[23] vitamin B_{12}[23] and BSA[23] were in a neutral state, and riboflavin[23] and cefazolin[24] showed ionized structures.

Table 6.4 shows the results of solute permeation experiment measured at 25 and 45 °C, respectively. The permeation of solutes through IPN hydrogels as a

Table 6.3 Characteristics of permeating solutes

Solutes	MW	Hydrodynamic radius (Å)	Ionic property at pH 7
Theophylline	180	3.5	Neutral
Riboflavin	376	5.8	Ionic
Cefazolin	476	6.5	Ionic
Vitamin B_{12}	1355	8.5	Neutral
BSA	65 000	36.1	Neutral

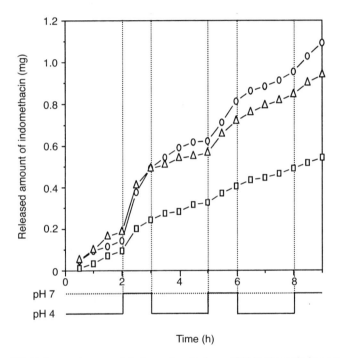

6.5 Release change of indomethacin from IPN hydrogels in response to stepwise pH change between pH 4 and 7; (○) IPN46, (△) IPN55, (□) IPN64.

function of temperature was in accordance with the swelling behaviours of the IPNs. IPN46 exhibited a positive swelling change with temperature, as described in the swelling experiments. Therefore, all the solutes showed higher diffusion coefficients at high temperature. This is explained by the fact that the increase of diffusion coefficients at high temperature does not alter the interaction between solutes and IPN structure, but only causes an expansion

Table 6.4 The permeation studies of various solutes through IPN46 hydrogel at pH 7 buffer solution ($I = 0.01$)

Solute	Temp. (°C)	P^a ($\times 10^6$ cm²/s)	K_d^b	D^c ($\times 10^6$ cm²/s)
Theophylline	25	7.63 ± 0.127	1.55 ± 0.035	4.94 ± 0.195
Riboflavin	25	0.99 ± 0.007	2.97 ± 0.028	0.33 ± 0.001
Cefazolin	25	0.42 ± 0.007	5.67 ± 0.064	0.07 ± 0.003
Vitamin B_{12}	25	1.80 ± 0.099	1.20 ± 0.064	1.51 ± 0.163
BSA	25	1.03 ± 0.134	2.75 ± 0.043	0.41 ± 0.052
Theophylline	45	9.90 ± 0.141	0.81 ± 0.014	12.22 ± 0.039
Riboflavin	45	1.52 ± 0.028	2.35 ± 0.014	0.65 ± 0.016
Cefazolin	45	1.38 ± 0.092	9.14 ± 0.042	0.15 ± 0.011
Vitamin B_{12}	45	4.06 ± 0.170	1.89 ± 0.099	2.15 ± 0.023
BSA	45	3.86 ± 0.134	4.74 ± 0.031	0.81 ± 0.034

[a]P: permeability coefficient, [b]K_d: partition coefficient, [c]D: diffusion coefficient
$n = 3$ for each sample.

of IPN networks. Higher swelling in IPN at high temperature increases the free volume through which the solutes permeate more freely.

The partition coefficients (K_d) of solutes between hydrogel membranes and surrounding solutions are also shown in Table 6.4. The value of K_d is driven by interactions between the solutes and the polymer molecular structure. Due to the hydrophilicity of PVA/PAAc IPN hydrogels, it is predicted that more hydrophilic solutes are well distributed throughout the entire hydrogel membranes. As can be seen in Table 6.4, the K_d values of ionic solutes such as riboflavin and cefazolin are greater than those of non-ionic solutes, obviously due to induced ionic interaction between ionic solutes and the carboxylic acid group. BSA showed the largest K_d value among non-ionic solutes. This is believed to be caused by the protein adsorption to the polymer surfaces. However, further study is needed to verify the interaction, which affects and determines the partition coefficients of solutes within the swollen hydrogels.

The size of the solute was an important factor in determining the diffusion and thus permeation through IPN hydrogels. The hydrodynamic sizes of three solutes is in the range from 3.5 Å to 36.1 Å, as seen in Table 6.3. For non-ionic solutes which had no ionic interaction with ionizable hydrogel membranes, theophylline, having the smallest molecular size of 3.5 Å, showed the most rapid diffusion through the hydrogel. The diffusion of vitamin B_{12} (hydrodynamic radius = 8.5 Å) was faster than that of BSA (hydrodynamic radius = 36.1 Å). The free volume caused by hydration may allow the pathway of solutes through which the larger molecules were excluded in the permeation. However, the diffusion of ionic solutes such as riboflavin and cefazolin were independent of their hydrodynamic size. Riboflavin and cefazolin, which are smaller in size than vitamin B_{12} and BSA, gave a drastic reduction of diffusion coefficients. The permeability of solutes through IPN46 exhibited a similar tendency with diffusion coefficients. These results are indicative of the fact that, in the ionizable hydrogel membranes, the solute diffusion is much affected by solute size, as well as by ionic interaction between solutes and membranes.

6.3.5 Electro-responsive properties of PVA/PAAc IPNs

To investigate the electric-responsive behaviour of PVA/PAAc IPN hydrogel under electric stimulus, we measured the bending degree of IPN by varying several factors. When PVA/PAAc IPN hydrogel in NaCl electrolyte solution is subjected to an electric field, the IPN bends toward the anode. The gradient slope in the plot of the bending angle vs. time became steeper with increasing applied voltage, and then levelled off at steady state. Note that the equilibrium bending angle (EBA) and the bending speed increase with applied voltage. This could be explained by the fact that there was an enhancement in the transfer rate of the counter ions of the immobile carboxylate groups of PAAc within PVA/PAAc IPN from IPN to external solution and free ion moved from external solution into IPN, as the potential gradient in the electric field increased.

We investigated the effect of PAAc content on the EBA of IPN in 0.1 M NaCl solution at 10 V. As the content of the PAAc network increased, the EBA increased linearly. The bending speed also increased in proportion to the content of PAAc in the IPN. This may indicate that the EBA is not a rate-limiting step for bending in an equilibrated state.

In addition, the EBA and bending kinetics of IPN37 were measured in NaCl solution with different ionic strength. There was a critical ionic strength for the bending of PVA/PAAc IPN. The bending degree increased with ionic strength when the latter was less than 0.1, while it decreased at ionic strength of greater than 0.1. Furthermore, when the ionic strength of NaCl solution was 1.0, we could not observe the bending behaviour of PVA/PAAc IPN under the 10 V of applied voltage. As described above, an increase of electrolyte concentration in the solution induces the increase of free ions moving from the surrounding solution toward their counter electrode or into the IPN. As a result, the bending degree and speed of the IPN could increase. However, if the ionic strength of the solution is greater than a critical concentration, the shielding effect of polyions by the ions in the electrolytic solute occurred, leading to a reduction in the electrostatic repulsion of polyions and a decrease in the degree of bending. As already shown in Table 6.2, the swelling ratio of PVA/PAAc IPN in solution with the lower concentration of electrolyte (ionic strength = 0.01) was greater than that of higher electrolyte concentration (ionic strength = 0.1). Therefore, if it is compared with the bending results depending on the ionic

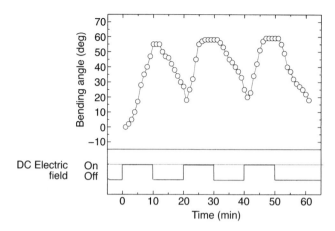

6.6 Reversible bending behaviour of IPN37 in NaCl solution (ionic strength = 0.1) with changes in applied voltage 10 volts every 10 min.

strength of the solution, we could consider that the swelling of PVA/PAAc IPN by ionic strength does not significantly affect the bending behaviours of the IPN. In addition, stepwise bending behaviour of PVA/PAAc IPN depending on electric stimulus was observed for IPN in 0.1 M NaCl solution, with changes in the applied voltage of 10 V every 10 minutes. When the electric stimulus was removed, the IPN displaced to its original position. Also, if the polarity of the electric field was altered, IPN bent in the opposite direction. As shown in Fig. 6.6, PVA/PAAc IPN exhibited reversible bending behaviour according to the application of the electric field.

6.3.6 Drug release behaviour under electric fields

To investigate the electrically induced release behaviours of drugs, we determined the amount of released drug from PVA/PAAc IPN into a release medium under various conditions by UV-spectrophotometer. The drug-loaded IPN disk was placed between two carbon electrodes in a release chamber filled with a release medium. Applied voltage was then altered every 30 min.

Figure 6.7 exhibits the release behaviours of drugs from IPN37 as a function of time and the voltage of applied electric field in 0.9 wt % NaCl solution at 37 °C. The release behaviours of drugs from IPN37 as a function of time and the voltage of applied electric field in 0.9 wt % NaCl solution at 37 °C were investigated. Figure 6.7 shows the plot of M_t/M_∞ vs. time (min) while altering electric stimulus. Here, M_t is the amount of drug released up to time t, and M_∞ is the amount of drug present in the IPN at time $t = 0$. Depending on the electric stimulus being turned on and off, the release of drug molecules loaded

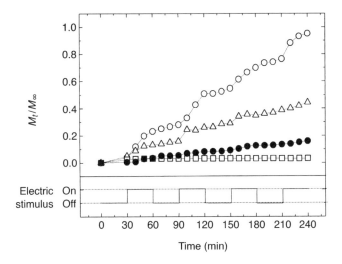

6.7 Release behaviours of cefazoline and theophylline from PVA–PAAc IPN55 as a function of applied voltage in 0.9% NaCl solution at 37 °C. Electric potential was altered every 30 min. —○— IPN55C, 10 V, —△— IPN55C, 5 V, —□— IPN55C, 0 V, —●— IPN55T, 5 V.

within PVA/PAAc IPN was switched on and off in a pulsatile fashion. A possible reason for this switching pattern of the release of drug molecules was that the electrically induced changes in osmotic pressure within the gel and local pH gradient attributed to water electrolysis could affect the swelling of IPN with charged groups.

In the case of theophylline (non-ionic solute), however, the magnitude of the 'ON–OFF' release pattern induced by the electrical stimulus was not so large as that of cefazolin (ionic solute). Particularly, the release rate of theophylline from IPN55T (theophylline-loaded IPN55) under 5 volts of electric potential was even smaller than that of cefazolin from IPN55C (cefazolin-loaded IPN55) without electric stimulus.

We conducted the release experiment of a drug from cefazolin- or theophylline-loaded IPN46 in NaCl solution with different ionic strength to examine the effect of the ionic strength of the release medium. Figure 6.8 shows the release behaviours of cefazolin and theophylline as a function of ionic strength in release medium at 5 V. In the case of IPN46C, the release rate of cefazolin increased in inverse proportion to the ionic strength of the release medium. Therefore, an increase of electrolyte concentration in the solution induces the movement of free ions from the surrounding solution toward their counter electrode or into IPN, resulting in an increase in release rate of cefazolin from charged PVA/PAAc IPN. The change of swelling in the PVA/PAAc IPN due to the ionic strength of the surrounding medium does not

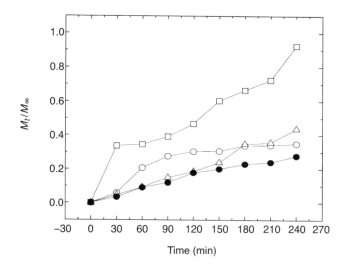

6.8 Effect of ionic strength in release medium on release behaviours of cefazoline and theophylline from IPN46 under 5 V applied voltage. —○— IPN46C, 0.01 M NaCl, —△— IPN46C, 0.1 M NaCl, —□— IPN46C, 0.2 M NaCl, —●— IPN46T, 0.1 M NaCl.

significantly affect the release behaviours of the drug. It rather indicates that the electrically induced changes of osmotic pressure within the gel, the local pH gradient attributed to water electrolysis, and the squeezing effect of IPN gel were more important than the decrease of the swelling ratio due to the ionic strength of the release medium. However, in view of the ionic property of drugs, we can see that the release rate of ionic drugs is much faster than that of non-ionic drugs. An IPN46T sample containing a non-ionic drug, theophylline, exhibited a slower release rate in 0.1 M NaCl at 5 V than the ionic drug, cefazolin, in 0.01 M NaCl. This could be attributed to the enhancement of the release by electrophoresis of charged drug molecules.

6.4 Conclusions

PVA/PAAc IPNs with various compositions were prepared by a sequential synthetic method composed of UV polymerization and freeze–thawing process. A PVA/PAAc IPN exhibited different swelling patterns as a function of pH, temperature, ionic strength of external solution, and the content of ionic group in IPN. The release of indomethacin from these IPN hydrogels was dominated by the magnitude of swelling in the IPN hydrogels. In the temperature-sensitive system, the positive release of drugs caused by the hydrogen bonding between PVA and PAAc was observed, showing an increase of the released amount with higher temperature. In the case of pH-dependent release, changing the pH controlled the release rate. At pH 7, higher release amounts were observed, since the water contents after swelling increased significantly. The permeability of non-ionic solutes changed according to the variation in swelling in the IPN hydrogel. Smaller solutes and higher swelling of the IPN hydrogel produced higher diffusion coefficients. However, the permeability of ionic solutes at pH 7 through IPN decreased, while the swelling of the IPN hydrogel increased. When a swollen PVA/PAAc IPN was placed between a pair of electrodes, the IPN exhibited bending behaviours upon the application of an electric field. The release behaviours of drug molecules from negatively charged PVA/PAAc IPN were switched on and off in a pulsatile pattern depending on the applied electrical stimulus. The release amount and the release rate of the drug were influenced by the applied voltage, the ionic group contents of the IPN, the ionic properties of the drug, and the ionic strength of the release medium. Therefore, the present IPN system was valid in pH/temperature/electrical-responsive drug delivery.

References

1 Kwon C, Bae Y H and Kim S W, 'Heparin release from polymer complex', *J. Controlled Release*, 1994, **30**, 155.
2 Yuk S H, Cho S H and Lee H B, 'Electric current-sensitive drug delivery systems using sodium alginate/polyacrylic acid composites', *Pharm. Res.*, 1992, **9**, 955.
3 Okahata Y and Seki T, 'pH-sensitive permeation of ionic fluorescent probes nylon capsule membranes', *Macromolecules*, 1984, **17**, 1880.
4 Okahata Y, Noguchi H and Seki T, 'Functional capsule membrane. 26. Permeability control of polymer-grafted capsule membrane responding to ambient pH changes', *Macromolecules*, 1987, **20**, 15.
5 Gudeman L F and Peppas N A, 'pH-sensitive membranes from poly(vinyl alcohol)/poly(acrylic acid) interpenetrating networks', *J. Mem. Sci.*, 1995, **107**, 239.
6 Feil H, Bae Y H, Feijen T and Kim S W, 'Mutual influence of pH and temperature on the swelling of ionizable and thermosensitive hydrogels', *Macromolecules*, 1992, **25**, 5228.
7 Yoshida M, Yang J S, Kumakaru M, Hagiwara M and Katakai R, 'Artificially intelligent hydrogels: responding to external stimuli such as temperature and pH', *Eur. Polym. J.*, 1991, **27**, 997.
8 Gutowska A, Bae Y H, Jacobs H, Feijen J and Kim S W, 'Thermosensitive interpenetrating polymer network: synthesis, characterisation, and macromolecular release', *Macromolecules*, 1994, **27**, 4167.
9 Okano T, Bae Y H, Jacobs H and Kim S W, 'Thermally on–off switching polymers for drug permeation and release', *J. Controlled Release*, 1990, **11**, 255.
10 Siegal R A and Firestone B A, 'pH-dependent equilibrium swelling properties of hydrophobic polyelectrolyte copolymer gels', *Macromolecules*, 1988, **21**, 3254.
11 Kim J H, Kim J Y, Lee Y M and Kim K Y, 'Properties and swelling characteristics of crosslinked poly(vinyl alcohol)/chitosan blend membrane', *J. Appl. Polym. Sci.*, 1992, **45**, 1711.

12 Kim J H, Kim J Y, Lee Y M and Kim K Y, 'Controlled release of riboflavin and insulin through crosslinked poly(vinyl alcohol)/chitosan blend membrane', *J. Appl. Polym. Sci.*, 1992, **44**, 1823.
13 Katono H, Maruyama A, Sanui K, Ogata N, Okano T and Sakurai Y, 'Thermo-responsive swelling and drug release switching of interpenetrating polymer networks composed of polycylamide-*co*-butyl methacrylate) and poly(acrylic acid)', *J. Controlled Release*, 1991, **16**, 215.
14 Schilld H G, 'Poly(*N*-isopropyl acrylamide): experiment, theory and application', *Prog. Polym. Sci.*, 1992, **17**, 163.
15 Brannon L and Peppas N A, 'Solute and penetrant deffusion in swellable polymers. IX. The mechanisms of drug release from pH-sensitive swelling-controlled systems', *J. Controlled Release*, 1989, **8**, 267.
16 Hoffman A S and Dong L C, 'A novel approach for preparation of pH-sensitive hydrogels for enteric drug delivery', *J. Controlled Release*, 1991, **15**, 141.
17 Kishi R, Hara M, Sawahata K and Osada Y, *Polymer Gels: Fundamentals and Biomedical Applications*, Plenum Press, New York, 1991, p. 205.
18 Osada Y, Okuzaki H and Hori H, 'A polymer gel with electrically driven motility', *Nature*, 1992, **355**, 242.
19 Chiarelli P, Umezawa K and Rossi D D, *Fundamentals and Biomedical Applications*, Plenum Press, New York, 1991, p. 195.
20 Ma J T, Liu L R, Yang X J and Yao K D, 'Blending behavior of gelatin/poly(hydroxyethyl methacrylate) IPN hydrogel under electric stimulus', *J. Appl. Polym. Sci.*, 1995, **56**, 73.
21 Kim S S and Lee Y M, 'Synthesis and properties of semi-interpenetrating polymer networks composed of β-chitin and poly(ethylene glycol) macromer', *Polymer*, 1995, **36**, 4497.
22 Lee Y M, Kim S H and Cho C S, 'Synthesis and swelling characteristics of pH and thermo-responsive interpenetrating polymer network hydrogel composed of poly(vinyl alcohol) and poly(acrylic acid)', *J. Appl. Polym. Sci.*, 1996, **62**, 301.
23 O'Neil M J, Smith A, Heckelman P E and Budavari S, in *The Merck Index*, 11th ed., Merck & Co., Rahway, NJ, 1989.
24 Florey K, *Analytical Profiles of Drug Substances, Vol. 4*, Academic Press, New York, 1975.

7
Permeation control through stimuli-responsive polymer membrane prepared by plasma and radiation grafting techniques

YOUNG MOO LEE AND JIN KIE SHIM

7.1 Introduction

Stimuli-sensitive polymer drug delivery, especially using temperature and pH-sensitive hydrogels and polymeric membranes, may be applied to enhance environment-sensitive properties. Temperature-sensitive polymers demonstrate lower critical solution temperatures (LCST), which provide reversible high swelling at low temperatures and low swelling at high temperatures.[1-4] Similarly, pH-sensitive polymers synthesized with either acidic or basic components demonstrate reversible swelling/deswelling in an acidic or basic medium.[5-14]

Hoffman and Dong[7] combined these two properties by synthesizing cross-linked pH/temperature-sensitive hydrogels. Positively charged hydrogels were prepared by copolymerizing in varying ratios of N-isopropylacrylamide and N,N'-dimethylaminopropylmethacrylamide. Kim et al.[14] made beads formed from linear pH/temperature polymers, poly(N-isopropylacrylamide-co-butylmethacrylate-co-acrylic acid) and reported on the release of insulin through the polymeric system.

There have been several studies reported on the surface modification of porous membranes by grafting acrylic monomers utilizing corona discharge, glow discharge and UV techniques.[15-21] Iwata et al. reported on the graft polymerization of acrylamide onto a polyethylene[15,16] using corona discharge and of N-isopropylacrylamide onto a poly(vinylidene fluoride)[17] using the glow discharge technique. Osada et al.[18] grafted poly(methacrylic acid) onto a porous poly(vinyl alcohol) membrane using plasma treatment, and their ion transport, albumin and poly(ethylene glycol) permeation behaviours were studied. X-ray photoelectronspectroscopy (XPS) was utilized to confirm the graft reaction.[15-18] Recently, Ito et al.[19] reported on the water permeation through porous polycarbonate membrane having poly(carboxylic acid) on the membrane surface by pH and ionic strength.

It is our objective to review the intelligent pH/temperature-sensitive membrane by grafting acrylic acid (AAc) and N-isopropylacrylamide (NIPAAm) onto the porous polymer membrane, which was prepared by the plasma or UV polymerization technique.

7.2 Experimental

7.2.1 Materials

Acrylic acid (AAc) was obtained from Junsei Chemical Co. and used as a grafting monomer. It was purified by using a glass column filled with an inhibitor remover (Aldrich Co., Milwaukee, WI) and stored in a dark and cold place. N-isopropylacrylamide (NIPAAm; Tokyo Kasei Chemical Co., Japan) was used after recrystallization in hexane and toluene (40:60 in vol%). Riboflavin was purchased from Junsei Chemical Co. and used without any further treatment. Porous polyamide membrane from Gelman Science Co. with 127 μm in thickness, 0.45 μm pores and 1,1-dimethyl-2-picrylhydrazyl (DPPH) (Aldrich Co.) was used for the determination of peroxides produced by UV irradiation.

7.2.2 Determination of peroxide produced by UV irradiation

The PSF membrane was irradiated by a medium-pressure mercury lamp (450 W, Ace Glass Inc.) at room temperature in air for a predetermined time. The membrane samples were prepared to an area of 8.675 cm^2 and irradiated at a distance of 5 cm below the UV lamp for all samples. Peroxides produced on the membrane surface were determined by DPPH. The DPPH (1×10^{-4} mol/l) solution in toluene was degassed by nitrogen blowing for 60 min. The UV-treated PSF membrane was quickly dipped into the DPPH solution at 60 °C for 2 hours in a shaking water bath. Using the reacted solutions, a consumed amount of DPPH due to peroxides produced on the surface was measured by a UV spectrophotometer (UV-2101PC, Shimadzu, Kyoto, Japan) at 520 nm. A calibration curve was obtained by using a DPPH solution of a known concentration.

7.2.3 Preparation of stimuli-responsive polymeric membranes

7.2.3.1 *Preparation of poly(amide-g-NIPAAm) and poly(amide-g-(AAc-NIPAAm))*

For the plasma polymerization, a bell jar type plasma reaction apparatus was used to accommodate the AAc and NIPAAm monomers. This low pressure glow discharge apparatus has a radio-frequency generator (13.56 MHz, 0–300 W

Table 7.1 Effect of feed composition on the graft yield of surface modified polyamide membranes using plasma polymerization method at 60 °C for 2 hours

Sample	Feed composition NIPPAm/AAc (w/w)	Concentration (wt%)	Graft yield ($\mu g/cm^2$)
PA	—	20	—
PNA1	100:0	20	202
PNA2	95:5	20	222
PNA3	90:10	20	297
PNA4	85:15	20	197

power) and an impedance matching circuit. After washing and drying, the porous polyamide membrane was applied to 50 mtorr and 30 W argon plasma for 30 seconds. The sample was exposed immediately to the air and immersed in the monomer solution. The AAc and NIPAAm monomer solutions were prepared by dissolving each monomer in deionized water (see Table 7.1) for sample designation. To remove the oxygen remaining in the solution, nitrogen gas was bubbled into the solution at room temperature for 30 min. The graft polymerization of AAc and NIPAAm onto the plasma-treated polyamide was performed for 2 h at 60 °C.

The grafting amount of treated polyamide membranes was calculated as follows:

$$\text{Graft yield } (\mu g/cm^2) = \frac{(W_t - W_0)}{(A)} \times 100,$$

where W_0, W_t and A represent the weight of the membrane before and after the graft reaction and membrane area, respectively.

7.2.3.2 Preparation of PSF-g-AAc membrane

The PSF membrane (17.35 cm²) was weighed precisely and irradiated at the same conditions as in the peroxide-determination experiments. 20 wt. % AAc aqueous solution was degassed by nitrogen blowing for 60 min. UV-treated PSF was quickly dipped into the solution at 60 °C for 2 h in a shaking water bath. The grafted membranes were washed with deionized water at 70 °C for 24 h and dried in a vacuum oven at 40 °C for 48 h. Details on analytical equipment and methods and the solute permeation experiments can be found in the references.[20,21]

7.3 Results and discussion

7.3.1 Characterization of stimuli-responsive polymeric membranes

7.3.1.1 *Poly(amide-g-NIPAAm) and poly(amide-g-(AAc-NIPAAm)) membranes*

The surface of AAc and NIPAAm grafted polyamide membrane was analysed by XPS[22–25] as shown in Fig. 7.1 and 7.2. The PNA membrane surface showed new peaks resulting from the incorporation of –C–O– at ~ 286.6 eV and ester carbon atoms at ~ 289.1 eV (O=C–O–). The poly(amide-g-NIPAAm) (PNA1) surface showed a new peak ~ 288.0 eV as O=C–N, indicating the presence of *N*-isopropyl groups on the polyamide surface. The oxygen peak in the C=O groups appeared at 529 eV for membranes. The peak at high binding energy region (533.5 eV) is a peak for the oxygen in the –OH groups, indicating the presence of –COOH groups on the grafted PNA membrane surface. As the AAc content increases, the –OH peak becomes larger relative to the peak of the C=O bond.

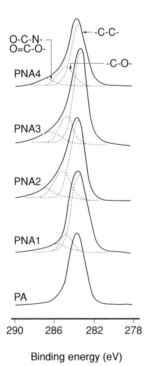

7.1 XPS carbon 1S core level spectra of polyamide and grafted polyamide surfaces.

In the present study, the graft yields of the polymer range between 197–297 μg/cm² (see Table 7.1) and increase with increasing AAc content, except for the PNA4 sample. When the AAc content was more than 10 wt % of the feed composition, the graft yield of the membrane decreased. These results also showed that upon increasing the pH, the reactivity ratio decreased and increased for acrylic acid and acrylamide, respectively. The pH of the feed solution in Table 7.1 decreased from 4.7 to 2.7 as acrylic acid increased. Therefore, the lower graft yield in PNA4 is probably due to the low reactivity of NIPAAm at above 10 wt % of acrylic acid in the feed composition because of the low pH.

7.3.1.2 PSF-g-AAc membranes

It is known that PSF and poly(ether sulfone) show strong absorption bands in the wavelength range between 250 and 300 nm.[26] During UV irradiation, chain scission, cross-linking and extensive yellowing occur. UV irradiation in a vacuum or in air yields several degraded products, such as gaseous products, oligomeric or polymeric sulfonic acids, and polymeric peroxides.

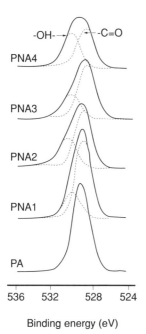

7.2 XPS carbon 1S core level spectra of polyanide and grafted polyamide surfaces.

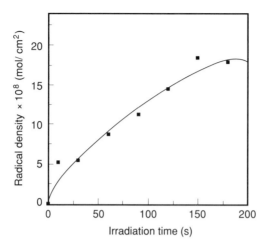

7.3 Effect of radical density as a function of UV irradiation time as determined by the DPPH method.

Kuroda et al.[26] reported the tendency of chain scission and cross-linking of poly(ether sulfone) films below and above the glass transition temperature (T_g). Chain scission and cross-linking occur simultaneously in all ranges of experimental temperatures. Cross-linking is dominant at above 170 °C, while chain scission is more important at room temperature. Yamashita et al.[27] reported on the photodegradation of poly(ether sulfone) and PSF in the presence and absence of oxygen over the temperature range from room temperature to 225 °C, and investigated the quantum yields for cross-linking and chain scission by gel permeation chromatography (GPC) measurements. They showed similar results on the temperature effect of chain scission and cross-linking from the degradation of poly(ether sulfone). However, in the case of PSF, chain scission occurs even at higher temperatures.

Figure 7.3 shows the radical density of a UV-irradiated PSF membrane with varying irradiation times. The radical density increased with increasing irradiation time up to 150 seconds, but decreased slightly with further irradiation. This indicates that the produced peroxides are partially converted into inactive species, which cannot generate radicals.

Table 7.2 summarizes the amount of PAAc grafted onto the surface of the PSF membrane. The graft amount of PAAc increased with irradiation time up to 150 s, exhibiting the same tendency as did the result from the radical density data. When we further irradiated the PSF samples, the graft amount and radical density of the modified PSF membrane decreased slightly. This means that long irradiation times do not always provide a merit in the amount of PAAc grafted onto the PSF membranes. Characterization and permeation experiments were done for grafted membranes irradiated up to 150 s.

Table 7.2 Effect of UV irradiation time on the amount of grafting of surface-modified PSF membranes

Sample[a]	Irradiation time (s)	Amount of grafting ($\mu g/cm^2$)
PSF	—	—
UA1	10	53
UA2	30	57
UA3	90	62
UA4	120	71
UA5	150	104
UA6	180	98

[a] A 20 wt% aqueous solution of AAc was used.

Table 7.3 XPS surface analysis of surface-modified PSF membranes by UV irradiation technique

Sample	Atomic (%)			
	C_{1S}	O_{1S}	S_{2P}	O_{1S}/S_{2P}
PSF	90.70	3.92	5.37	0.73
UA1	90.81	4.18	5.01	0.83
UA2	91.69	4.00	4.11	0.97
UA3	87.93	6.28	5.60	1.08
UA4	90.20	5.43	4.37	1.24
UA5	91.08	5.08	3.85	1.32

To investigate the chemical composition of the membrane surface, XPS analyses of the PSF and PSF-g-AAc membranes were performed and are summarized in Table 7.3. The atomic concentration of S_{2p} for the unmodified PSF membrane is 5.37% and can be used as a reference on the basis that it does not change after UV irradiation. The atomic concentration of grafted membranes, the ratio of O_{1S} to S_{2P}, gradually increases upon prolonging the UV irradiation time up to 150 s (UA5). This indicates that the atomic concentration of O_{1S} in the surface region increases as AAc is grafted further.

An inert gas ion beam was used to ablate the sample surface, and the chemical composition of the new surface was determined by XPS surface analysis. XPS ion-sputter depth profiling is a valuable means of determining the thickness of the graft layer. The effective thickness of the graft layer was calculated and the thickness of the PAAc grafted onto the PSF membrane surface in UA1 and UA3 was determined to be around 80–100 nm. Moreover, we can also determine the thickness of the graft layer from the relation

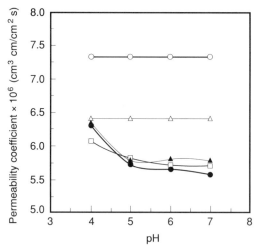

7.4 Effect of pH on the permeation of riboflavin through PA and PNA membranes measured at 37 °C. —○— PA, —△— PNA1, —□— PNA2, —●— PNA3, —▲— PNA4.

$l = m/\rho A$, where l is the thickness of the graft layer, m the amount of grafting, ρ the density of the graft PAAc and A the unit area. Here, the PAAc density is approximately 1. The thickness of the UA1 and UA3 samples can be calculated to be approximately 530–620 nm, indicating a large discrepancy between the thickness values determined by the two methods. The thickness determined by an XPS method is much smaller than that by the amount of grafting. This result suggests that PAAc was grafted not only onto the surface of the PSF membrane but also onto the inside of the PSF membrane. In this case, an aqueous AAc solution was expected to diffuse into the PSF membrane, and the vinyl monomer was grafted not only at the surface but also within the membrane. Tazuke[28] reported that acrylamide was grafted more from the bulk of the hydrophobic-oriented polypropylene membranes than on the surface of the membranes when the solvent had strong interactions with the base polymer.

7.3.2 Riboflavin permeation through stimuli-responsive membranes

7.3.2.1 Poly(amide-g-NIPAAm) and poly(amide-g-(AAc-NIPAAm)) membranes

The riboflavin release patterns of PA and PNA membranes at different pH regions and at 37 °C are demonstrated in Fig. 7.4. The permeation of riboflavin was also measured at various temperatures and at neutral pH, as shown in

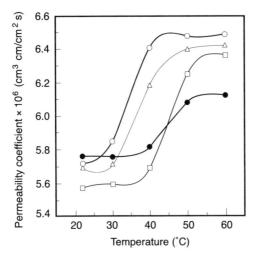

7.5 Effect of temperature on the permeation of riboflavin through PA and PNA membranes measured at pH 7. —○— PNA1, —△— PNA2, —□— PNA3, —●— PNA4.

Fig. 7.5. In PA and PNA1 membranes that did not have acrylic acid on the polyamide surface, the release profile of riboflavin was constant as pH varied. However, the PNA2, PNA3 and PNA4 membranes showed a pH-dependent release of riboflavin. The permeability of riboflavin decreased from 6.3×10^{-6} cm^3 cm/cm^2 s at pH 4 to around 5.7×10^{-6} cm^3 cm/cm^2 s at pH 7. The absolute permeability value was one order of magnitude higher than that of the PVDF membrane[29] in our previous study due to the larger pore radius of the PA membrane.

However, the magnitude of the decrease of permeability at pH 4 to pH 7 was somewhat lower than that of the PVDF membrane. This may be caused by two possible factors: the large pore size of the PA membrane and the lower acrylic acid content in the feed solution.

The effective graft chain for permeation is present on the inner pore or on the edge of the pore, and may shrink or enlarge upon pH changes. Large pore size may reduce the effectiveness of the pH-dependent permeation of small solutes unless the graft chain is long enough to cover the pores. Competitive AAc and NIPAAm reaction reduced the number of the AAc content in the chain and lowered the pH sensitivity, which is defined as the ratio of permeability at pH 4 and pH 7.

The effective pore size of the PNA membranes at any pH and temperature can be calculated using a simple Hagen–Poiseuille's law and the ratio of flux or permeability coefficient of the virgin and graft membrane.

Table 7.4 Determination of effective pore radii and effective areas for permeation at pH 4 and 7 through pH-sensitive polyamide membranes estimated by Eq. (7.1)

Sample	pH 4		pH 7		$\dfrac{A_{pH\,4}}{A_{pH\,7}}$ (%)
	r(Å)	A^* (× 10^6 Å^2)	r(Å)	A^* (× 10^6 Å^2)	
PA	2250	15.90	2250	15.90	100.0
PNA1	2176	14.88	2176	14.88	100.0
PNA2	2147	14.48	2115	14.05	103.1
PNA3	2167	13.75	2103	13.89	106.2
PNA4	2172	14.82	2121	14.13	104.9

*A = Effective membrane area for permeation of solute.

$$r_0/r_1 = [J_0/J_i]^{\frac{1}{4}} \qquad [7.1]$$

Here, J_0 and J_i are the fluxes of solute through the PA and PNA membranes respectively, at 37 °C in varying pH ranges.

The effective pore radius was reduced from 2250 Å to 2103 Å upon grafting and expansion of the graft chain at pH 7 (Table 7.4). The above results show the possibility of controlling the pH region, in which the riboflavin permeability changes most sensitively with pH, by choosing the nature of the polymers to be grafted.

As NIPAAm is grafted on the porous polyamide membrane using the plasma grafting technique, the permeability changes with temperature. Poly(NIPAAm)[1,2] is known to have LCST at around 31–33 °C. Below LCST, poly(NIPAAm) forms hydrogen bonds with water and exists in solution form. However, above LCST, inter- and intramolecular interaction in poly(NIPAAm) is much stronger, resulting in an undissolved state. In this state, the hydrogen bonding between grafted poly(NIPAAm) (PNA1) and water breaks down and the mobility of the polymer chain, inter- and intramolecular interactions, and the hydrophobic interaction due to the presence of the alkyl groups in the polymer chain increase.[30] The grafted poly(NIPAAm) chain shrinks, leading to an enlargement of the effective pore size in the porous polyamide membrane. Using the same Hagen–Poiseuille's equation and Eq. (7.1), the effective pore radius was calculated to be 2113 Å at 30 °C and the graft chains shrink to 2175 Å at 50 °C for PNA2 membrane, resulting in an expansion of the effective area for permeation, as indicated in increase in $A_{50°C}/A_{30°C}$ in Table 7.5.

PNA2, PNA3 and PNA4 used the acrylic acid as a comonomer with NIPAAm. In this case, the transition temperatures of riboflavin permeation change from 35 °C to 50 °C, as illustrated in Fig. 7.5.

Permeation control through stimuli-responsive polymer membrane 119

Table 7.5 Determination of effective pore radii and effective areas for permeation at 30 °C and 50 °C through temperature-sensitive polyamide membranes estimated by Eq. (7.1)

Sample	Temperature 30 °C		Temperature 50 °C		$\dfrac{A_{50°C}}{A_{30°C}}$ (%)
	$r(\text{Å})$	A^* (× 10^6 Å^2)	$r(\text{Å})$	A^* (× 10^6 Å^2)	
PA	2250	15.90	2250	15.90	100.0
PNA1	2126	14.20	2182	14.96	100.3
PNA2	2113	14.03	2175	14.86	105.9
PNA3	2102	13.88	2162	14.69	105.9
PNA4	2118	14.09	2147	14.48	102.8

*A = Effective membrane area for permeation of solute.

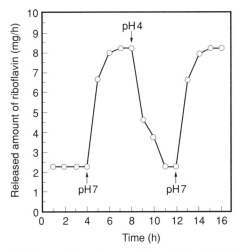

7.6 Reversible release pattern of riboflavin from PNA3 membrane with step-wise changing of the pH between 7 and 4 measured at 37 °C. —O— PNA3.

In order to study the pH/temperature-dependent change of the permeability of riboflavin, the permeation of riboflavin through the PNA3 membrane was investigated by alternating pH from 7 to 4 (Fig. 7.6), and temperature from 30 to 50 °C (Fig. 7.7). It was observed that a discontinuous change in concentration of riboflavin was brought about by stepwise changing of the temperature or pH. When the permeation experiment of riboflavin through the PNA3 membrane was conducted by changing pH or temperature, the riboflavin release increased rapidly but reverted to the same permeability within an hour.

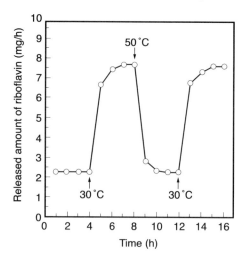

7.7 Reversible concentration control of riboflavin from PNA3 membrane with step-wise changing of the temperature between 30 °C and 50 °C measured at pH 7. —O— PNA3.

7.3.2.2 PSF-g-AAc membranes

Riboflavin permeability through the original PSF membrane and the PSF-g-AAc membranes is shown in Fig. 7.8, measured at different pH values. Note that the virgin PSF membrane exhibits no response to the change in pH, whereas PSF-g-AAc membranes are responsive to the pH change. A remarkable decline in permeability is noted in the range of pH 4–5. From these data, it is clear that the grafted PAAc is responsible for the permeability decline in riboflavin permeation.

It has been reported that the pK_a value of poly(acrylic acid) (PAAc) is 4.8.[20,30] Above pH 4.8, the carboxylic acid groups of the grafted PAAc chains are dissociated into carboxylate ions and have an extended conformation because of the electrostatic repulsion forces between the chains. Extended chains block the pores of the PSF membrane, causing a decrease in the permeability. At below pH 4.8, carboxylic acid groups do not dissociate: the grafted PAAc chains will shrink and be precipitated on the surface. Thus, the pores become open and permeability increases sharply. These conformational changes are obviously due to both intra- and intermolecular interactions between the grafted PAAc chains.

As the amount of grafted PAAc increases further, pore blocking overwhelms the conformational changes of the grafted chains due to the interactions of the polymer chains, causing only small changes in the permeability of riboflavin in response to the pH. Therefore, for a UA5 sample, the extension and

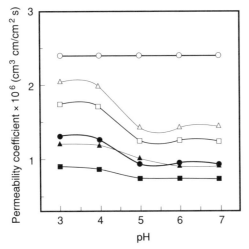

7.8 Effect of pH on the riboflavin permeation through (a) PSF (○), (b) UA1 (△), (c) UA2 (□), (d) UA3 (●), (e) UA4 (▲), (f) UA5 (■) membranes.

shrinkage of the grafted chains are hindered, and the extent of the change in the permeability is reduced, meaning that the permeability depends on the amount of grafting and the environmental pH.

7.4 Conclusions

We have prepared pH/temperature-sensitive polymer membranes by grafting AAc or NIPAAm or AAc and NIPAAm on the surface of polyamide and polysulfone membranes, utilizing plasma and UV radiation techniques. The structure of the graft chain was confirmed by XPS and ATR-FTIR spectra. Through this study, we investigated the morphology and permeability of the riboflavin of the unmodified and modified polymer membranes. AAc or NIPAAm or AAc and NIPAAm-grafted polymer membranes showed the pH or temperature and pH-dependent permeation behaviours of riboflavin, respectively.

Therefore, the present stimuli-responsive membrane system was valid in pH/temperature-responsive drug delivery or intelligent separation membranes. This study envisioned the possibility of controlling the permeability of the membrane by varying the pH of the external solution. These pH/temperature-responsive membranes will be useful for sensing and modulating external chemical signals, and also for drug-delivery applications, because they change their chain conformation according to the surrounding circumstances.

Acknowledgement

Financial support from the Korea Institute of Science & Technology Evaluation and Planning through National Research Laboratory Programme is greatly appreciated.

References

1. Okano T, Bae Y H, Jacobs H and Kim S W, 'Thermally on–off switching polymers for drug permeation and release' *J. Controlled Release*, 1990, **11**, 255.
2. Okano T, Bae Y H and Kim S W, 'On–off thermocontrol of solute transport: II. Solute release from thermosensitive hydrogels' *Pharmaceut. Res.*, 1991, **8**, 624.
3. Nozawa I, Suzuki Y, Sato S, Sugibayashi K and Morimoto Y, 'Preparation of thermo-responsive polymer membranes', *J. Biomed. Mater. Res.*, 1991, **25**, 243.
4. Okahata Y, Lim H J, Nakamura G and Hachiya S, 'A large nylon capsule coated with a synthetic bilayer membrane permeability control of NaCl by phase transition of the dialkylammonium bilayer coating', *J. Amer. Chem. Soc.*, 1983, **105**, 4855.
5. Kitano H, Akatsuka Y and Ise N, 'pH-responsive liposomes which contain amphiphiles prepared by using lipophilic radical initiator', *Macromolecules*, 1991, **24**, 42.
6. Siegel R A, Falamarzian M, Firestone B A and Moxley B C, 'pH-controlled release from hydrophobic/polyelectrolyte copolymer hydrogels', *J. Controlled Release*, 1988, **8**, 179.
7. Hoffman S and Dong L, 'A novel approach for preparation of pH-sensitive hydrogels for enteric drug delivery', *J. Controlled Release*, 1991, **15**, 141.
8. Park T G and Hoffman A S, 'Synthesis and characterization of pH- and/or temperature-sensitive hydrogels', *J. Appl. Polym. Sci.*, 1992, **46**, 659.
9. Iwata H and Matsuda T G, 'Preparation and properties of novel environment-sensitive membranes prepared by graft polymerization onto a porous membrane', *J. Membrane Sci.*, 1988, **38**, 185.
10. Klumb L A and Horbett T A, 'Design of insulin delivery devices based on glucose sensitive membranes', *J. Controlled Release*, 1992, **18**, 59.
11. Kim J H, Kim J Y, Lee Y M and Kim K Y, 'Controlled release of riboflavin and insulin through crosslinked poly(vinyl alcohol)/chitosan blend membrane', *J. Appl. Polym. Sci.*, 1992, **44**, 1823.
12. Chung J, Ito Y and Imanishi Y, 'An insulin-releasing membrane system on the basis of oxidation reaction of glucose', *J. Controlled Release*, 1992, **18**, 45.
13. Ito Y, Kotera S, Ibana M, Kono K and Imanishi Y, 'Control of pore size of polycarbonate membrane with straight pores by poly(acrylic acid)', *Polymer*, 1990, **31**, 2157.
14. Kim Y H, Bae Y H and Kim S W, 'pH/temperature-sensitive polymers for macromolecular drug loading and release', *J. Controlled Release*, 1994, **28**, 143.
15. Iwata H, Kishida A, Suzuki M, Hata Y and Ikada Y, 'Oxidation of polyethylene surface by corona discharge and the subsequent graft polymerization', *J. Polym. Sci. Polym. Chem.*, 1988, **26**, 3309.

16 Suzuki M, Kishida A, Iwata H and Ikada Y, 'Graft copolymerization of acrylamide onto a polyethylene surface prepared with a glow discharge', *Macromolecules*, 1986, **19**, 1804.
17 Iwata H, Oodate M, Uyama Y, Amemiya H and Ikada Y, 'Preparation of temperature-sensitive membranes by graft polymerization onto a porous membrane', *J. Membrane Sci.*, 1991, **55**, 119.
18 Osada Y, Honda K and Ohta M, 'Control of water permeability by mechanochemical contraction of poly(methacrylic acid)-grafted membranes', *J. Membrane Sci.*, 1986, **27**, 327.
19 Ito Y, Inaba M, Chung D J and Imanishi I, 'Control of water permeation by pH and ionic strength through a porous membrane having poly(carboxylic acid) surface-grafted', *Macromolecules*, 1992, **25**, 7313.
20 Kim J H, Lee Y M and Chung C N, 'Permeation of drug through porous polyurethane membranes grafted with poly(acrylic acid)', *J. Korean Ind. Eng. Chem.*, 1992, **3**, 233.
21 Lee Y M, Ihm S Y, Shim J K, Kim J H, Cho C S and Sung Y K, 'Preparation of surface-modified stimuli-responsive polymeric membranes by plasma and ultraviolet grafting methods and their riboflavin permeation', *Polymer*, 1994, **36**, 81.
22 Palit S R and Ghosh P, 'Quantitative determination of carboxyl endgroups in vinyl polymers by the dye-interaction method, a theoretical investigation of molecular core binding and relaxation', *J. Polym. Sci.*, 1962, **58**, 1225.
23 Beamson G and Briggs D, *High Resolution XPS of Organic Polymers*, Wiley, UK, 1992, pp. 110–11, 188–201.
24 Clark D T, Cromarty B J and Dilks A H R, 'A theoretical investigation of molecular core building and relaxation energies in a series of oxygen-containing organic molecules of interest in the study of surface oxidation of polymer', *J. Polym. Sci. Polym. Chem.*, 1978, **16**, 3173.
25 Rabek J F, *Polymer Degradation*, 1st. edn., Chapman & Hall, London, 1995, p. 317.
26 Kuroda S I, Mita I, Obata K and Tanaka S, 'The tendency of chain scission and crosslinking of poly(ether sulfone) films below and above the glass transition temperature (Tg)', *Polym. Degrad. Stabil.*, 1990, **27**, 257.
27 Yamashita T H, Tomitaka T, Kudo K, Horie K and Mita I, 'The photodegradation of poly(ether sulfone) and PSF in the presence and absence of oxygen', *Polym. Degrad. Stabil.*, 1993, **9**, 47.
28 Tazuke S, Matoba T, Kimura H and Okada T, ACS Symposium Series 121, American Chemical Society, Washington, DC, 1980, p. 217.
29 Lee Y M and Shim J K, 'Plasma surface graft of acrylic acid onto porous poly(vinylidene fluoride) membrane and its riboflavin permeation', *J. Appl. Polym. Sci.*, 1996, **61**, 1245.
30 Mandel M, 'Energies in a series of oxygen-containing organic molecules of interest in the study of surface oxidation of polymers', *Eur. Polym. J.*, 1970, **6**, 807.

8
Mechanical properties of fibre Bragg gratings

XIAOGENG TIAN AND XIAOMING TAO

8.1 Introduction

Self-organized Bragg grating in optical fibres, or changes in refractive index induced by radiation was first observed[1,2] in 1978. Since permanent Bragg gratings were first written by Meltz et al.[3] in 1989 by the transverse holographic method, they have attracted great attention in the fields of telecommunication, sensing, smart materials and structures. Fibre Bragg grating (FBG)-based sensors have some distinct advantages. First, they give an absolute measurement insensitive to any fluctuation in the irradiance of the illuminating source, as the information is directly obtained by detecting the wavelength-shift induced by strain or temperature change. Secondly, they can be written into a fibre without changing the fibre size, making them suitable for a wide range of situations where small diameter probes are essential, such as in advanced composite materials, the human body, etc. Thirdly, they can be mass produced with good repeatability, making them potentially competitive with traditional strain gauges from the fabrication cost point of view.[4,5] Finally, they can be multiplexed in a way similar to methods used on conventional fibre-optic sensors, such as wavelength-, frequency-, time- and spatial-division-multiplexing and their combination,[6,7] making quasi-distributed sensing feasible in practice.

Because of these advantages, FBGs have been applied as embedded sensors to monitor or measure the internal strain of composite structures.[8–13] Friebele et al.[9,10] reported internal distribution strain sensing with fibre grating arrays embedded in continuous resin transfer moulding™ (CRTM™) composites. Bullock and his colleagues[11] used a translaminar embedded fibre grating sensor system to monitor the structure of composites. Du et al.[14] embedded fibre Bragg gratings in a glass woven fabric laminated composite beam for internal strain measurements. FBGs are also used in mine operating accurate stability control[15] and medical monitoring.[16]

With FBG sensors being prepared for field applications, growing interest is focused on the mechanical reliability of the fibre gratings. The influence of various parameters like humidity,[17,18] chemical composition of coating,[19] chemical agent,[20] stripping methods[21,22] and fibre splicing[23] have been investigated. At the same time, more than ten different models have been developed to estimate the lifetime from static and dynamic fatigue measurement under different environmental conditions.[24–26] In this chapter, we will examine the influence of the fabrication procedure and the mechanical properties of fibre Bragg grating sensors.

8.2 Fabrication techniques

8.2.1 Optical fibres

As shown in Fig. 8.1, an optical fibre consists of a core surrounded by a cladding whose refractive index is slightly smaller than that of the core. The optical fibre is coated during the fibre drawing process with a protective layer. Inside the fibre core, light rays incident on the core-cladding boundary at angles greater than the critical angle undergo total internal reflection and are guided through the core without refraction. Silica glass is the most common material for optical fibres, where the cladding is normally made from pure fused silica, and the core from doped silica containing a few mol% of germania. Other dopants, such as phosphorus, can also be used. Extra-low absorption occurs in a germanosilicate fibre, with a local minimum attenuation coefficient $\alpha = 0.3\,dB/km$ at 1.3 µm and an absolute minimum $\alpha = 0.16\,dB/km$ at 1.55 µm. Therefore, light in the two windows travels down tens of kilometres of fibre without strong attenuation in a correctly guided mode condition.

8.2.2 Fibre Bragg gratings

A fibre Bragg grating consists of a modulation in the refractive index, $n(z)$, along a short length (z) of the core in germania-doped silica fibre with a period of Λ, which is given by:

$$n(z) = n_0 + \Delta n_0 \cdot \cos[(2\pi/\Lambda) \cdot z + \phi_0] \quad [8.1]$$

where n_0 is the linear refractive index for a guided mode in the fibre core, Δn_0 is

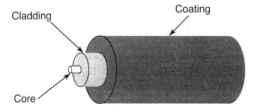

8.1 Structure of optical fibre.

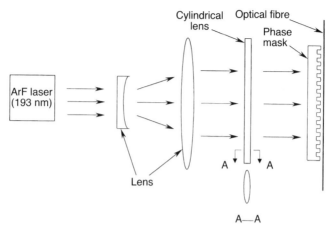

8.2 The set-up of FBG fabrication.

the modulation amplitude of the refractive index and ϕ_0 is the initial phase of the grating.

The inscription of permanent Bragg grating can be achieved in the core of an optical fibre by interference of two coherent ultra-violet light beams.[3] The grating thus has a holographically induced refractive index modulation with a period $\Lambda = \lambda_{uv}/2\sin(\theta/2)$, where θ is the angle between the two writing beams of wavelength λ_{UV}. When a broadband light source is coupled into the fibre, those components with wavelengths that satisfy the Bragg condition are strongly reflected, but all other components pass through the grating with negligible insertion loss and no change in signal. The central reflecting wavelength, referred to as the Bragg wavelength (λ_B) of a fibre grating, is determined by the Bragg condition:

$$\lambda_B = 2n_0\Lambda \qquad [8.2]$$

The central reflectivity is given by:

$$R = \tanh^2(\pi \Delta n_0 L/\lambda_B) \qquad [8.3]$$

where L is the grating length. The modulation amplitude of the refractive index, Δn_0, represents the grating strength and is dependent on the time of exposure to UV illumination. The width of the FBG reflecting band is determined by L and Δn_0. The higher the values of Δn_0 and L, the narrower the bandwidth.

Nowadays, the gratings are often fabricated by phase mask technology. Figure 8.2 shows an experimental set-up for writing Bragg gratings on single mode germanium-doped silica fibres. The light is spread by using a lens after it is guided from an ArF excimer laser, then paralleled by passing through a cylindrical lens and focused on the optical fibre by using another cylindrical lens, which is normal to the other lens.

8.3 Mechanisms of FBG sensor fabrication

Why can UV irradiation cause a change in the refractive index? To date, there appear to be three possible mechanisms by which a photo-induced refractive index change can occur in germanosilica optical fibres: (i) through the formation of colour centres (GeE'), (ii) through densification and increase in tension, and (iii) through formation of GeH. Broadly speaking, all three mechanisms prevail in germanosilica optical fibres. The relative importance of each contribution depends on the type of optical fibre and the photosensitization process used.

8.3.1 The formation of colour centres (GeE') and GeH

There are a lot of defects in the germanium-doped core. The paramagnetic Ge(n) defects, where n refers to the number of next-nearest-neighbour Ge atoms surrounding a germanium ion with an associate unsatisfied single electron, were first identified by Friebele et al.[27] These defects are shown schematically in Fig. 8.3.

The Ge(1) and Ge(2) have been identified as trapped-electron centres.[28] The characteristic absorption of Ge(1) and Ge(2) are 280 nm and 213 nm. The GeE', previously known as the Ge(0) and Ge(3) centres, which is common in oxygen-deficient germania, is a hole trapped next to a germanium at an oxygen vacancy,[29] and has been shown to be independent of the number of next-neighbour Ge sites. Here, an oxygen atom is missing from the tetrahedron, while the germanium atom has an extra electron as a dangling bond. Other defects include the non-bridging oxygen hole centre (NBOHC), which is claimed to have absorption at 260 nm and 600 nm, and the peroxy radical, believed to have absorption at 163 nm and 325 nm.[30] Both are shown in Fig. 8.3.

If the optical fibre is treated with high temperature hydrogen, germania will decrease and the concentration of GeO molecules will be enhanced. The reduction process may occur as follows:

$$-\text{Ge}-\text{O}-\text{Si}- + 1/2\,\text{H}_2 \longrightarrow -\text{Ge}^{e^-} + \text{OH} - \text{Si}-$$

Most fibres, if not all, show an increase in the population of the GeE' centre (trapped hole with an oxygen vacancy) after UV exposure. This is formed by

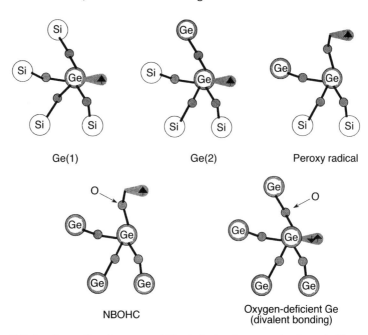

8.3 A schematic of proposed Ge defects of germania-doped silica.

the conversion of the electron-trapped Ge(I) centre, which absorbs at ~ 5 eV, and the GeO defect. GeO defect is shown in Fig. 8.4. It has a germanium atom coordinated with another Si or Ge atom. This bond has the characteristic 240 nm absorption peak that is observed in many germanium-doped photosensitive optical fibres.[31] On UV illumination, the bond readily breaks, creating the GeE' centre. It is thought that the electron from the GeE' centre is liberated and is free to move within the glass matrix via hopping or tunnelling, or by two-photon excitation into a conduction band. The change in the population of the GeE' centres causes changes in the UV absorption spectra, which lead to a change in the refractive index of the fibre at a wavelength λ directly through the Kramers–Kronig relationship:[32]

$$\Delta n = \frac{1}{(2\pi)^2} \sum_i \int_{\lambda_1}^{\lambda_2} \frac{\Delta\alpha_i(\lambda') \cdot \lambda^2}{(\lambda^2 - \lambda'^2)} d\lambda' \qquad [8.4]$$

where the summation is over discrete wavelength intervals around each of the i changes in measured absorption, $\Delta\alpha_i$. Therefore, a source of photoinduced change in the absorption at $\lambda_1 \leq \lambda' \leq \lambda_2$ will change the refractive index at wavelength λ. This process is common to all fibres. The colour centre model, originally proposed by Hand and Russel,[33] only explains part of the observed refractive index changes of ~ 2×10^{-4} in non-hydrogenated optical fibres.[34]

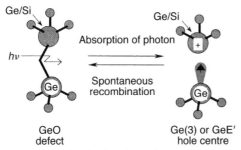

8.4 The capital GeO defect of germania-doped silica, in which the atom adjacent to germanium is either a silica or another germanium.

For molecular hydrogen, the suggested reaction is the formation of GeH and OH ions from a Ge(2) defect. The possible route may be as follows:

$$-\overset{|}{\underset{|}{Ge}} - O - \overset{|}{\underset{|}{Ge}} - + H_2 \longrightarrow -\overset{H}{\underset{|}{Ge}}\overset{|}{\underset{|}{\bullet}}\overset{e^-}{} + H - O - \overset{|}{\underset{|}{Ge}}-$$

After the treatment, the concentration of GeE' and GeH will increase greatly.

8.3.2 Densification and increase in tension

The refractive index of glass depends on the density of the material also,[35,36] so that a change in the volume through thermally induced relaxation of the glass will lead to a change Δn. The refractive index n is shown as:

$$\frac{\Delta n}{n} \approx \frac{\Delta V}{V} \approx \frac{3n}{2}\varepsilon \qquad [8.5]$$

where the volumetric change ΔV as a fraction of the original volume V is proportional to the fractional change ε in linear dimension of the glass.

A possible effect of the irradiation is a collapse of a higher-order ring structure leading to densification.[37] The densification of silica under UV irradiation is well documented.[38] The process of densification has been shown to occur in fibres, as evidenced by scans using an atomic force microscope of the surface of D-shape fibres and in etched fibres,[39] and in preform samples that were drawn into a D-shaped fibre.[35] These observations are on the surface of the sample and are unable to replicate the stress profiles within the core of the fibre directly. Direct optical measurement of in-fibre stress has indicated that, rather than the relief of the stress, tensile stress actually increases with an associated reduction in the average refractive index by $\sim$ 30% of the observed UV induced refractive index change in non-hydrogen loaded, high germania-content fibre. The changes in the stress profile of the fibre are consistent with the shift in the Bragg wavelength of a grating during inscription.[40]

8.4 Mechanical properties

8.4.1 Strength

Telecommunication optical fibres are made from silica glass with ring structures of Si–O tetrahedral. Defects or small imperfections which have been created on the surface of optical fibre during the fibre fabrication processing lead to a reduced fibre strength. Griffith suggested that these defects, called flaws, act as stress intensifiers and cause fracture, when they induce large enough stress to break chemical bonds mechanically.[41] The stress intensity factor K_I at a crack tip as a function of time, t, can be written as:[24]

$$K_I(t) = Y\sigma_\alpha(t)\sqrt{\alpha(t)} \qquad [8.6]$$

where Y is the dimensionless crack geometry shape parameter which is assumed to be constant for a given flaw type, σ_α is the applied stress and α is the flaw depth, i.e. the flaw size normal to the direction of the applied stress.

When the applied stress, σ_α, increases to its allowable maximum given by the intrinsic strength, S, of the crack, a catastrophic failure occurs. Equation (8.6) becomes:

$$K_{IC} = YS\sqrt{\alpha} \qquad [8.7]$$

K_{IC} is an intrinsic constant of silica and has been empirically determined to be: $K_{IC} = 0.79 \text{ MPa m}^{1/2}$.

If $n(\sigma)$ denotes a flaw distribution, then the number of the flaws which will fail between σ and $\sigma + d\sigma$ is $n(\sigma)d\sigma$. The total number of flaws whose strengths are less than σ is:

$$N(\sigma) = \int_0^\sigma n(\sigma)d\sigma \qquad [8.8]$$

According to the Weibull distribution, the empirical form of the cumulative flaw distribution is given by:

$$N(\sigma) = \left(\frac{\sigma}{\sigma_0}\right)^m \qquad [8.9]$$

where σ is the applied stress to the fibre during the fatigue test, m is the scaling factor and σ_0 is a constant characteristic of the material. The probability of a breakage between stress level σ and $\sigma + d\sigma$ is given by the joint probability that the sample contains a flaw whose strength falls into this interval and that the sample has survived a stress level σ:

$$dF(\sigma) = F(\sigma + d\sigma) - F(\sigma) = (1 - F(\sigma))n(\sigma)d\sigma \quad [8.10]$$

$F(\sigma)$, commonly referred to as the cumulative failure probability, is defined as the probability of breakage below a stress level σ. Consequently $1 - F(\sigma)$ is the survival probability. Assuming a group of M samples, $F(\sigma)$ is calculated as below:

$$F(\sigma) = \frac{i - 0.5}{M} \quad i = 1, 2, \ldots, M \quad [8.11]$$

The failure stresses are listed with increasing magnitude as $\sigma_1, \sigma_2, \ldots, \sigma_i, \ldots, \sigma_M$ so that the cumulative failure probability $F(\sigma_i)$ can be determined. In the Weibull plot, the breaking stress σ is plotted versus the cumulative failure probability F. To compare different Weibull distributions, we use the median breaking stress, σ_m, which corresponds to 50% failure probability. Combining Eqs. (8.9) and (8.10) we obtain:

$$\log\left(\ln\frac{1}{1 - F(\sigma)}\right) = m \log\left(\frac{\sigma}{\sigma_0}\right) \quad [8.12]$$

The slope of the curve is the Weibull scaling parameter m, which indicates the shape of the distribution. Low m-values indicate a large range of surface flaw sizes, while high m-values indicate high surface quality and a close approach to the tensile strength of a flaw-free fibre under normal condition.[42] m-Values for pristine fibres are normally of the order of 80–150.

8.4.2 Young's modulus and hardness

The tensile test can provide information on the average Young's modulus and mechanical strength. A nano-indenter can measure the Young's modulus and hardness locally on a much smaller scale. Interest in load and displacement-sensing indentation testing as an experimental tool for measuring elastic modulus began in the early 1970s.[43] The nano-indenter is made of very hard material. When it is pressed on the surface of the tested material, deformation occurs on both the indenter and the tested material; according to the deformation of the material and the acted force, the material properties, such as Young's modulus and hardness, can be obtained.

The effects of the non-rigid indenter on the load-displacement behaviour can be effectively accounted for by defining a reduced modulus, E_r, through the equation:

$$\frac{1}{E_r} = \frac{1 - v_s^2}{E_s} + \frac{1 - v_i^2}{E_i} \quad [8.13]$$

where E_i and v_i are the Young's modulus and Poisson's ratio for the indenter and E_s and v_s are those of the material being tested. Therefore, the Poisson's

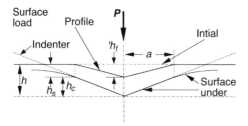

8.5 A schematic representation of load vs. indenter displacement showing quantities used in the analysis as well as the graphical interpretation of the contact depth.

ratio must be estimated for the material being tested to determine the modulus.

Using instrumented microhardness testing machines, the indentation load-displacement curve can be obtained as shown in Fig. 8.5. They are analysed according to the equation:

$$S = \frac{dP}{dh} = \frac{2}{\pi} E_r \sqrt{A} \qquad [8.14]$$

Here, $S = dP/dh$ is the experimentally measured stiffness of the upper portion of the unloading data, E_r is the reduced modulus (previously defined) and A is the projected area of the elastic contact.

The contact area at peak load can be determined by the geometry of the indenter and the depth of contact, h_c, assuming that the indenter geometry can be described by an area function $G(h)$ which relates the cross-sectional area of the indenter to the distance from its tip, h. Given that the indenter does not itself deform significantly, the projected contact area at peak load can then be computed from the relation $A = G(h_c)$. The functional form of G must be established experimentally prior to analysis.[43]

At any time during loading, the displacement, h, can be written as:

$$h = h_c + h_s \qquad [8.15]$$

where h_s is the displacement of the surface at the perimeter of the contact, which can be ascertained from the load-displacement data. At the peak, the load and displacement are P_{max} and h_{max}, respectively. We have $h_s = \varepsilon \frac{P_{max}}{S}$, ε is the geometrical constant, which relates to the geometry of the indenter.[44] ε for different geometries of indenter are listed in Table 8.1. The contact depth, h_c, can be expressed as:

$$h_c = h_{max} - \varepsilon \frac{P_{max}}{S} \qquad [8.16]$$

Table 8.1 ε for different shapes of indenter

Indenter	Conical indenter	Triangular pyramid	Flat punch
ε	0.72	0.75	1.00

So the contact area, A, can be obtained from Eq. (8.14) and combining it with Eq. (8.13), Young's modulus of the tested material can be expressed as:

$$E_s = \left[\frac{2}{\sqrt{\pi}}\frac{\sqrt{A}}{S} - \frac{1-v_i^2}{E_i}\right]^{-1} \cdot (1-v_s^2) \qquad [8.17]$$

The data obtained using the current method can be used to determine the hardness, H. The hardness is defined as the mean pressure the material will support under load. The hardness is computed from:

$$H = \frac{P_{max}}{A} \qquad [8.18]$$

8.5 Influence of the UV-irradiation on mechanical properties

8.5.1 The influence on the mechanical strength

8.5.1.1 The mechanical strength of gratings not affected

Although type I grating has a reflectivity of only a few per cent, it can easily be written at any arbitrary position along the fibre.[45] Such FBG arrays are ideally suited for distributed sensing, where low reflectivity is not a major drawback. Askins et al.[46] and Hagemann et al.[47] have studied the mechanical strength of optical fibre containing FBG fabricated in this method. In their experiment, laser exposures occurred only immediately before the protective jacketing was applied during fibre drawing.

FBG sensors produced on-line during the process using an excimer laser pulse suffer no measurable loss of strength relative to unirradiated fibre. This emphasizes the fact that it is favourable to use draw-tower FBGs as sensor elements to overcome the problem related to fibre stripping and high fluence UV irradiation.

8.5.1.2 Mechanical strength degradation of gratings

Feced et al.[48] has studied the influence of different wavelength UV irradiation on the strength of optical fibres with a single pulse radiation. The single mode

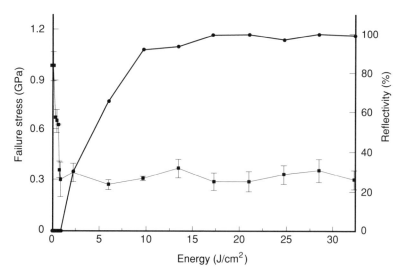

8.6 Mechanical strength and reflectivity: —■— strength, —●— reflectivity.

optical fibres used in their investigations were Corning's SMF-28 (∼ 3 mol % of germanium), and a highly photosensitive germanium and boron (Ge–B) co-doped fibre (10 mol % of germanium) from the University of Southampton. Their results suggest that the fabrication of Bragg gratings using 193 nm irradiation will yield higher-strength gratings with enhanced reliability. A similar analysis shows that the difference in strength between Ge–B doped fibre and SMF-28 when exposed to 248 nm radiation is significantly higher.

It has previously been reported that the processes induced in the fibre by 248 nm radiation are different from those induced by 193 nm radiation.[49] In addition, it is thought that the exposure of optical fibre to UV increases the stress in the core. The results reinforce these concepts; they suggest that the 248 nm mechanism increases the internal core stress of the fibre significantly more than the 193 nm mechanism, such that 248 nm radiation causes a larger degradation of the fibre strength.

We have tested the fracture strength of FBG sensors fabricated by ourselves. In order to eliminate the mechanical strength degradation of the coating stripping, we decoated the coating of the fibre with acetone prior to UV laser exposure. The UV radiation was a laser operating at 193 nm. The light focused with a cylindrical lens then passed through a phase mask before irradiating the fibre. The frequency of the UV irradiation is 5 Hz.

Figure 8.6 shows the mechanical strength and the reflectivity of the FBG sensors. In the fracture strength test, the length of the gauge is 0.25 m and the strain rate is 1 mm/min. It can be seen that the mechanical strength of the

Mechanical properties of fibre Bragg gratings 135

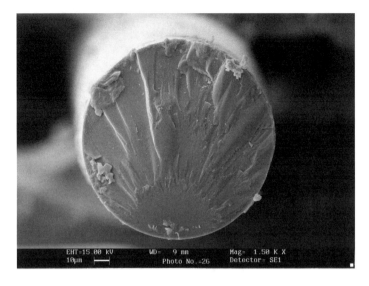

8.7 SEM of the cross-section of the broken grating. The start of the break is at the lower side. The UV laser is radiated from bottom to top.

FBG sensor degraded drastically with the increase in UV irradiation, and after a certain dose the mechanical strength will not degrade with the growth of UV irradiation. The strength of the FBG sensor is only about one-third of the optical fibre without UV irradiation. We can also see that the reflectivity increases sharply once the reflected peak appears, and then flattens as the UV irradiation increases. The strength degradation has been completed even before the reflected peak appears in our experiment, as if there were an obvious relation between the strength degradation and the reflectivity of the grating.

The SEM of the cross-section of the broken FBG sensor is shown in Fig. 8.7. The UV light irradiated the optical fibre from bottom to top. The beginning of the fracture was not always at the UV irradiation side, and the lateral side of optical fibre with UV irradiation was observed carefully with SEM; no flaw could be seen. This means the mechanical strength degradation takes place not only because of the flaws on the fibre surface induced by UV radiation. The mechanism of the mechanical strength degradation of the FBGs is not clear. Possibly, it is related to the structural change of the optical fibre molecule after UV irradiation.

Figure 8.8 is the comparison of the strength degradation with different irradiation energy/pulse at the same frequency. The strength degradation is larger with higher irradiation energy/pulse than with lower energy/pulse. This means that, in order to increase the mechanical strength of the FBGs, the

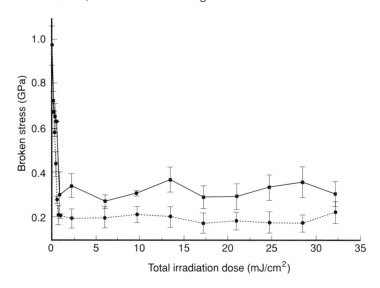

8.8 Strength degradation with different energy/pulse: —■— lower energy per pulse, ···●··· higher energy per pulse.

irradiation energy/pulse should be lowered. We also observed that, not only does the reflectivity increase slowly with the growth of the total dose of UV irradiation after certain reflectivity (99%), but also the line-width of the reflected peak gets wider. The wide reflected peak is disadvantageous for the accuracy of the test. Thus, in FBG fabrication, the UV irradiation time should not be too long.

Because FBG fabrication is a complex procedure and glass fibre is brittle, the mechanical strength degradation is related not only to the UV irradiation but also to the coating stripping method. We have stripped coating with different methods in our FBG fabrication and compared their influence on the mechanical strength of the optical fibre. Optical fibres were divided into two groups, each with 20 samples. One group had the coating stripped with acetone, and the other with sulfuric acid. Coating stripped with acetone was a physical procedure. The stripped part of the optical fibre was soaked into acetone for several minutes, the acetone caused the costing to soften and swell; the coating then separated from the cladding. The coating can be pulled off gently. When the optical fibre was immersed in sulfuric acid (98%) at room temperature ($\sim 20\,°C$), the coating was ablated by sulfuric acid, which took about 5 minutes. The optical fibre was then rinsed in water to remove the residual acid and other materials. No mechanical shock occurred to fibres in these two coating stripping procedures. The mechanical fracture during stripping was reduced to as little as possible. The mechanical strength of these stripped

Mechanical properties of fibre Bragg gratings   137

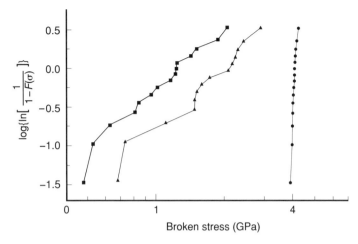

8.9 Weibull plot of optical fibre with different coating stripping method. —●— pristine fibre, —■— decoated with acetone, —▲— decoated with H_2SO_4.

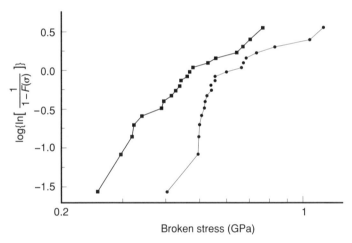

8.10 Influence of UV irradiation frequency on mechanical strength of FBGs. —■— 26 kV, 5 Hz, —●— 26 kV, 8 Hz.

optical fibres was tested. The Weibull plot is shown in Fig. 8.9. Table 8.2 gives the parameters for the distribution. The mechanical strength degraded notably when the coating was stripped with acetone. This may be related to the properties of these two solvents.

We have also compared the mechanical strength of FBGs fabricated with different frequencies of UV irradiation. The energy per pulse was the same

138 Smart fibres, fabrics and clothing

Table 8.2 Mechanical strength of optical fibre with different stripping method

	Stripped with acetone	Stripped with sulfuric acid	Pristine fibre
Mean of broken stress	1.14	1.76	4.09
Median stress	1.16	1.67	4.09
m value	2.8	2.8	52.0

Table 8.3 The influence of UV irradiation frequency

	Without UV irradiation	Grating 5 Hz	Grating 8 Hz
Mean of broken stress	1.76	0.47	0.63
Median stress	1.67	0.44	0.55
m value	2.8	4.0	4.1

Coating stripped with H_2SO_4.

with different frequencies. Figure 8.10 is the Weibull plot of FBGs with different UV irradiation. We can ascertain that the mechanical strength degradation is larger with lower frequency irradiation. As shown in Table 8.3, the m-values are almost the same in both situations and, obviously, the median broken strength of gratings with 8 Hz irradiation is higher than that of gratings fabricated with 5 Hz irradiation. This may be because irradiation with high frequency can increase the surface temperature of the fibre, resulting in the thermal annealing of some of the flaws produced in the first phase, thereby increasing the median break stress of the fibre.

We have also studied the influence of spot size on the mechanical strength. In FBG fabrication, we have found that when UV lasers pass through the phase mask, the density of light is not uniform. There was a notable change in the area of the edge of the phase mask. We have fabricated FBGs under different fabrication conditions and compared their mechanical strength. Group 1 was fabricated under normal fabrication conditions. The spot size is 27 mm × 1 mm. When Group 2 was fabricated, part of the UV light was covered so that the spot size was only within the mask area. The spot size was 9 mm × 1 mm. The edge effect of the phase mask was avoided. The Weibull plot of the FBG is shown in Fig. 8.11.

Mechanical properties of fibre Bragg gratings

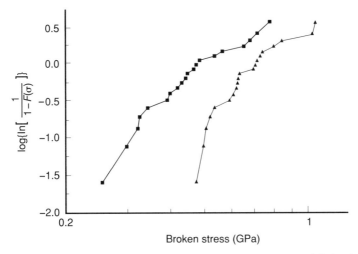

8.11 Weibull plot of FBGs with different spot size during fabrication: —■— 27 mm × 1 mm, —▲— 9 mm × 1 mm.

8.5.1.3 Enhancing the mechanical performance of FBG sensors

Because the strength degradation of the FBG sensors has important consequences for the reliability of the UV-induced Bragg gratings which are used as strain, temperature, and pressure sensors in a variety of fields, Varelas et al.[50,51] have examined the influence of homogeneous irradiation, using a continuous wave (CW) laser in the mechanical degradation of the fibre and compared it with the equivalent irradiation dose delivered by a pulsed laser source. Imamura et al.[52] have developed a new method for fabricating Bragg gratings with direct writing. The strength of Bragg gratings has been enhanced notably.

CW laser irradiation results in the mechanical resistance of a fibre being similar to that of a pristine fibre. This is in contrast to the case where the fibre undergoes pulse excimer irradiation. The total dose dependence is also less pronounced in the case of CW irradiation. This phenomenon has important consequences for Bragg grating fabrication, where high mechanical resistance is required for strain applications.

Although high strength Bragg gratings with higher reflectivity can be fabricated by CW UV exposure after chemical stripping of the protective polymer coating, this is not practical as it requires recoating and packaging. It has the potential for reducing the fibre strength due to exposure of the bare fibre to air. A direct fibre Bragg grating writing method through the coating has been demonstrated.[52,53] Imamura et al.[54] directly fabricated fibre gratings through the coating and have succeeded in producing a high strength fibre

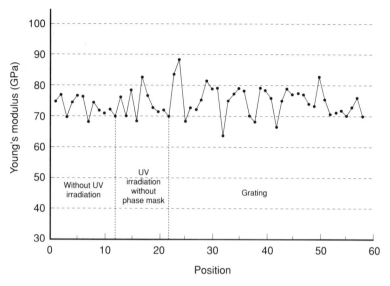

8.12 Young's modulus of the optical fibre.

grating. Tin-codoped germanosilicate fibre is more photosensitive than germanium-doped fibre.[55] The germanosilicate single mode fibre was codoped with 15 000 ppm Sn and 900 ppm Al in the core. To further enhance the fibre sensitivity, the fibres were treated under low temperature hydrogen for 2 weeks at 20 MPa prior to the UV exposure. The silica cladding had a diameter of 125 μm and was single coated with a UV-transparent UV-curable resin to an outer diameter of 200 μm.

8.5.2 Influence of UV irradiation on Young's modulus and hardness of fibre

The influence of UV irradiation on the mechanical strength of optical fibres is notable. How about the influence of UV irradiation to the other mechanical properties of optical fibres? Young's modulus and hardness of the optical fibre with and without UV irradiation have been tested using the Nano Indenter IIp.

In the test, the surface of the sample should be a smooth plane. The optical fibre is so thin that the lateral side has a large curvature. In order to get a plane in the lateral side of the optical fibre, the tested fibres were put into a polymer, then the polymer was polished carefully with lapping film (0.2 μm) so that a very smooth plane along the fibre was obtained. The sample contained three parts: optical fibre without UV irradiation, with UV irradiation but without phase mask, and grating. The result of Young's modulus of the cladding is

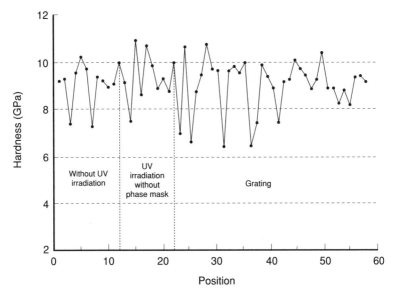

8.13 Hardness of the optical fibre.

shown in Fig. 8.12. There is no notable difference between the three parts. Statistical analysis[56] was performed. Suppose the measured data belong to normal distribution, with a 95% confidence level, the confidence interval of three parts are 71.29 GPa $< \mu <$ 75.30 GPa, 66.00 GPa $< \mu <$ 77.97 GPa and 73.28 GPa $< \mu <$ 76.76 GPa, respectively to optical fibre without UV irradiation, with UV irradiation without phase mask, and grating. They are almost in the same range. That is to say, the UV irradiation cannot influence Young's modulus of the optical fibre. Figure 8.13 is the hardness of the optical fibre. The UV irradiation cannot influence the hardness of the optical fibre either. Changing the depth of the plane (by polishing the tested fibre so that the test plane is closer to the core of the fibre), the results are the same.

From the experiment above we can confirm that it is reasonable to assume that the deformation of optical fibre contains uniform Bragg grating during the tension and the strain measured is believable.

8.6 Polymeric fibre

8.6.1 Properties of polymer optical fibre

Polymer optical fibre (POF) is a circular optical waveguide. POF is receiving increasing interest in short-distance communication systems because of its large power-carrying capacity, ease of joining and light weight. The structure

142 Smart fibres, fabrics and clothing

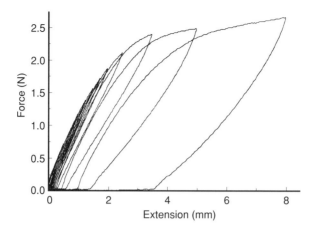

8.14 Extension properties of plastic optical fibre with ten cycles.

of POF is as the same as silica optical fibre. It consists of a central part, the core, with refractive index n_1 and of cylindrical shape, surrounded by a second part, the cladding, with a lower refractive index n_2. These two areas, which are essential for the light guiding, are composed of two different but transparent organic materials. In order to avoid external damage or the deterioration of this surface during the handling or use of the optical fibre, it is often covered with a protective coating which plays no part in the light guiding.

The core diameter varies according to the type of fibre, but is of the order of a millimetre, except in special cases. For economic reasons, the thickness of the cladding is much less, of the order of tens of micrometres. The surface of the core/cladding separation must be as even as possible.

A possible classification of POF can be based on the principal core material. The principal types of POF are fibres with PMMA (polymethyl methacrylate) core, fibres with PS (polystyrene) core, fibres with PC (polycarbonate) core and fibres with deuterated core.[57]

Physical properties, such as tensile resistance, compression resistance, shock resistance, torsion resistance, the effect of permanent winding, bending resistance, etc., temperature property and the chemical properties of POF have been studied systematically.[57]

8.6.1.1 The mechanical property

Figure 8.14 shows the elongation property of the measured fibre. In the measurement, the gauge length was 100 mm. The measurement was controlled by a programme, in which ten cycles were set. Each cycle contained loading and unloading; the speed was 0.5 mm/min. The elongation of the cycles was

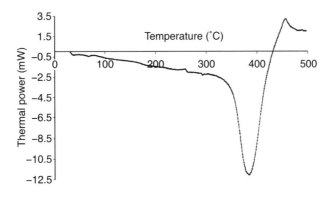

8.15 DSC of the sample of the plastic optical fibre.

0.5%, 1.0%, 1.25%, 1.5%, 1.75%, 2%, 2.5%, 3.5%, 5% and 8%, respectively. In this figure, the unloading curves were not the same as the loading curves. Like plastic material, the fibre has the hysteresis property. In the meantime, we find that, if the elongation of the fibre is larger than 1.5%, the sample cannot return to its original length even when the external force is released: this means that permanent deformation was induced in the sample. Sonic modulus was measured as 6.3 GPa.

8.6.1.2 DSC results

The heat flow to a sample can be measured in thermally controlled conditions with differential scanning calorimetry (DSC) by Mettler TA4000 Thermal Analysis System. The DSC measurement result is shown in Fig. 8.15. At 67.5 °C, some physical properties of the fibre were changed. This temperature is called the glass transition temperature. At 330 °C, the fibre began to melt. At 384.3 °C, the fibre began to be oxygenated, and it was completely oxygenated at 456.9 °C.

8.6.2 Bragg grating of polymer optical fibre

Although silica optical fibre Bragg gratings can be used to measure many physical parameters, the changes in the Bragg wavelength due to the change of these measured values are small. The measured range is limited, typically a few nanometres.[58] This is because silica glass has a small thermal effect and large Young's modulus. In the case of polymer optical fibre, the situation is different. Its Young's modulus (1 GPa), for example, is more than 70 times smaller than that of glass (72.9 GPa), and its elongation is much larger than that of glass optical fibre. The refractive index changes that can be induced by photoreaction are relatively high ($\Delta n \sim 10^{-2}$). Therefore, it is expected that a Bragg grating written into a polymer optical fibre would be tunable over a very wide wavelength range. In addition, polymer optical fibre gratings are ideal devices for inclusion in an organic solid-state fibre laser,[59] which in turn provides a compact and low-cost optical source over a broad range of wavelengths throughout the visible spectrum.[60]

Xiong et al. have fabricated a photosensitive polymer optical fibre, whose refractive index in the core region can be changed under irradiation from a pulse UV laser beam, and have shown that it is possible to write gratings in these fibres.[61] They have fabricated several types of grating, including surface gratings on the flat surface of a preform, gratings in preform core, and gratings in the core of a multimode polymer optical fibre.

The polymer preform was prepared according to the method reported in Peng et al.,[62] except that the core monomer used contained a much lower level of lauryl peroxide and chain transfer agent so as to increase its photosensitivity. The preform was then cut into a square block with the core exposed and polished. A grating was fabricated by means of a phase mask with a period of 1.06 μm. The illumination period was 10 ns, and the pulse energy was about 5 mJ. The grating created at wavelength 248 nm was a surface grating formed by period removal of the core polymer. An atomic force microscope (AFM) was used to obtained a surface profile which has a grating period of 1.06 μm, coinciding with that of the pulse mask. The diffraction efficiency is 15%.

The mechanisms responsible for the formation of periodic structures in polymers have been explained.[63–65] They can be classified into photolysis, oxidization and laser ablation for the surface relief gratings, while chain scission, cross-link and photopolymerization are considered to be responsible for bulk or volume gratings and in-fibre gratings. The cross-link in Bragg grating fabrication can be shown as:

$$\left[H_2C - \underset{\underset{\bigcirc}{|}}{\overset{\overset{COOCH_3}{|}}{C}} \right] \quad \longrightarrow \quad \left[H_2C - \underset{\underset{\bigcirc}{|}}{\overset{\overset{COOCH_3}{|}}{C}} \right]$$

$$-CH-CH_2-M-H_2C-HC-$$

Peng et al. have written Bragg gratings in single-mode polymer optical fibre[66] and have shown that these gratings can be readily tuned over a wavelength range of 12 nm without changing their reflection spectra.

8.7 Conclusions

The mechanism and procedure of FBG fabrication are discussed, and the influences of UV irradiation on the mechanical strength of optical fibres are studied. The mechanical strength of optical fibres is unaffected by the on-line writing of a single pulse during fibre drawing. The strength degrades drastically with UV irradiation after the coating is stripped, and will degrade further with 248 nm single pulse UV irradiation than with 193 nm. The strength degradation is more serious with higher radiation energy per pulse than lower radiation energy. Lower UV irradiation frequency degrades the mechanical strength more than higher frequency. The mechanical strength of the optical fibre will not be lowered significantly after certain doses of UV irradiation. The mechanical strength of the FBG sensor can be increased notably with CW UV irradiation and the direct writing method without stripping the coating. Young's modulus and the hardness of the optical fibre will not change with UV irradiation.

Because of its large elongation property and relatively high refractive index changes induced by photoreaction, a Bragg grating written into a plastic optical fibre would be tunable over a very wide wavelength range. Plastic FBGs can be used to measure large strain and other physical properties with a large change range of structures.

Acknowledgements

The authors wish to acknowledge Hong Kong Research Grants Council for supporting this work (Grant No Polyu5112/98E) and Dr Dongxiao Yang for conducting the measurement of PMMA optical fibres.

References

1 Hill K, Fujii Y, Johnson D C and Kawasaki B S, 'Photosensitivity in optical fibre wave guide applications to reflection filter fabrication', *Appl. Phys. Lett.*, 1978, **32**, 647-9.
2 Kawasaki B S, Hill K O, Johnson C D and Fujii Y, 'Narrow-band Bragg reflectors in optical fibres', *Opt. Lett.*, 1978, **3**, 66-8.
3 Meltz G, Morey W W and Glenn W H, 'Forming of Bragg grating in optical fibres by a transverse holographic method', *Opt. Lett.*, 1989, **14**, 823-5.
4 Askins C C, Putnam M A, Williams G M and Friebele E J, 'Stepped-wavelength optical-fibre Bragg grating arrays fabricated in line on a draw tower', *Opt. Lett.*, 1994, **14**, 823-4.
5 Archambault J L, Reekie L and Russell, 'High reflectivity and narrow bandwidth fibre grating written by single excimer pulse', *Electron. Lett.*, 1993, **29**, 28-9.

6 Kersey A D, 'Interrogation and multiplexing techniques for fibre Bragg grating strain-sensors', *Proc. SPIE*, 1993, **2071**, 30–48.
7 Rao Y J, 'In fibre Bragg grating sensors', *Meas. Sci. Technol.*, 1997, **8**, 355–75.
8 Mather M H, Tabrizi K, Prohaska J D and Snitzer E, 'Fibre Bragg gratings for civil engineering applications', *Proc. SPIE*, 1996, **2682**, 298–302.
9 Friebele E J, Askins C G, Putnam M A et al., 'Distributed strain sensing with fibre Bragg grating arrays embedded in CRTM™ composite', *Electron. Lett.*, 1994, **30**, 1783–1784.
10 Friebele E J, Askins C G, Putnam M A et al., 'Demonstration of distributed strain sensing in production scale instrumented structures', *Proc. SPIE*, 1996, **2721**, 118–24.
11 Bullock D E, Dunphy J R, Hufstetler G H, 'Embedded Bragg grating fibre optic sensor for composite flexbeams', *Proc. SPIE*, 1993, **1789**, 253–61.
12 Chen B X, Mather M H and Nawy E G, 'Fibre-optic Bragg grating sensor for nondestructive evaluation of composite beam', *J. Struct. Eng.*, 1994, **12**(12), 3456–70.
13 Measures R M, Alavie A T, Maaskant R, Ohn M, Karr S and Huang S, 'Structurally integrated Bragg grating laser sensing system for a carbon fibre pre-stressed concrete highway bridge', *Smart Mater. Struct.*, 1995, **4**(1), 20–30.
14 Du W C, Tao X M, Tam H Y and Choy C L, 'Optical Bragg grating sensors in smart textile structural composite', *Proceeding of the 4th International Conference on Composite Engineering, ICCE/4*, Hawaii, USA, 1997, 289–90.
15 Ferdinand P, Ferragu O, Lechien J L et al. 'Mine operating accurate stability control with optical fibre sensing and Bragg grating technology: the European brite/euram stabilos project', *J. Lightwave Technol.*, 1995, **13**, 1303–12.
16 Rao Y J, Jackson D A, Zhang L and Bennion I, 'In-fibre Bragg grating temperature sensor system for medical applications', *J. Lightwave Technol.*, 1997, **15**, 779–85.
17 Griffioen W, 'Mechanical life time model for optical fibre in water', *Proc. SPIE*, 1994, **2074**, 2–10.
18 Haslove P, Jensen K B and Skovgaard N H, 'Degradation of stressed optical fibres in water: new worst case lifetime estimation model', *J. Am. Ceram. Soc.*, 1994, **77**, 1531–40.
19 Bouten P C P and Broer D J, 'Coating composition and fibre life time', *Proc. SPIE*, 1994, **2074**, 59–64.
20 Inniss D, Brownlow D L and Kurkjian C R, 'Chemical effects on the fatigue of light-guide fibres', *Proc. SPIE*, 1992, **1791**, 3–10.
21 Wei T, Yuee H H, Hasz C H and Key P L, 'Degradation of fibre strength during coating stripping', *Int. Wire and Cable Symp. Proc.*, 1989, 199–204.
22 Rondinella V V and Matthewson M J, 'Effect of chemical stripping on the strength and surface morphology of fused silica fibre', *Proc. SPIE*, 1994, **2074**, 52–8.
23 Damsgaard H and Hansen O, 'Factory spliced fibres: technology, performance and field experience', *Int. Wire and Cable Symp. Proc.*, 1990, 284–90.
24 Olshansky R and Maurer D R, 'Tensile strength and fatigue of optical fibres', *J. Appl. Phys.*, 1976, **47**(10), 4497–9.
25 Mitsunaga Y, Katsuyama Y and Ishida Y, 'Reliability assurance for long-length optical fibre based on proof testing', *Electron. Lett.*, 1981, **17**, 568–71.
26 Griffioen W, Breuls T, Cocito G, Dodd S, Ferri G, Haslov P, Oksanen L, Stockon D and Svensson T, 'COST 218 evaluation of optical fibre lifetime models', *Proc. SPIE*, 1992, **1791**, 190–5.

27 Friebele E J, Griscom D L and Siegal G H Jr, 'Defect centers in germania doped silica optical fibre', *J. Appl. Phys.*, 1974, **45**, 3424–8.
28 Kawazoe H, 'Effects of modes of glass formation on the structure of intrinsic or photoinduced defects centered on III, IV, or V cations in oxide glasses', *J. Non-cryst. Solids*, 1985, **71**, 213–34.
29 Tsai T E, Griscom D L, Friebele E J and Fleming J W, 'Radiation induced defect centers in high purity GeO_2 glass', *J. Appl. Phys.*, 1987, **62**, 2262–8.
30 Griscom D L and Mizuguchi M, 'Determination of the visible range optical absorption spectrum of peroxy radicals in gamma irradiation fused silica'. In *Bragg Gratings, Photosensitivity and Poling in Glass Fibres and Waveguides: Fundamentals and Applications*, 1997 OSA Technical Digest Series, **17**, Paper JMD2, 139–41.
31 Honso H, Abe Y, Kinser D L, Weeks R A, Muta K and Kawazoe H, 'Nature and origin of the 5eV band in SiO_2: GeO_2 glasses', *Phys. Rev. B*, 1995, **46**, II-445–II-451.
32 Tsai T E and Friebele E J, 'Kinetics of defects centers formation in Ge–SiO_2 fibre of various compositions'. In *Bragg Gratings, Photosensitivity and Poling in Glass Fibres and Waveguides: Fundamentals and Applications*, 1997, OSA Technical Digest Series, **17**, Paper JMD2, 101–3.
33 Hand D P and Russel P St J, 'Photoinduced refractive index changes in germanosilicate optical fibres', *Opt. Lett.*, 1990, **15**(2), 102–4.
34 Williams D L, Davey S T, Kashyap R, Armitage J R and Ainslie B J, 'Direct observation of UV-induced bleaching of 240 nm absorption band in photosensitive germanosilicate glass fibres', *Electron. Lett.*, 1992, **28**(4), 369–71.
35 Riant I, Borne S and Sansonetti P, 'Evidence of densification in UV-write Bragg gratings in fibres'. In *Photosensitivity and Quadratic Nonlinearity in Waveguides: Fundamentals and Applications*, 1995, **22**, 51–5.
36 Poumellec B, Guenot P, Riant I et al., 'UV induced densification during Bragg grating inscription in Ge:SiO_2 preforms', *Opt. Mater.*, 1995, **4**, 441–9.
37 Bernardin J P and Lawandy N M, 'Dynamics of the formation of Bragg gratings in germanosilicate optical fibres', *Opt. Commun.*, 1990, **79**, 194–8.
38 Rothschild M, Erlich D J and Shaver D C, 'Effects of excimer irradiation on the transmission, index of refraction, and density of ultraviolet grade fused silica', *Appl. Phys. Lett.*, 1989, **55**(13), 1276–8.
39 Douay M, Ramecourt D, Tanuay T et al. 'Microscopic investigations of Bragg gratings photo-written in germanosillicate fibres'. In *Photosensitive and Quadratic Nonlinearity in Waveguides: Fundamentals and Applications*, 1995, OSA Technical Digest Series, **22**, 48–51.
40 Fonjallaz P Y, Limberger H G, Salathe, Cochet F and Leuenberger B, 'Tension increase correlated to refractive index change in fibres containing UV written Bragg gratings', *Opt. Lett.*, 1995, **20**(1), 1346–8.
41 Griffith A A, 'The phenomena of rupture and flaw in solids', *Phil. Trans. Roy. Soc.*, 1920, **221A**, 163–9.
42 Kurkjian C R and Paek U C, 'Single value strength of perfect silica fibres', *Appl. Phys. Lett.*, 1983, **42**, 251–3.
43 Oliver W C and Pharr G M, 'An improved technique for determining hardness and elastic modulus using load and displacement sensing indentation experiments', *J. Mater. Res.*, 1992, **7**(6), 1564–83.
44 Sneddon I N, 'The relation between load and penetration in the axisymmetric

boussinesq problem for a punch of arbitrary profile', *Int. J. Engng. Sci.*, 1965, **3**, 47–57.
45 Askins C G, Putnam M A, Williams G M and Friebele E J, 'Stepped-wavelength optical fibre Bragg grating arrays fabricated in line on a draw tower', *Opt. Lett.*, 1994, **19**(2), 147–9.
46 Askins C G, Putnam H J, Patrick H J and Friebele E J, 'Fibre strength unaffected by on-line writing of single-pulse Bragg grating', *Electron. Lett.*, 1997, **33**(15), 1333–4.
47 Hagemann V, Trutzel M N, Staudigel L, Rothhardt M, Muller H-R and Krumpholz O, 'Mechanical resistance of draw-tower-Bragg grating sensors', *Electron. Lett.*, 1998, **34**(2), 211–12.
48 Feced R, Roe-Edwards M P, Kanellopoulos S E, Taylor N H and Handerick V A, 'Mechanical strength degration of UV exposed optical fibres', *Electron Lett.*, 1997, **33**(2), 157–9.
49 Albert J, Malo B, Hill K O, Bilodeau F, Johnson D C and Theriault S, 'Comparison of one-photon and two-photon effects in photosensitivity of germanium-doped silica optical fibres exposed to intense ArF excimer laser pulses', *Appl. Phys. Lett.*, 1995, **67**(24), 3529–31.
50 Varelas D, Limberger H G, Salathe R P and Kotrotsios C, 'UV-induced mechanical degradation of optical fibres', *Electron. Lett.*, 1997, **33**(9), 804–5.
51 Varelas D, Limberger H G and Salathe R P, 'Enhanced mechanical performance of single mode optical fibres irradiated by a CW UV laser', *Electron. Lett.*, 1997, **33**(8), 704–5.
52 Imamura K, Nakai T, Moriura K, Sudo Y and Imada Y, 'Mechanical strength characteristics of tin-codoped germanosilicate fibre Bragg gratings fabricated by writing through UV-transparent coating', *Electron. Lett.*, 1998, **34**(10), 1016–17.
53 Chao L, Reekie L and Ibsen M, 'Grating writing through fibre coating at 244 and 248 nm', *Electron. Lett.*, 1999, **35**(11), 924–5.
54 Imamura K, Nakai T, Sudo Y and Imada Y, 'High reliability of tin-codoped germanosilicate fibre Bragg gratings fabricated by direct writing method', *Electron. Lett.*, 1998, **34**(18), 1772–4.
55 Dong L, Cruz J L, Reekie L, Xu M G and Payne D N, 'Enhanced photosensitivity in tin-codoped germansilicate optical fibres', *IEEE Photon. Technol. Lett.*, 1995, **7**, 1048–50.
56 Sanders D H, *Statistics: A Fresh Approach*, McGraw-Hill, Inc., 1990.
57 Marcou J (ed.), *Plastic Optical Fibres: Practices and Applications*, John Wiley, New York, 1997.
58 Hill K O and Meltz G, 'Fibre Bragg grating technology fundamental and overview', *J. Lightwave Technol.*, 1997, **15**(8), 1263–76.
59 Kaminov I P, Weber H P and Chandross E A, 'Poly(methyl methacrylate) dye laser with internal diffraction grating resonator', *Appl. Phys. Lett.*, 1971, **18**, 497–9.
60 Dodabalapur A, Chandross E A, Berggren M and Slusher R E, 'Organic solid-state lasers: past and future', *Science*, 1997, **277**, 1787–8.
61 Peng G D, Xiong Z and Chu P L, 'Photosensitivity and gratings in dye-doped polymer optical fibres', *Optical Fibre Technol.*, 1999, **5**(2), 242–51.
62 Peng G D, Chu P L, Xiong Z, Whitbread T and Chaplin R P, 'Dye-doped step-index polymer optical for broadband optical amplification', *J. Lightwave Technol.*, 1996, **14**, 2215–23.
63 Tomlinson W J, Kaminow I P, Chandross A, Fork R L, and Silfvast W T,

'Photoinduced refractive index increase in poly(methylmethacrylate) and its applications', *Appl. Phys. Lett.*, 1970, **16**, 486–9.
64 Bolle M, Lazare S, Blanc M L and Wilmes A, 'Submicron periodic structures produced on polymer surfaces with polarized excimer laser ultraviolet radiation', *Appl. Phys. Lett.*, 1992, **60**, 674–6.
65 Decker C, 'The use of UV irradiation in polymerization', *Polymer Int.*, 1998, **45**, 133–7.
66 Xiong Z, Peng G D, Wu B and Chu P L, 'Highly tunable Bragg gratings in single-mode polymer optical fibres', *IEEE Photon. Technol. Lett.*, 1999, **11**(3), 352–4.

9
Optical responses of FBG sensors under deformations

DONGXIAO YANG, XIAOMING TAO AND
APING ZHANG

9.1 Introduction

Many sensors[1-16] can be used to measure the strain and deformations, such as traditional strain-meter sensors, piezoelectric sensors, piezo-resistive sensors, optic sensors based on fluorescence spectroscopy or Raman spectroscopy, fibre optic sensors based on Brillouin scattering, fibre optic sensors based on interferometer or/and polarimeter, fibre optic sensors based on photography and fibre Bragg grating (FBG) sensors. FBG sensors have been considered excellent sensing elements, suitable for measuring static and dynamic fields, like temperature, strain and pressure. Since the measured information is wavelength encoded,[13-22] FBG makes the sensor self-referencing, rendering it independent of fluctuating light levels and the system immune to source power and connector losses that plague many other types of fibre optic sensors. The advantages of FBG sensors include light weight, flexibility, stability, potential low cost, longer lifetime, higher temperature capacity, unique wavelength-multiplexing capacity, suitable size for embedding into composites without introducing significant perturbation to the characteristics of the structure, good invulnerability to electro-magnetic interference, and even durability against high radiation environments, making reproducible measurements possible. Therefore, FBG sensors seem to be ideal for realizing fibre optic smart structures where sensors are embedded in or attached to structures for achieving a number of technical objectives, such as health monitoring, impact detection, shape control and vibration damping, via the provision of real-time sensing information, such as strain, temperature and vibration. FBG sensors have been used for measurements of a wide variety of parameters: some FBG sensor systems have been installed in large-scale practical applications, and some are commercially available.[16] FBG sensors have been used in bridges, highways, textiles, mines, marine vehicles, medical therapies and aircrafts.[16,23-35] They[13,14,36-74] can be used for quasi-static strain monitoring, dynamic strain sensing, time- and wavelength-division multiplexing, and temperature/strain discrimination. The combination of FBG and long period fibre grating can be used to simultaneously determine strain and temperature. FBGs can also be used in laser sensors and interferometer sensors.

Most of the sensing applications of FBG sensors focus on its reflection spectra, which depend on the relationship between the Bragg wavelength of FBG and its physical quantities.[13,14] The key detection issue is the determination of the often small measurand-induced Bragg wavelength shift. In addition to fibre grating spectra, the polarization optics of FBG sensors will also be discussed in this chapter. The contents include optical methodologies, optical responses under tension, torsion, lateral compression and bending.

9.2 Optical methodology for FBG sensors

An optical fibre is a cylindrical dielectric waveguide made of low-loss materials such as silica glass. It has a central core in which the light is guided, embedded in an outer cladding of slightly lower refractive index. Light rays incident on the core-cladding boundary at angles greater than the critical angle undergo total internal reflection and are guided through the core without refraction. Rays of greater inclination to the fibre axis lose part of their power into the cladding at each reflection and are not guided.

Light in fibre propagates in the form of modes. In mathematics, a mode in optic fibre is a solution of the Maxwell's equations under the boundary conditions. The guided modes are electric and magnetic fields that maintain the same transverse distribution and polarization at all distances along the fibre axis. Each mode travels along the axis of the fibre with a distinct propagation constant and group velocity, maintaining its transverse spatial distribution and its polarization. There are two independent configurations of electric and magnetic vectors for each mode, corresponding to two states of polarization. When the core diameter is small, only a single mode is permitted and the fibre is said to be a single mode fibre. Fibres with large core diameters are multimode fibres.

9.2.1 Electric field in single mode fibre

In a single mode fibre, the weakly guiding approximation $n_{co} - n_{cl} \ll 1$ is satisfied, where n_{co} and n_{cl} are the refractive index of the core and the cladding, respectively. The fundamental mode is hybrid mode HE_{11}, which can be simplified in the analysis by linear polarized mode LP_{01}. This fundamental mode consists of two orthogonal modes HE_{11}^x and HE_{11}^y in accordance with their polarization directions, so the electric fields in a single mode fibre which is not under Bragg condition can be well represented as:

$$E(x, y, z; t) = \sum_{m=1}^{2} c_m(z)E_m(x, y)\exp[-i(\omega t - \beta_m z)] \quad [9.1]$$

where c_1 and c_2 denote the slowly varying amplitudes of the orthogonal HE_{11} modes, ω, β_1, β_2, E_1 and E_2 are the angular frequency, the propagation constants of HE_{11}^x, and HE_{11}^y, the transverse spatial distribution of electric field HE_{11}^x, and the transverse spatial distribution of electric field HE_{11}^y, respectively, and

$$E_1(x, y) = F_0(r)e_x + (i/\beta_0)\cos\theta[dF_0(r)/dr]e_z \quad [9.2]$$

$$E_2(x, y) = F_0(r)e_y + (i/\beta_0)\sin\theta[dF_0(r)/dr]e_z \quad [9.3]$$

where $F_0(r)$ represent the zero-order Bessel functions J_0 in the core, and modified Bessel functions K_0 in the cladding, respectively, β_0, e_x, e_y and e_z are the propagation constant of the mode in an ideal single mode fibre, the unit vectors in x, y and z directions, respectively.

The coupled mode theory is often used to describe the polarization characteristics, or modes coupling in optical fibre, and the reflection spectra of FBG sensors. When the optical response of an optical fibre is analysed based on the coupled mode theory, the electric field is normalized as $\iint E_{nt} \cdot E_{nt}^* dxdy = 1$. Based on the perturbation approach, the slowly varying amplitudes c_m are determined by the following coupled mode equations:

$$dc_1(z)/dz = i\{\kappa_{11}c_1(z) + \kappa_{12}c_2(z)\exp[(\beta_2 - \beta_1)z]\} \quad [9.4]$$

$$dc_2(z)/dz = i\{\kappa_{21}c_1(z)\exp[-(\beta_2 - \beta_1)z] + \kappa_{22}c_2(z)\} \quad [9.5]$$

where the subscripts 1 and 2 denote the modes HE_{11}^x and HE_{11}^y, respectively. The amplitude coupling coefficient from mode m to mode n is given by:

$$\kappa_{nm} = (k_0^2/2\beta_0)\iint(\tilde{\varepsilon}_{ij}E_m) \cdot E_n^* dxdy, \quad n, m = 1, 2 \quad [9.6]$$

where $\tilde{\varepsilon}_{ij}$ is the dielectric permittivity perturbation tensor.

9.2.2 Polarization optics[75–81]

When the polarization behaviour of FBGs under various deformations is discussed, one has to consider three types of perturbation, that is, those induced by the fibre making process (intrinsic), UV side exposure and strain caused by deformation.

An evolution velocity Ω is usually introduced to describe qualitatively the polarization characteristics of specific polarization behaviour. Ω in a generalized Poincaré sphere can be expressed by the coupled coefficients as:[77]

$$|\Omega| = [(\kappa_{11} - \kappa_{22})^2 + 4\kappa_{12}\kappa_{21}]^{1/2} \quad [9.7]$$

Optical responses of FBG sensors under deformations 153

$$2\chi = \arg(\kappa_{11} + \kappa_{12} - \kappa_{21} - \kappa_{22}) \quad [9.8]$$

$$2\psi = \arctan[(\kappa_{11} - \kappa_{22})/(4\kappa_{12}\kappa_{21})]^{1/2} \quad [9.9]$$

where 2χ and 2ψ are the latitude and longitude of the generalized Poincaré sphere, respectively. Hence the general approach of the polarization analysis based on the coupled mode theory is used, by analysing the permittivity perturbation first, then inserting it into the coupled mode equations, obtaining the solutions of c_1 and c_2, and qualitatively presenting the polarization characteristics by the evolution velocity.

The Poincaré sphere,[75] which has a unit radius and the spherical angular coordinates 2ψ (longitude) and 2χ (latitude), is often used to represent the different states of polarization geometrically. According to the traditional terminology which is based on the apparent behaviour of the electric vector when 'viewed' face on by the observer, we say that the polarization is right-handed when to an observer looking in the direction from which the light is coming, the end point of the electric vector would appear to describe the ellipse in the clockwise sense. Hence right-handed polarization is represented by points on the Poincaré sphere which lie below the equatorial plane, and left-handed by points on the Poincaré sphere which lie above this plane. Linear polarization is represented by points on the equator. Right- and left-handed circular polarization are represented by the south and north pole on the Poincaré sphere, respectively.

9.2.3 Reflection spectra

A fibre Bragg grating (FBG) consists of a periodic modulation of refractive index in the core of a single mode fibre. The electric fields in FBGs are represented by the superposition of the ideal modes travelling in both forward and backward directions:

$$E(x, y, z; t) = \sum_k [a_k(z)\exp(i\beta_k z)$$
$$+ b_k(z)\exp(-i\beta_k z)] E_k(x, y)\exp(-i\omega t) \quad [9.10]$$

where a_k and b_k are slowly varying amplitudes of the kth mode travelling in the $+z$ and $-z$ directions. The transverse mode fields $E_k(x, y)$ might describe the LP modes, or the cladding modes. The modes are orthogonal in an ideal waveguide, hence do not exchange energy. The presence of a dielectric perturbation causes them to be coupled. Based on the slowly varying envelope approximation, the coupled mode equations for amplitudes a_k and b_k can be obtained by:

$$\frac{da_j(z)}{dz} = i\sum_k (\kappa^t_{jk} + \kappa^z_{jk})a_k(z)\exp[i(\beta_k - \beta_j)z]$$

$$+ i\sum_k (\kappa^t_{jk} - \kappa^z_{jk})b_k(z)\exp[-i(\beta_k + \beta_j)z] \quad [9.11]$$

$$\frac{db_j(z)}{dz} = -i\sum_k (\kappa^t_{jk} - \kappa^z_{jk})a_k(z)\exp[i(\beta_k + \beta_j)z]$$

$$- i\sum_k (\kappa^t_{jk} + \kappa^z_{jk})b_k(z)\exp[-i(\beta_k - \beta_j)z] \quad [9.12]$$

where the transverse coupling coefficients between modes k and j are given by:

$$\kappa^t_{jk} = \frac{k_0^2}{2\beta_j}\int\!\!\int_\infty (\tilde{\varepsilon}_r \mathbf{E}_k)_t \cdot \mathbf{E}^*_{jt}\,dxdy \quad [9.13]$$

where $\tilde{\varepsilon}_r$ is the permittivity perturbation. It is often treated as a scalar term for FBG sensors, and can be approximately expressed by index perturbation as $\tilde{\varepsilon}_r \approx 2n \cdot \delta n$. However, in the cases where there is significant linear birefringence in an optical fibre, such as in polarization maintaining fibre, or under high lateral compression, it should be corrected. The longitudinal coefficient κ^z_{jk} can usually be neglected, since generally $\kappa^z_{jk} \ll \kappa^t_{jk}$ for modes. In most FBGs, the induced index change is non-existent outside the core, and the core refractive index changes can be expressed as:

$$\delta n = \delta n_{\text{eff}}(z)\left\{1 + \eta \cdot \cos\left[\frac{2\pi}{\Lambda}z + \phi(z)\right]\right\} \quad [9.14]$$

where $\delta n_{\text{eff}}(z)$ is the DC index change spatially averaged over a grating period, η is the fringe visibility of the index change, Λ is the nominal period, and $\phi(z)$ describes the grating chirp, respectively. Figure 9.1 shows some reflection spectra of normal FGBs with different fringe visibility.

If we insert the perturbation term Eq. (9.14) into Eq. (9.13), we can obtain quantitative information about the reflection spectra of fibre gratings. One of the most important results of FBGs is the phase-matching condition, or Bragg condition. The phase-matching condition of FBGs with period Λ is given by $\beta_1 - \beta_2 = 2\pi/\Lambda$, where β_1 and β_2 are the propagation constants of forward and backward propagation modes. $\beta_1 = -\beta_2 = \beta$ for directional coupling between the same modes. The phase-matching condition can then be simplified to $\beta = \pi/\Lambda$. If the effective refractive index is used to represent the propagation characteristics as $\beta = 2\pi \cdot n_{\text{eff}}/\lambda$, it can then be expressed as:

$$\lambda_B = 2n_{\text{eff}}\Lambda \quad [9.15]$$

Optical responses of FBG sensors under deformations 155

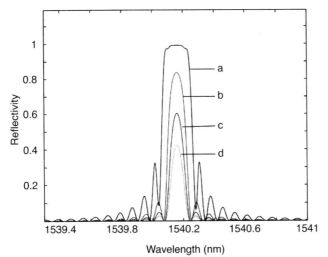

9.1 Reflection spectra of an FBG sensor with different 'AC' modulation dose (a) 2.0×10^{-4}; (b) 1.0×10^{-4}; (c) 0.67×10^{-4}; (d) 0.5×10^{-4}).

where the Bragg wavelength λ_B is the free space centre wavelength of the input light that will be back-reflected from the Bragg grating. Equation (9.15) is the first-order Bragg condition of the grating.

This first-order Bragg condition is simply the requirement that satisfies both energy and momentum conservation. Energy conservation requires that the frequencies of the incident radiation and the scattered radiation are the same, $\hbar\omega_s = \hbar\omega_i$, where $\hbar = h/2\pi$, h is Planck's constant, and ω_i and ω_s are radian frequencies of incident radiation and scattered radiation, respectively. Momentum conservation requires that the incident wave-vector $\mathbf{k}_i$, plus the FBG wave-vector $\mathbf{k}_g$, equal the wave-vector of the scattered radiation $\mathbf{k}_s = \mathbf{k}_i + \mathbf{k}_g$, where $\mathbf{k}_g$ has a direction normal to the FBG planes with a magnitude $2\pi/\Lambda$, Λ is the spacing period of the FBG, $\mathbf{k}_i$ has a direction along the propagation direction of the incident radiation with a magnitude $2\pi n_{\text{eff}}/\lambda_i$, λ_i is the free space wavelength of the incident radiation, n_{eff} is the modal index (the effective refractive index of the fibre core) at the free space wavelength, $\mathbf{k}_s$ has a direction along the propagation direction of the scattered radiation with a magnitude $2\pi n_{\text{eff}}/\lambda_s$, and λ_s is the free space wavelength of the scattered radiation. If the scattered radiation is the reflected radiation of incident radiation, that is, $\mathbf{k}_s = -\mathbf{k}_i = -2\pi n_{\text{eff}}/\lambda_i \mathbf{e}_z$, the momentum conservation condition becomes the first-order Bragg condition Eq. (9.15). The guided light along the core of an optical fibre will be scattered by each FBG plane. If the Bragg condition is not satisfied, the reflected light from each of the subsequent planes becomes progressively out of phase and will eventually cancel out. It will experience very weak reflection at each of the FBG planes because of the index mismatch. This reflection accumulates over the length of the FBG.

A very important advantage of FBG sensors is that they are wavelength encoded. Shifts in the spectrum, seen as a narrow-band reflection or dip in transmission, are independent of the optical intensity and uniquely associated with each FBG, provided no overlap occurs in each sensor stop-band. FBG sensors have achieved significant applications in monitoring or inspecting the mechanical or temperature response in smart materials and structures. Most of these applications focus on the axial deformation (or strain) and temperature measurements, because the sensitivities to axial deformation and temperature are much higher than those to other modes of deformation.

It is known from Eq. (9.15) that the Bragg wavelength is proportional to the modal index and the FBG spacing period. Both the index and period may change with external conditions which can be divided to temperature and applied disturbances, such as deformations. The induced Bragg wavelength shift of mode j (including polarization mode) can be expressed by:

$$\Delta \lambda_{Bj} = 2 \left[\left(n_{effj} \frac{\partial \Lambda}{\partial T} + \Lambda \frac{\partial n_{effj}}{\partial T} \right) \Delta T \right. \\ \left. + \sum_i \left(n_{effj} \frac{\partial \Lambda}{\partial u_i} + \Lambda \frac{\partial n_{effj}}{\partial u_i} \right) \Delta u_i \right] \quad [9.16]$$

where n_{effj} and u_i are the effective refractive index of mode j and a perturbation, respectively. Measurement of the perturbation-induced Bragg wavelength shift from a single FBG does not facilitate the discrimination of the response to these variables. The simplest approach is to isolate the unwanted perturbations. In applications, sensors must be embedded with minimal intrusion. In case the deformation sensing is considered, temperature is the main unwanted perturbation. Temperature-compensating methods may be classified as intrinsic or extrinsic. The elimination of cross-sensitivity may be achieved by measurements at two different wavelengths or two different optical modes, in which the strain and the temperature sensitivity are different. The sensor schemes can be constructed by the combination of FBGs with different grating types, such as FBGs with different diameter, different Bragg wavelength, different codope, hybrid FBGs and long period fibre grating, Fabry–Perot cavity, stimulated Brillouin scattering or fibre polarization rocking filter. The measurands may be Bragg wavelength, intensity, Brillouin frequency or polarization rocking resonant wavelength.[54–67] In this chapter, only the optical responses of FBG sensors under deformations are included.

9.3 Optical responses under tension

No significant polarization signals can be observed for FBG sensors under tension. But an FBG sensor has good linear characteristics when it is applied

Optical responses of FBG sensors under deformations

to measure the axial strain by reflection spectra. According to Eq. (9.16), the relative shift of the Bragg wavelength due to strain is given by:[22]

$$\frac{\Delta \lambda_B}{\lambda_B} = \frac{1}{n_{eff}} \sum_i \frac{\partial n_{eff}}{\partial \xi_i} \xi_i + \frac{1}{\Lambda} \sum_i \frac{\partial \Lambda}{\partial \xi_i} \xi_i \quad [9.17]$$

where ξ_i is the applied strain field to the FBG sensor. According to the photoelastic effect, the first term on the right side of Eq. (9.17) can be written as:

$$\frac{1}{n_{eff}} \sum_i \frac{\partial n_{eff}}{\partial \xi_i} \xi_i = -\frac{n_{eff}^2}{2} \sum_i p_{ij} \xi_i \quad [9.18]$$

where p_{ij} is the strain-optic tensor. For a homogeneous isotropic medium,

$$p_{ij} = \begin{bmatrix} p_{11} & p_{12} & p_{12} & 0 & 0 & 0 \\ p_{12} & p_{11} & p_{12} & 0 & 0 & 0 \\ p_{12} & p_{12} & p_{11} & 0 & 0 & 0 \\ 0 & 0 & 0 & p_{44} & 0 & 0 \\ 0 & 0 & 0 & 0 & p_{44} & 0 \\ 0 & 0 & 0 & 0 & 0 & p_{44} \end{bmatrix} \quad [9.19]$$

where $p_{44} = (p_{11} - p_{12})/2$. For the case of tension, the strain response arises due to both the change in fibre index due to photoelastic effect, and the physical elongation including the corresponding fractional change in grating pitch of the FBG. If a uniaxial longitudinal stress σ_z is applied to the FBG sensor in the z-direction, the resulting strain from first-order elastic theory has three principal components:

$$\xi_i = [-v\xi_z \quad -v\xi_z \quad \xi_z \quad 0 \quad 0 \quad 0]^T \quad [9.20]$$

where the superscript T denotes the transpose of a matrix, $\xi_z = \sigma_z/E$ is the longitudinal strain, E is Young's modulus and v is Poisson's ratio. The index-weighted strain-optic coefficient is then given by:

$$p_{eff} = n_{eff}^2 [p_{12} - (p_{11} + p_{12})v]2 \quad [9.21]$$

Hence the right side of Eq. (9.18) can be written as $-p_{eff}\xi_z$.

If it can be assumed that the strain-induced change in the period of the FBG is only dependent on the axial strain, the second term on the right side of Eq. (9.17) will be equal to ξ_z. Equation (9.17) is then expressed as:

$$\Delta \lambda_B / \lambda_B = (1 - p_{eff})\xi_z \quad [9.22]$$

If the fibre is extended at both ends and no body force applied, the fibre core and cladding can be regarded as homogeneous and isotropic, the strain response of the germanosilicate FBG sensor at constant temperature under tension is then given by:

$$(\Delta\lambda_B/\lambda_B)/\xi_z \approx 0.78 \qquad [9.23]$$

In a silica fibre there are three transmission wavelength windows in which the losses are very low, such as 0.15 dB/km at 1550 nm. According to Eq. (9.23), typical values for the sensitivity to an applied axial strain in these windows are 1.2 nm/milli-strain at 1550 nm, 1 nm/milli-strain at 1300 nm, and 0.66 nm/milli-strain at 850 nm. The resolutions of FBG are 8.3 micro-strain at 1550 nm, 9.8 micro-strain at 1300 nm and 15 micro-strain at 850 nm by an optical spectrum analyser with a wavelength resolution of 10 pm.

9.4 Optical responses under torsion

Unlike the optical responses of an FBG sensor under tension, the sensitivity of the Bragg wavelength shift of an FBG sensor under torsion is very small. However, the polarization response to torsion is significant. The coupled mode theory can be used to analyse the polarization behaviour of an FBG sensor under torsion.[82]

9.4.1 Shear strain-induced polarization behaviour

The torsion introduces shearing stress in the cross-section of the fibre. If the twisted length of a fibre is L, and the angle of twist is $\phi = \tau L$, where τ is torsion ratio that is the angle of twist per unit length along the axis of the fibre, the matrix for the strain due to this torsion is:

$$\xi_i = [0 \ 0 \ 0 \ \tau x \ -\tau y \ 0]^T \qquad [9.24]$$

so that the matrix for ΔD_i, perturbation in optical impermeability, is:

$$\Delta D_i = p_{ij}\xi_i = [0 \ 0 \ 0 \ p_{44}\tau x \ -p_{44}\tau y \ 0]^T \qquad [9.25]$$

The relationship between the dielectric permittivity perturbation and the optical impermeability perturbation can be expressed as:

$$\tilde{\varepsilon}_{\text{torsion } ij} = -n_{co}^4 \Delta D_{ij} \qquad [9.26]$$

The induced polarization behaviour can be analysed by using Eq. (9.26) and Subsection 9.2.2. The induced circular birefringence in a single mode optical fibre is given by:[83]

$$B_c = n_{co}\tau\lambda(p_{11} - p_{12})/(2\pi) \qquad [9.27]$$

9.4.2 UV-induced polarization behaviour

The polarization of pulse UV beam and asymmetric geometry associated with the side-exposure of UV light during the FBG fabrication process[84,85] will induce linear birefringence. The peak birefringence of the FBG can be calculated from the expression:[84]

$$\Delta n = \frac{\lambda}{2\pi \cdot L} |\text{phase}(Q_1) - \text{phase}(Q_2)| \qquad [9.28]$$

where L is the length of the FBG, Q_1 and Q_2 are the eigenvalues of the corresponding Jones matrix:

$$\det[T(t)T^{-1}(0) - QI] = 0 \qquad [9.29]$$

where T, t and I are the Jones matrix, the time from the beginning of the UV exposure and the identity matrix, respectively. The symbol det and the superscript $^{-1}$ denote the determinant and the inverse of a matrix, respectively. Based on the birefringence, a permittivity perturbation tensor can be used to represent its polarization behaviour:

$$\tilde{\varepsilon}_{UV} = \begin{bmatrix} \Delta\varepsilon_x & 0 & 0 \\ 0 & \Delta\varepsilon_y & 0 \\ 0 & 0 & 0 \end{bmatrix} \qquad [9.30]$$

where $\Delta\varepsilon_y - \Delta\varepsilon_x = 2n_{\text{eff}} \cdot \Delta n$. By considering the azimuth ϕ of the faster or slow axis of the FBG sensor and the tilted angle ϑ, Eq. (9.30) becomes:

$$\tilde{\varepsilon}'_{UV} = T_z T_x \tilde{\varepsilon}_{UV} T_x^T T_z^T \qquad [9.31]$$

where T_z and T_x are the rotation matrix around z, x axes, respectively. The induced polarization behaviour can be analysed by using Eq. (9.31) and Subsection 9.2.2.

In the case of an FBG sensor under torsion, both shear strain-induced birefringence and UV-induced birefringence are considered:

$$\tilde{\varepsilon}'_{ij} = \tilde{\varepsilon}_{\text{torsion } ij} + \tilde{\varepsilon}'_{UV} \qquad [9.32]$$

There are UV-induced linear birefringence and shear strain-induced circular birefringence in the fibre.

9.4.3 Simulated and experimental results

In numeric simulations, the parameters of the FBG sensor are the same as those used in the corresponding experiments. The core radius, cladding radius and the effective refractive index are 4.25 µm, 62.5 µm and 1.46, respectively. The strain-optic coefficients p_{11} and p_{12} are taken as 0.113 and 0.252, respectively.

Based on the preceding theoretical analysis, the FBG sensors can be treated as the wave-plates. Apparently, when an FBG sensor is under torsion, these

160 Smart fibres, fabrics and clothing

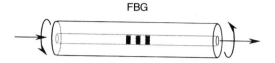

9.2 FBGs under torsion. The length of the FBG and the torsion gauge of the fibre are 1 cm and 17.5 cm, respectively.

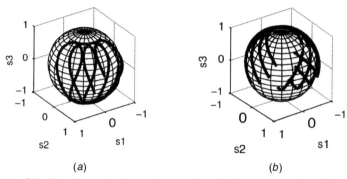

9.3 Simulated polarization signals of an FBG under torsion. The orientation is 0, $\pi/12$, $\pi/6$, $\pi/4$, respectively, and the angle of ellipticity is zero in (a), and $\pi/8$ in (b).

wave-plates will be rotated. This means the geometric parameter, the azimuth ϕ, will be changed during the twisting of an optical fibre. Then the output polarization signals will represent the combination effects of the torsion-induced circulation birefringence and the UV irradiation-induced linear birefringence.

The torsion model under investigation is shown in Fig. 9.2, where the FBG is located at the middle point of an optical fibre. The experimental setup is similar to that of T. Erdogan and V. Mizrahi,[84] and the UV-induced birefringence of a FBG is approximately 2.5×10^{-5}. The wavelength of the incident laser is set at 1525 nm, and the loading rotation angle is from 0 to 360°. The simulated results are presented in Fig. 9.3. According to these simulations, the initial orientations do not appear to affect the shape of the output polarization signals. However, the ellipticity of the input light will affect the output polarization signals significantly.

The position of the FBG can influence the output polarization signals significantly. On one side, it will change the direction of the angular velocity vector of the FBG sensor, and on the other side, it will provide a different initial state of polarization from the FBG segment. Although both their directions vary along the equator in a generalized Poincaré sphere, their respective variation velocities are different. These differences will result in different output polarization signals for various FBG sensor positions.

Figure 9.4 shows one measured polarization signal of an FBG sensor under

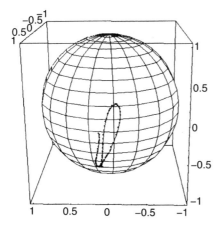

9.4 Measured polarization signal of an FBG under torsion.

torsion. The twisted length was 17.5 cm, the Bragg wavelength without external strain was 1555 nm, the wavelength of incident laser was 1525 nm, the torsion ratios were 0–0.36 rad/cm with a step of 0.005 rad/cm. The states of polarization were measured by a commercial polarimeter. The simulated result coincides with the experimental result.

The reflective spectrum of the FBG sensor was measured by an optical spectrum measuring system composed of a tunable laser, an optical power meter with GP-IB interface, and a computer. The wavelength resolution is 10 pm. The measured wavelength shift under the torsion ratio of 1 rad/cm is smaller than the resolution of the optical spectrum measuring system.

9.5 Optical responses under lateral compression

In the case of optical fibres under lateral compression, linear birefringence will be induced based on the strain–optic relationship. Thus a fibre can be used to sense the lateral compression by measuring the change of state of polarization of travelled light. For general FBG sensors the wavelength-compression sensitivity is relatively small compared with the axial strain sensitivity, but spectrum bifurcation, or spectrum split, will occur with the compression-induced birefringence in optical fibres.[86] The transverse strain can also be sensed by an FBG written into a polarization-maintaining fibre (PMF) by measuring the Bragg wavelength shifts of two reflective peaks associated with the orthogonal polarization modes.[87-95] In addition, transverse strain can be measured by a long-period fibre grating.[96]

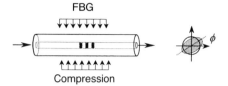

9.5 FBG under lateral compression. The length of FBG and compressed length of fibre are 1 cm and 1.8 cm, respectively.

9.5.1 Polarization responses under lateral compression

The strain of an optical fibre under lateral compression will induce linear birefringence. Since the glass optical fibre is a rigid medium, the birefringence induced by the geometric property, such as tilted angle for tilted FBG, is small. The polarization behaviour of the FBG under lateral compression can be determined by the combination effects of both the compression-induced linear birefringence and the UV-induced linear birefringence.

The case that an uncoated silica optical fibre is laminated by two rigid plates is considered, as shown in Fig. 9.5. Since the silica optical fibre is a rigid medium, this contact problem can be simplified to line force loading. Then the close form of the stress solutions can be obtained with the help of the plane stress assumption. Since the electric field in a single mode fibre is concentrated in the core region, the solution in the centre is approximately selected to represent the stress of a single mode fibre under lateral compression. The stress in the core can be expressed as:

$$\sigma_{11} = f/(\pi \cdot r_{cl}) \qquad [9.33]$$

$$\sigma_{22} = -3f/(\pi \cdot r_{cl}) \qquad [9.34]$$

$$\sigma_{33} = 0 \qquad [9.35]$$

$$\tau_{12} = \tau_{23} = \tau_{31} = 0 \qquad [9.36]$$

where f is the lateral loading, and r_{cl} is the radius of optical fibre cladding. Based on the strain–optic relationship, the permittivity perturbation $\tilde{\varepsilon}_{ij}$ can be derived. It can be inserted into the coupled mode equations, Eqs. (9.4) and (9.5), and the coupled mode coefficients can be obtained by coupling integrals, Eq. (9.6). The strain induces a linear birefringence:

$$B_1 = \left(\frac{4n_{co}^3 \cdot f}{\lambda E \cdot r_0}\right)(1 + v)(p_{12} - p_{11}) \qquad [9.37]$$

In the simulated analysis, the lateral loading is from 0 to 1000 N/m, and the

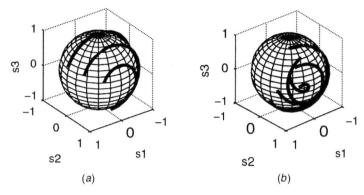

9.6 Polarization signals of FBG under lateral compression. The normal compression is shown in (a), and slant compression with 45° in (b); the angle of ellipticity is 0. The orientations are 0, π/12, π6 and π/4, respectively.

permittivity perturbation $\tilde{\varepsilon}_{ij}$ induced by the transverse strain and UV-induced birefringence are considered together. For a normal lateral compression, the UV irradiation-induced birefringence axis coincides with the strain-induced birefringence. Since the silica optical fibre is a rigid medium, the compression-induced displacement is small, and it has little effect on the geometric parameters of the UV-induced perturbation. Thus, the output polarization signal is the same as the single mode fibre. The output polarization signals are shown in Fig. 9.6(a). They are insensitive to the FBG position.

For a slant compression, there is an angle between the strain- and UV irradiation-induced birefringence axis. Though the UV irradiation-induced component change is small during the loading, it still affects the output polarization signals during the compression. Apparently, the effects are related to the angle and the position of the FBG. The output polarization signals are shown in Fig. 9.6(b).

9.5.2 Spectrum responses under lateral compression

As shown in the previous section, lateral compression will induce linear birefringence in a single mode optical fibre. This means that a difference of the propagation constants in the two orthogonal polarization modes in the fibre is induced. Based on the phase-matching condition Eq. (9.15), these two propagation constants correspond to two different Bragg wavelengths. Thus, with the increasing of compression loading, the spectrum bifurcation or spectrum split will be observed. An extended multidimension measurement is applying the double wavelength measurement in the birefringent FBG. Since the effective indices of the orthogonal LP modes are different from each other in birefringent optical fibre, every FBG in a birefringent fibre has two corresponding Bragg wavelengths. If we write two different FBGs in a single optical fibre, then it is expected to obtain four different Bragg wavelengths, which can establish four equations as Eq. (9.16), and to measure not only the axial strain and temperature, but also two transverse strain components.

In the case of lateral compression, the propagated fundamental mode HE_{11} perfectly degenerates into x- and y-polarized modes with different propagation constants. According to the photoelastic effect and the stress–strain relationship, the effective index changes of the two polarized modes can be expressed as:[86]

$$\Delta n_{\text{eff}x}(x, y, z) = -\frac{n_{\text{eff}}^3}{2E}\{(p_{11} - 2vp_{12})\sigma_x(x, y, z)$$

$$+ [(1 - v)p_{12} - vp_{11}]\sigma_y(x, y, z)\} \quad [9.38]$$

$$\Delta n_{\text{eff}y}(x, y, z) = -\frac{n_{\text{eff}}^3}{2E}\{(p_{11} - 2vp_{12})\sigma_y(x, y, z)$$

$$+ [(1 - v)p_{12} - vp_{11}]\sigma_x(x, y, z)\} \quad [9.39]$$

Using the stress–strain relationship and Eq. (9.17), the relative shifts of the Bragg wavelength of two polarized modes at any point of the FBG due to lateral compression are then obtained by:

$$\frac{\Delta \lambda_{Bx}}{\lambda_B} = -\frac{n_{\text{eff}}^2}{2E}\{(p_{11} - 2vp_{12})\sigma_x(x, y, z) + [(1 - v)p_{12} - vp_{11}]\sigma_y(x, y, z)\}$$

$$- v[\sigma_x(x, y, z) + \sigma_y(x, y, z)]/E \quad [9.40]$$

$$\frac{\Delta \lambda_{By}}{\lambda_B} = -\frac{n_{\text{eff}}^2}{2E}\{(p_{11} - 2vp_{12})\sigma_y(x, y, z) + [(1 - v)p_{12} - vp_{11}]\sigma_x(x, y, z)\}$$

$$- v[\sigma_x(x, y, z) + \sigma_y(x, y, z)]/E \quad [9.41]$$

The spectrum split of the polarized modes at any point is then given by:

$$\frac{\Delta \lambda_{By}(x, y, z) - \Delta \lambda_{Bx}(x, y, z)}{\lambda_B} = -\frac{n_{\text{eff}}^2}{2E}(1 + v)(p_{11} - p_{12})[\sigma_y(x, y, z)$$

$$- \sigma_x(x, y, z)] \quad [9.42]$$

Figure 9.7 represents the reflection spectra of an FBG under different lateral loads. The spectrum of the FBG splits in more than one peak at the Bragg wavelength. For higher values of applied force, the peaks become totally separate.

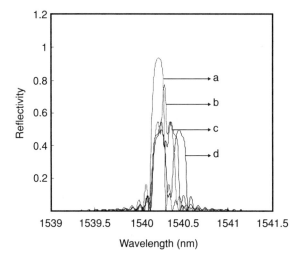

9.7 Reflection spectra of an FBG sensor under lateral compression (loadings a–d are 0.0, 1.5, 2.0 and 3.0 kN/m, respectively), the angle between the orientation of input linear polarized light and the compression direction is 45°.

9.6 Optical responses under bending

The major effects of FBG under bending (see Fig. 9.8) may come from the geometrically bent structure and non-uniform lateral stress. Using the perturbation method, the scalar propagation characteristics of a geometrically bent optical waveguide can be analysed. Corrected electric fields or Gaussian approximations have been proposed to describe the field deformations.[97,98] The equivalent straight waveguide is often used to describe its electric field in the bent waveguide. More rigorous description adopted a full-vectorial wave propagation theory.[99] Based on the photoelastic relationship, the non-uniform stress distribution-induced birefringence has been presented.[100-102] The induced linear birefringence can be expressed as:[83]

$$B_1 = 0.25 n_{co}^2 (p_{11} - p_{12})(1 + v)(r_{cl}/r_{bent})^2 \qquad [9.43]$$

Assuming that the mode holds and the phase-matching condition works well in the bent FBG, the effects under bending can be qualitatively analysed. Then the shift of Bragg wavelength is determined by the changes of propagation constants or effective refractive index, and grating period. In the case of stress-induced birefringence, only the effects of non-uniform lateral compressive stress are considered, since it is only sensitive to the difference of that in two orthogonal directions. But, in order to obtain an absolute change of refractive index, the effect of the axis component is required. According to the

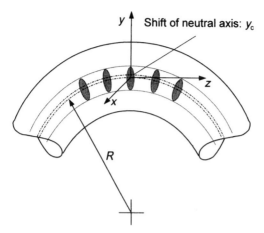

9.8 Fibre Bragg gratings under bending.

photoelastic relationship, the index of the optical fibre is decreased by tension and is increased by compression. Since the electric field in the bent optical fibre shifts towards the outer core/cladding boundary, and the outer side of the bent optical fibre is under tension, the field deformation will induce a decrement of effective refractive index. Apparently due to the birefringence of the bending optical fibre, two independent Bragg wavelengths occur. The difference of these two Bragg wavelengths is small, and only a broader bandwidth of reflection spectrum can be observed.[103]

Another characteristic of the silica fibre is the non-linear stress–strain relationship. It is known that the neutral axis is also shifted toward the outer side of the fibre under bending.[104] Its direct effect on the FBG is that the grating period will be compressed, so these effects should also be considered, together with the change of effective refractive index of the FBG under bending.

Other fibre gratings can also be used in bend sensing, such as long-period fibre grating[105,106] and multicore fibre Bragg gratings.[107]

9.7 Conclusions

This chapter has given the optical responses of FBG sensors under different modes of deformation. Both the polarization behaviour and optical spectrum of FBGs have been used to analyse the deformation perturbations. In the case of an FBG under tension, the wavelength sensitivity to tension is significant. In the case of torsion, the wavelength sensitivity is very small, and the polarization behaviour is sensitive to it. In the case of lateral compression, the wavelength sensitivity is much smaller than that under tension, but the deformation induced birefringence can broaden and split the reflection peaks.

Table 9.1 Optical responses of normal germanosilicate FBG under different deformations

Mode of deformation	Optical response		
	Bragg wavelength shift [14,85,102]	UV induced linear birefringence[83] B_l	Deformation induced linear/circular birefringence[82,85] B_l/B_c
Tension	1200 pm/m ε, at wavelength 1550 nm	2.5×10^{-5}	0
Torsion	< 10 pm, at torsion ratio 0.072 rad/mm, from experiment	2.5×10^{-5}	$B_l = 9.7 \times 10^{-7}\,\tau$, $[\tau]$ = rad/mm, at wavelength 1550 nm
Lateral compression	2.0 pm/N (parallel), 8.6 pm/N (perpendicular), with compression length 10 mm	2.5×10^{-5}	$B_l = 5.1 \times 10^{-4}$, at 10 N/mm
Pure bending	< 260 pm, at curvature 2.43 cm^{-1}	2.5×10^{-5}	$B_l = 1.5 \times 10^{-5}\,(1/r_{bent})^2$, $[r_{bent}]$ = cm

The lateral compression can be sensed by the spectrum behaviour in some sensing schemes. Both the polarization behaviour and the optical spectrum of FBGs can be used to analyse lateral compression. Bending of FBGs can be analysed for both the polarization behaviour and the optical spectrum, but it is difficult to measure the polarization evolution of fibre under bending. The optical responses of normal germanosilicate FBG under different individual modes of deformations are summarized in Table 9.1.

In practical FBG sensors, temperature must be considered in the spectrum because of their high sensitivity to temperature. Some approaches include isolating or compensating or simultaneously measuring the temperature perturbation.

Acknowledgements

The authors wish to thank Hong Kong Research Grants Council for supporting this work (Grant No. PolyU5112/98E).

References

1 Gardner J W, *Microsensors: Principles and Applications*, John Wiley & Sons, Chichester, UK, 1994.
2 Yang X and Young R J, 'Determination of residual strains in ceramic-fibre

reinforced composites using fluorescence spectroscopy', *Acta Metallurgica et Materialia*, 1995, **43**(6), 2407–16.
3 Grattan K T V and Sun T, 'Fiber optic sensor technology: an overview', *Sensors and Actuators*, A-Physical, 2000, **82**(1–3), 40–61.
4 Posey R and Vohra S T, 'An eight-channel fiber-optic Bragg grating and stimulated Brillouin sensor system for simultaneous temperature and strain measurements', *IEEE Photonics Technol. Lett.*, 1999, **11**(12), 1641–3.
5 Brown A W, DeMerchant M D, Bao X et al., 'Spatial resolution enhancement of a Brillouin-distributed sensor using a novel signal processing method', *IEEE J. Lightwave Technol.*, 1999, **17**(7), 1179–83.
6 Kim S H, Lee J J, Lee D C et al., 'A study on the development of transmission-type extrinsic Fabry–Perot interferometric optical fiber sensor', *IEEE J. Lightwave Technol.*, 1999, **17**(10), 1869–74.
7 Udd E, *Fiber Optic Sensors: An Introduction for Engineers and Scientists*, John Wiley & Sons, New York, 1991.
8 Van Steenkiste R J, *Strain and Temperature Measurement with Fiber Optic Sensors*, Technomic Publishing Company, Lancaster, 1997.
9 Johnson P, 'Strain field measurements with dual-beam digital speckle photography', *Optics Lasers Eng.*, 1998, **30**(3–4), 315–26.
10 Johnson P, 'Strain field measurements in industrial applications using dual-beam digital speckle photography', *Optics Lasers Eng.*, 1998, **30**(5), 421–31.
11 Hack E and Broennimann R, 'Electronic speckle pattern interferometry deformation measurement on lightweight structures under thermal load', *Optics Lasers Eng.*, 1999, **31**(3), 213–22.
12 Rae P J, Goldrein H T, Bourne N K et al., 'Measurement of dynamic large-strain deformation maps using an automated fine grid technique', *Optics Lasers Eng.*, 1999, **31**(2), 113–22.
13 Kersey A D, Davis M A, Patrick H J et al., 'Fiber grating sensors', *IEEE J. Lightwave Technol.*, 1997, **15**(8), 1442–63.
14 Othonos A and Kalli K, *Fiber Bragg Gratings: Fundamentals and Applications in Telecommunications and Sensing*, Artech House, Boston, 1999.
15 Rao Y J, 'In-fiber Bragg grating sensors', *Measurement Sci. Technol.*, 1997, **8**(4), 355–75.
16 Rao Y J, 'Recent progress in applications of in-fiber Bragg grating sensors', *Optics Lasers Eng.*, 1999, **31**(4), 297–324.
17 Friebele E J, 'Fiber Bragg grating strain sensors: present and future applications in smart structures', *Optics Photon. News*, 1999, **9**(1), 33–7.
18 Hill K O and Meltz G, 'Fiber Bragg grating technology fundamentals and overview', *IEEE J. Lightwave Technol.*, 1997, **15**(8), 1263–76.
19 Kashyap R, *Fiber Bragg Gratings*, Academic Press, San Diego, 1999.
20 Meltz G, 'Overview of fiber grating-based sensors'. In *Proc. SPIE–Distributed and Multiplexed Fiber Optic Sensors VI*, 1996, **2838**, 2–22.
21 Goyal A and Muendel M, 'Optical fiber Bragg gratings have a wide variety of uses', *Photonics Spectra*, 1998, **32**(9), 116–22.
22 Udd E, *Fiber Optic Smart Structures*, John Wiley, New York, 1995.
23 Morey W W, Ball G A and Singh H, 'Applications of fiber Bragg grating sensors'. In *Proc. SPIE – Fiber Optic and Laser Sensors XIV*, 1996, **2839**, 2–7.

24 Measures R M, Maaskant R, Alavie T et al., 'Fiber-optic Bragg gratings for bridge monitoring', *Cement Concrete Composites*, 1997, **9**(1), 21–3.
25 Meissner J, Nowak W, Slowik V et al., 'Strain monitoring at a prestressed concrete bridge'. In *Proc. 12th Int. Conf. Optical Fiber Sensors*, Williamsburg, 48–51, 1997.
26 Schulz W L, Udd E, Seim J M et al., 'Advanced fiber grating strain sensor systems for bridges, structures, and highways'. In *Proc. SPIE – Smart Structures and Materials 1998: Smart Systems for Bridges, Structures, and Highways*, 1998, **3325**, 212–21.
27 Prohaska J D, Snitzer E, Chen B et al., 'Fiber optic Bragg grating strain sensor in large scale concrete structures'. In *Proc. SPIE – Fiber Optic Smart Structures and Skins V*, 1992, **1798**, 286–94.
28 Du W C, Tao X M, Tam H Y et al., 'Fundamentals and applications of optical fiber Bragg grating sensors to textile structural composites', *Composite Structures*, 1998, **42**(3), 217–29.
29 Jin X D, Sirkis J S, Venkateswaran V S et al., 'Simultaneous measurement of two strain components in composite structures using embedded fiber sensors'. In *Proc. 12th Int. Conf. Optical Fiber Sensors*, Williamsburg, 44–7, 1997.
30 Ferdinand P, Ferragu O, Lechien J L et al., 'Mine operating accurate stability control with optical fiber sensing and Bragg grating technology: the European BRITE/EURAM STABILOS project', *IEEE J. Lightwave Technol.*, 1995, **13**(7), 1303–13.
31 Hjelme D R, Bjerkan L, Neegard S et al., 'Application of Bragg grating sensors in the characterization of scaled marine vehicle modes', *Appl. Optics*, 1997, **36**(1), 328–36.
32 Friebele E J, Askins C G, Bosse A B et al., 'Optical fiber sensors for spacecraft applications', *Smart Mater. Struct.*, 1999, **8**(6), 813–38.
33 Ezbiri A, Kanellopoulos S E and Handerek V A, 'High resolution instrumentation system for demodulation of Bragg grating aerospace sensors', In *Proc. 12th Int. Conf. Optical Fiber Sensors*, Williamsburg, 456–9, 1997.
34 Tang L Q, Tao X M and Choy C L, 'Effectiveness and optimization of fiber Bragg grating sensor as embedded strain sensor', *Smart Mater. Struct.*, 1999, **8**(1), 154–60.
35 Rao Y J, Webb D J, Jackson D A et al., 'In-fiber Bragg grating temperature sensor system for medical applications', *IEEE J. Lightwave Technol.*, 1997, **15**(5), 779–85.
36 Arie A, Lissak B and Tur M, 'Static fiber-Bragg grating strain sensing using frequency-locked lasers', *IEEE J. Lightwave Technol.*, 1999, **17**(10), 1849–55.
37 Kersey A D, Berkoff T A and Morey W W, 'Multiplexed fiber Bragg grating sensor system with a fiber Fabry–Perot wavelength filter', *Optics Lett.*, 1993, **18**(16), 1370–2.
38 Davis M A, Bellemore D G and Kersey A D, 'Design and performance of a fiber Bragg grating distributed strain sensor system'. In *Proc. SPIE – Smart Structures and Materials 1995: Smart Systems for Bridges, Structures, and Highways*, 1995, **2446**, 227–35.
39 Kersey A D, Davis M A and Bellemore D G, 'Development of fiber sensors for structural monitoring'. In *Proc. SPIE – Nondestructive Evaluation of Aging Bridges and Highways*, 1995, **2456**, 262–8.
40 Dunphy J R, Ball G, Amato F D et al., 'Instrumentation development in support of fiber grating sensor arrays', In *Proc. SPIE – Distributed and Multiplexed Fiber Optic Sensors III*, 1993, **2071**, 2–11.

41 Jackson D A, Ribeiro A B L, Reekie L et al., 'Simple multiplexing scheme for a fiber optic grating sensor network', *Optics Lett.*, 1993, **18**(14), 1192–4.
42 Davis M A and Kersey A D, 'A fiber Fourier transform spectrometer for decoding fiber Bragg sensors', *IEEE J. Lightwave Technol.*, 1995, **13**(7), 1289–95.
43 Kersey A D and Berkoff T A, 'Fiber optic Bragg differential temperature sensor', *IEEE Photon. Technol. Lett.*, 1992, **4**(10), 1183–5.
44 Ferreira L A, Diatzikis E V, Santos J L et al., 'Frequency-modulated multimode laser diode for fiber Bragg grating sensors', *IEEE J. Lightwave Technol.*, 1998, **16**(9), 1620–30.
45 Moreira P J, Ferreira L A, Santos J L et al., 'Dynamic range enhancement in fiber Bragg grating sensors using a multimode laser diode', *IEEE Photon. Technol. Lett.*, 1999, **11**(6), 703–5.
46 Tanaka Y and Ogusu K, 'Tensile-strain coefficient of resonance frequency of depolarized guided acoustic-wave Brillouin scattering', *IEEE Photonics Technol. Lett.*, 1999, **11**(7), 865–7.
47 Lissak B, Arie A and Tur M, 'Highly sensitive dynamic strain measurements by locking lasers to fiber Bragg gratings', *Optics Lett.*, 1998, **23**(24), 1930–2.
48 Ogawa O, Kato T and Kamikatano M, 'Technique for measuring the dynamic strain on an optical fiber based on Brillouin ring amplification', *IEEE J. Lightwave Technol.*, 1999, **17**(2), 234–42.
49 McGarrity C and Jackson D A, 'A network for large numbers of interferometric sensors and fiber Bragg gratings with high resolution and extended range', *IEEE J. Lightwave Technol.*, 1998, **16**(1), 54–65.
50 Berkoff T A, Davis M A, Bellemore D G et al., 'Hybrid time and wavelength division multiplexed fiber grating array', In *Proc. SPIE – Smart Structures and Materials 1995: Smart Sensing, Processing, and Instrumentation*, 1995, **2444**, 288–94.
51 Friebele E J, Putnam M A, Patrick H J et al., 'Ultrahigh-sensitivity fiber-optic strain and temperature sensor', *Optics Lett.*, 1998, **23**(3), 222–4.
52 Chan C C, Jin W, Rad A B et al., 'Simultaneous measurement of temperature and strain: an artificial neural network approach', *IEEE Photon. Technol. Lett.*, 1998, **10**(6), 854–6.
53 James S W, Dockney M L and Tatam R P, 'Simultaneous independent temperature and strain measurement using in-fiber Bragg grating sensors', *Electron. Lett.*, 1996, **32**(12), 1133–4.
54 Xu M G, Archambault J L, Reekie L et al., 'Discrimination between strain and temperature effects using dual-wavelength fiber grating sensors', *Electron. Lett.*, 1994, **30**(13), 1085–7.
55 Cavaleiro P M, Araújo F M, Ferreira L A et al., 'Simultaneous measurement of strain and temperature using Bragg gratings written in germanosilicate and boron-codoped germanosilicate fibers', *IEEE Photon. Technol. Lett.*, 1999, **11**(12), 1635–7.
56 Patrick H J, Williams G M, Kersey A D et al., 'Hybrid fiber Bragg grating/long period fiber grating sensor for strain/temperature discrimination', *IEEE Photon. Technol. Lett.*, 1996, **8**(9), 1223–5.
57 Du W C, Tao X M and Tam H Y, 'Temperature independent strain measurement with a fiber grating tapered cavity sensor', *IEEE Photon. Technol. Lett.*, 1999, **11**(5), 596–8.

Optical responses of FBG sensors under deformations 171

58 Du W C, Tao X M and Tam H Y, 'Fiber Bragg grating cavity sensor for simultaneous measurement of strain and temperature', *IEEE Photon. Technol. Lett.*, 1999, **11**(1), 105–7.
59 Lo Y L, 'In-fiber Bragg grating sensors using interferometric interrogations for passive quadrature signal processing', *IEEE Photon. Technol. Lett.*, 1998, **10**(7), 1003–5.
60 Koo K P, LeBlanc M, Tsai T E et al., 'Fiber-chirped grating Fabry–Perot sensor with multiple-wavelength-addressable free-spectral ranges', *IEEE Photon. Technol. Lett.*, 1998, **10**(7), 1006–8.
61 Smith J, Brown A, DeMerchant M et al., 'Simultaneous strain and temperature measurement using a Brillouin scattering based distributed sensor'. In *Proc. SPIE – Smart Structures and Materials 1999: Sensory Phenomena and Measurement Instrumentation for Smart Structures and Materials*, 1999, **3670**, 366–73.
62 Kanellopoulos S E, Handerek V A and Rogers A, 'Simultaneous strain and temperature sensing with photogenerated in-fiber gratings', *Optics Lett.*, 1995, **20**(3), 333–5.
63 Kang S C, Kim S Y, Lee S B et al., 'Temperature-independent strain system using a tilted fiber Bragg grating demodulator', *IEEE Photon. Technol. Lett.*, 1998, **10**(10), 1461–3.
64 Sinha P G and Yoshino T, 'Acoustically scanned low-coherence interrogated simultaneous measurement of absolute strain and temperature using highly birefringent fibers', *IEEE J. Lightwave Technol.*, 1998, **16**(11), 2010–15.
65 Yun S H, Richardson D J and Kim B Y, 'Interrogation of fiber grating sensor arrays with a wavelength-swept fiber laser', *Optics Lett.*, 1998, **23**(11), 843–5.
66 Putnam M A, Dennis M L, Duling III I N et al., 'Broadband square-pulse operation of a passively mode-locked fiber laser for fiber Bragg grating interrogation', *Optics Lett.*, 1998, **23**(2), 138–40.
67 LeBlanc M, Huang S, Ohn M et al., 'Distributed strain measurement based on a fiber Bragg grating and its reflection spectrum analysis', *Optics Lett.*, 1996, **21**(17), 1405–7.
68 Huang S, Ohn M M and Measures R M, 'A novel Bragg grating distributed-strain sensor based on phase measurements'. In *Proc. SPIE – Smart Structures and Materials 1995: Smart Sensing, Processing, and Instrumentation*, 1995, **2444**, 158–69.
69 Volanthen M, Geiger H and Dakin J P, 'Distributed grating sensors using low-coherence reflectometry', *IEEE J. Lightwave Technol.*, 1997, **15**(11), 2076–82.
70 Bhatia V and Vengsarkar A M, 'Optical fiber long-period grating sensors', *Optics Lett.*, 1996, **21**(9), 692–4.
71 Patrick H, Kersey A D, Pedrazzani J R et al., 'Fiber Bragg grating demodulation system using in-fiber long period grating filters'. In *Proc. SPIE – Distributed and Multiplexed Fiber Optic Sensors VI*, 1996, **2838**, 60–5.
72 Kersey A D and Morey W W, 'Multiplexed Bragg grating fiber laser strain sensor system with mode-locked interrogation', *Electron. Lett.*, 1993, **29**(1), 112–14.
73 Kersey A D and Morey W W, 'Multi-element Bragg grating based fiber laser strain sensor', *Electron. Lett.*, 1993, **29**(11), 964–66.
74 Koo K P and Kersey A D, 'Bragg grating based laser sensor systems with interferometric interrogation and wavelength division multiplexing', *IEEE J. Lightwave Technol.*, 1995, **13**(7), 1243–9.

75 Born M and Wolf E, *Principles of Optics*, 7th (Expanded) edn, Cambridge University Press, Cambridge, 1999.
76 Huard S, *Polarization of Light*, John Wiley, New York, 1997.
77 Ulrich R and Simon A, 'Polarisation optics of twisted single-mode fibers', *Appl. Optics*, 1979, **18**(13), 2241–51.
78 Buck J A, *Fundamentals of Optical Fibers*, John Wiley, New York, 1995.
79 Zhou W, Chen X F and Yang D X, *Fundamentals of Photonics*, Zhejiang University Press, Hangzhou, 2000.
80 Marcuse D, *Theory of Dielectric Optical Waveguides*, 2nd edn, Academic Press, Boston, 1991.
81 Wangsness R K, *Electromagnetic Fields*, John Wiley, New York, 1979.
82 Zhang A P, Tao X M and Tam H Y, Prediction of polarization behavior of twisted optical fibres containing Bragg grating sensors, *J. Textile Inst.*, 2000, Part 3, 105–16.
83 Jeunhomme L B, *Single-Mode Fiber Optics: Principles and Applications*, 2nd edn, Marcel Dekker, New York, 1990.
84 Erdogan T and Mizrahi V, 'Characterization of UV-induced birefringence in photosensitive Ge-doped silica optical fiber', *J. Opt. Soc. Amer. B*, 1994, **11**(10), 2100–5.
85 Vengarkar A M, Zhong Q, Inniss D et al., 'Birefringence reduction in side-written photoinduced fiber devices by a dual-exposure method', *Optics Lett.*, 1994, **19**(16), 1260–2.
86 Gafsi R and El-Sherif M A, 'Analysis of induced-birefringence effects on fiber Bragg gratings', *Opt. Fiber Technol.*, 2000, **6**(3), 299–323.
87 Kawase L R, Valente L C G, Margulis W et al., 'Force measurement using induced birefringence on Bragg grating'. In *Proc. 1997 SBMO/MTT-S International Microwave and Optoelectronics Conference*, (1) 394–6, 1997.
88 Lawrence C M, Nelson D V, Makino A et al., 'Modeling of the multi-parameter Bragg grating sensor', In *Proc. SPIE – Third Pacific Northwest Fiber Optic Sensor Workshop*, 1997, **3180**, 42–9.
89 Udd E, Nelson D V and Lawrence C M, 'Multiple axis strain sensing using fiber gratings written onto birefringent single mode optical fiber', In *Proc. 12th Int. Conf. Optical Fiber Sensors*, Williamsburg, 48–51, 1997.
90 Udd E, Lawrence C M and Nelson D V, 'Development of a three axis strain and temperature fiber optical grating sensor'. In *Proc. SPIE – Smart Structures and Materials 1997: Smart Sensing, Processing, and Instrumentation*, 1997, **3042**, 229–36.
91 Udd E, Schulz W L, Seim J M et al., 'Transverse fiber grating strain sensors based on dual overlaid fiber gratings on polarization preserving fibers'. In *Proc. SPIE – Smart Structures and Materials 1998: Sensory Phenomena and Measurement Instrumentation for Smart Structures and Materials*, 1998, **3330**, 253–63.
92 Udd E, Schulz W L and Seim J M, 'Measurement of multidimensional strain fields using fiber grating sensors for structural monitoring'. In *Proc. SPIE – Fiber Optic Sensor Technology and Applications*, 1999, **3860**, 24–34.
93 Udd E, Schulz W L and Seim J M, 'Multi-axis fiber grating strain sensor applications for structural monitoring and process control'. In *Proc. SPIE – Process Monitoring Applications of Fiber Optic Sensors*, 1999, **3538**, 206–14.
94 LeBlanc M, Vohra S T, Tsai T E et al., 'Transverse load sensing by use of pi-phase-shifted fiber Bragg gratings', *Optics Lett.*, 1999, **24**(16), 1091–3.

95 Canning J and Sceats M G, 'π-phase-shifted periodic distributed structures in optical fibers by UV post-processing', *Electron. Lett.*, 1994, **30**(16), 1344–5.
96 Liu Y, Zhang L, Bennion I, 'Fibre optic load sensors with high transverse strain sensitivity based on long-period gratings in B/Ge co-doped fibre', *Electron. Lett.*, 1999, **35**(8), 661–3.
97 Gambling W A, Matsumura H and Ragdale C M, 'Field deformation in a curved single-mode fiber', *Electronics Lett.*, 1978, **14**(5), 130–2.
98 Garth S J, 'Modes on a bent optical wave-guide', *IEE Proc. J: Optoelectronics*, 1987, **134**(4), 221–9.
99 Lui W W, Xu C L, Hirono T et al., 'Full-vectorial wave propagation in semiconductor optical bending waveguides and equivalent straight waveguide approximations', *J. Lightwave Technol.*, 1998, **16**(5), 910–14.
100 Ulrich R, Rashleigh S C and Eichoff W, 'Bending-induced birefringence in single mode fibers', *Optics Lett.*, 1980, **5**(6), 273–5.
101 Garth S J, 'Birefringence in bent single-mode fibers', *J. Lightwave Technol.*, 1988, **6**(3), 445–9.
102 Sakai J I and Kimura T, 'Birefringence and polarization characteristics of single-mode optical fibers under elastic deformations', *IEEE J. Quantum Electron.*, 1981, **17**(6), 1041–51.
103 Dai C, Yang D X, Tao X M et al., 'Effects of pure bending on the sensing characteristics of fiber Bragg gratings', In *Proc. SPIE – International Conference on Sensors and Control Techniques*, 2000, **4077**, 92–6.
104 Suhir E, 'Effect of the nonlinear stress–strain relationship on the maximum stress in silica fibers subjected to two-point bending', *Appl. Optics*, 1993, **32**(9), 1567–72.
105 Liu Y, Zhang L, Williams J A R et al., 'Optical bend sensor based on measurement of resonance mode splitting of long-period fiber grating', *IEEE Photon. Technol. Lett.*, 2000, **12**(5), 531–3.
106 Ye C C, James S W and Tatam R P, 'Simultaneous temperature and bend sensing with long-period fiber gratings', *Optics Lett.*, 2000, **25**(14), 1007–9.
107 Gander M J, MacPherson W N, McBride R et al., 'Bend measurement using Bragg gratings in multicore fibre', *Electron. Lett.*, 2000, **36**(2), 120–1.

10
Smart textile composites integrated with fibre optic sensors

XIAOMING TAO

10.1 Introduction

The rapid development in textile structural composites (TSCs) has created new marketing and research opportunities for the textile industry and textile scientists.[1] According to their structural integrity, three-dimensional textile composites have a network of yarn bundles in an integrated manner, thereby giving rise to significant increases in inter- and intralaminar strength, higher feasibility of complex structural shape formation, and greater possibility of large-scale manufacturing at reduced cost as compared with traditional laminated composites. Their higher strength and stiffness with lighter weight have led to increasing applications in aerospace, automobile industries and civil engineering. It has been predicted that the improvement of TSC fabrication technologies and their combination with smart structure technologies will lead to major industrial growth in the next century by challenging the position of metals and other conventional engineering materials.

A success in developing TSC fabrication technology relies on a better understanding of the processing structure–properties relationship. An important step in this direction involves the monitoring of the internal strain/stress distributions in realtime during the fabrication of textile preforms and the subsequent solidification to their final structures. Another important issue in the application of TSCs is to make them sensitive to their internal health conditions and external environments. Integration of sensing networks inside fabric-reinforced structures is considered to be the first step to make the materials smart. Furthermore, the complexity of the TSC structure, for instance, the skin–core effect of three-dimensional braid composites, makes characterization of the material a very difficult task. In the past, it has been almost impossible to measure the internal stress/strain distribution of such a complex material by using conventional methods such as strain gauges and ultrasonic sensors. In addition, there is a need for some kind of sensing network embedded in the structures to offer a means to (1) monitor the

Smart textile composites integrated with fibre optic sensors 175

internal stress distribution of TSCs *in situ* during the manufacturing process, (2) allow health monitoring and damage assessment of TSCs during services, and (3) enable a control system to actively monitor and react to the changes in the working environment.

Fibre optic technologies offering both sensing and signal transmission functions have attracted considerable attention in recent years, especially in smart concrete structures including highways, bridges, dams and buildings.[2,3] A number of researchers have applied fibre optic sensor (FOS) technologies to manufacture process monitoring and structure health assessment for fibre-reinforced composites.[4-6] Since optical fibres are of small size and lightweight, compatible with textile yarns, and readily embedded or even woven inside TSCs, they are the most promising medium to form the sensing network mentioned above.

This chapter provides a review of various types of fibre optic sensors, major issues of smart textile composites integrated with fibre Bragg grating sensors, that is temperature and strain coupling, sensitivity, multiaxial strain measurement, measurement effectiveness and reliability issues, as well as various measurement systems for smart textile composites integrated with fibre optic sensors.

10.2 Optical fibres and fibre optic sensors

Normally, an optical fibre consists of a core surrounded by a cladding whose refractive index is slightly smaller than that of the core. The optic fibre is coated during the fibre drawing process with a protective layer of polymer. Inside the fibre core, light rays incident on the core-cladding boundary at angles greater than the critical angle undergo total internal reflection and are guided through the core without refraction. Silica glass is the most common material for optical fibres, where the cladding is normally made from pure fused silica, and the core from doped silica containing a few mol% of germanium. Other dopants, such as phosphorus, can also be used. Extra-low absorption occurs in a germanosilicate fibre with a local minimum attenuation coefficient $\alpha = 0.3\,\text{dB/km}$ at 1.3 μm and an absolute minimum $\alpha = 0.16\,\text{dB/km}$ at 1.55 μm. Therefore, light in the two windows travels down tens of kilometres of fibre without strong attenuation in a correct guided mode condition. This is why optical fibre has been replacing copper coaxial cable as the preferred transmission medium for electromagnetic waves, revolutionizing the world's communication today.

Parallel with the rapid development in the area of optical fibre communications, fibre optical sensors have also attracted much attention and experienced a fast growth in the recent years. They are lightweight, small and flexible; thus they will not affect the structure integrity of the composite materials and can be integrated with the reinforcing fabrics to form the backbones in structures. They are based on a single common technology that enables devices to be developed for sensing numerous physical perturbations of a mechanical, acoustic, electric, magnetic and thermal nature. A number of sensors can be multiplexed along a single optical fibre using wavelength-, frequency-, time- and polarization-division techniques to form one-, two- or three-dimensional distributed sensing systems. They do not provide a conducting path through the structure and do not generate additional heat that could potentially damage the structure. They do not require electrical isolation from the structural material and do not generate electromagnetic interference; this could be a crucial advantage in some applications.

For the applications in smart structures, FOSs can be divided according to whether sensing is distributed, localized (point) or multiplexed (multipoint).[6] If sensing is distributed along the length of the fibre, the measurand distribution as a function of position can be determined from the output signal, so a single fibre can effectively monitor the changes in the entire object into which it is embedded. A localized sensor detects measurand variation only in the vicinity of the sensor. Some localized sensors can lead themselves to multiplexing, in which multiple localised sensors are placed at intervals along the fibre length. Each sensor can be isolated by wavelength, time or frequency discrimination, thereby allowing the real-time profiling of parameters throughout the structure.

Before the invention of fibre Bragg gratings (FBGs), FOSs could be classified into two broad categories, intensiometric and interferometric, according to the sensing scheme. Intensiometric sensors are simply based on the amount of light detected through the fibre. In its simplest form, a stoppage of transmission due to breakage of a fibre embedded in the structure indicates possible damage. Interferometric sensors have been developed for a range of high-sensitivity applications, such as acoustic and magnetic field sensors, and are usually based on single-mode fibres. For example, the Mach–Zehnder interferometer, as shown in Fig. 10.1, is one of the most common configurations. With this type of device, strain can be monitored directly by embedding the sensing fibre arm in the structure, while the referencing arm of equal length is isolated from the environment. Although such a configuration is highly sensitive to strain, the whole fibre length in the signal arm responds to strain, and thus localization of the sensing region is difficult. An alternative interferometric sensor, more suitable for localized sensing, is based on interference between light reflected from two closely spaced surfaces, which form a short gauge length Fabry–Perot (FP)-type interferometer (Fig. 10.2). The strain or stress applied on the gauge inside the structure can be determined by measuring the reflected spectrum or reflected light signal from the FP cavity, which is a function of the distance between the two reflected surfaces. The disadvantage of such devices is that it is difficult to perform

Smart textile composites integrated with fibre optic sensors

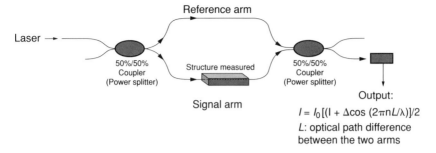

10.1 Measurement by interferometric sensors with Mach–Zehnder interferometer.

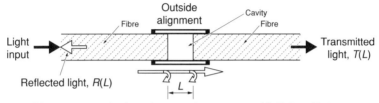

10.2 Measurement by interferometric sensors with Fabry–Perot-type interferometer.

absolute measurements, and hard to form a multiplexing sensor array along a single fibre length due to the large loss of the discontinuing structure of an FP cavity. Detailed review and analysis have been given by Udd[7] and Measures.[8]

10.3 Principal analysis of embedded fibre Bragg grating sensors

10.3.1 Principle of FBGS

As FBGs have many advantages over the other two groups, and show great promise, we will concentrate on FBGs in this section. The FBG is fabricated by modulation of the refractive index of the core in a single mode optic fibre, which has been described in detail in Chapters 8 and 9. Assume that the change in the index modulation period is independent of the state of polarization of the interrogating light and only dependent on the fibre axial strain, differentiating the Bragg wavelength in Eq. (9.15) yields:

$$\frac{\Delta \lambda}{\lambda} = \varepsilon_1 + \frac{\Delta n}{n} \qquad [10.1]$$

where ε_1 is the total axial strain of the optic fibre. Generally speaking, λ and n will have different values in the directions of polarization. Subscript $i = 1, 2, 3$ is denoted for λ and n as their values in the defined polarized direction. A local

Cartesian coordinate system is used, with 1, 2, 3 representing the three principal directions respectively. Eq. (10.1) can be rewritten as:

$$\frac{\Delta \lambda_i}{\lambda_i} = \varepsilon_1 + \frac{\Delta n_i}{n_i} \qquad [10.2]$$

For strain, subscript ($j = 1, 2, 3, 4, 5, 6$) is used. The first three represent the normal strains in the first (fibre axis), second, third directions, respectively, the latter three being the three shear strains, respectively. The strain ε of an optic fibre may be contributed by either thermal expansion or stress, hence the symbol ε^* is used for the optic fibre strain induced by stress only. The refraction index n is related to both temperature T and strain ε^*, therefore:

$$\frac{\Delta n_i}{n_i} = \frac{1}{n_i} \sum_j \frac{\partial n_i}{\partial \varepsilon_j^*} \varepsilon_j^* + \frac{\partial n_i}{\partial T} \Delta T \qquad [10.3]$$

According to the strain optic theory:[9]

$$\frac{1}{n_i} \sum_j \frac{\partial n_i}{\partial \varepsilon_j^*} \varepsilon_j^* = -\frac{n_i^2}{2} \sum_j P_{ij} \varepsilon_j^* \qquad [10.4]$$

where P_{ij} is the strain-optic coefficient matrix. For a homogeneous isotropic medium,

$$P_{ij} = \begin{bmatrix} P_{11} & P_{12} & P_{12} & & & \\ P_{12} & P_{11} & P_{12} & & 0 & \\ P_{12} & P_{12} & P_{11} & & & \\ & & & P_{44} & & \\ & 0 & & & P_{44} & \\ & & & & & P_{44} \end{bmatrix} \qquad [10.5]$$

where $P_{44} = (P_{11} - P_{12})/2$.

For a homogeneous isotropic medium, it can be assumed that the index of refraction n has a linear relation with temperature T:

$$\xi = \frac{\partial n_i}{\partial T} \qquad [10.6]$$

where ξ is regarded as a thermo-optic constant.

Because lights are transverse waves, only the transverse (2, 3 directions) deviations of the reflective index can cause the shift of the Bragg wavelength. Substituting Eqs. (10.4), (10.5) and (10.6) into Eq. (10.3), the peak wavelength shifts for the light linearly polarized in the second and third directions are given below:

$$\frac{\Delta\lambda_2}{\lambda_2} = \varepsilon_1 - \frac{n_2^2}{2}[P_{11}\varepsilon_2^* + P_{12}(\varepsilon_1^* + \varepsilon_3^*)] + \xi\Delta T \qquad [10.7]$$

and

$$\frac{\Delta\lambda_3}{\lambda_3} = \varepsilon_1 - \frac{n_3^2}{2}[P_{11}\varepsilon_3^* + P_{12}(\varepsilon_1^* + \varepsilon_2^*)] + \xi\Delta T \qquad [10.8]$$

In many cases, the wavelength shift for the Bragg grating sensor observed for each polarization eigen-mode of the optical fibre depends on all the three principal strain components within the optical fibre. Sirkis and Haslach[10] extended Butter and Hocker's model[9] and have shown that their results are closer to those observed in transverse loading experiments[11] for the interferometric optical fibre sensor.

The general cases will be discussed in Section 10.4.2. Here we will only discuss the axisymmetric problem where $\varepsilon_2^* = \varepsilon_3^*$. If the optic fibre is a thermal isotropic material with a constant expansion coefficient α, then $\varepsilon_j^* = \varepsilon_j - \alpha\Delta T$ ($j = 1, 2, 3$). Eqs. (10.7), (10.8) can be written into the same form:

$$\frac{\Delta\lambda}{\lambda} = \varepsilon_1\left\{1 - \frac{n^2}{2}\left[P_{12} + (P_{11} + P_{12})\frac{\varepsilon_2}{\varepsilon_1}\right]\right\}$$
$$+ \alpha\frac{n^2}{2}(P_{11} + 2P_{12})\Delta T + \xi\Delta T$$
$$= f\varepsilon_1 + \xi^*\Delta T \qquad [10.9]$$

where

$$f = 1 - \frac{n^2}{2}\left[P_{12} + (P_{11} + P_{12})\frac{\varepsilon_2}{\varepsilon_1}\right] \qquad [10.10]$$

and

$$\xi^* = \xi + \frac{\alpha n^2}{2}(P_{11} + 2P_{12}) \qquad [10.11]$$

f is defined as the sensitivity factor, ξ^* as the revised optic-thermal constant.

10.3.2 Sensitivity factor

10.3.2.1 Axial strain measurement alone

When the temperature change is so small that its effect can be ignored, the FBG can be considered as a strain sensor. Let us define $\upsilon^* = (-\varepsilon_2/\varepsilon_1)$ as the effective Poisson's ratio (EPR) of optic fibre. From Eq. (10.11), it is obvious that the sensitivity factor f is not a constant but a function of υ^*,

Table 10.1 Material parameters of single mode silica optic fibres

Strain-optic coefficient		Index of refraction (n)	Elastic modulus E (GPa)	Poisson's ratio v
P_{11}	P_{12}			
0.113	0.252	1.458	70	0.17

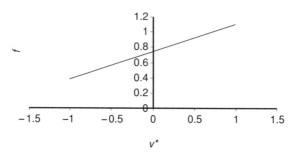

10.3 Sensitivity factor plotted against effective Poisson's ratio.

$$f = 1 - \frac{n^2}{2}[P_{12} - (P_{11} + P_{12})v^*] \qquad [10.12]$$

Figure 10.3 shows a typical curve of the sensitivity factor as a function of the effective Poisson's ratio, calculated by using the material parameters of optic fibre, provided in Table 10.1.

Let us consider the following special cases:

1. $v^* = 0.17$, $f = 0.798$: which implies that the EPR is equal to the fibre material Poisson's ratio and meets the requirement of Butter and Hocker's assumption.[9] The value of sensitivity factor, $f = 0.798$, is recommended by many FBGS manufacturers;
2. $v^* = -1$, $f = 0.344$: which implies that the strains in three principal directions of the fibre are equal, which corresponds to the case of static uniform stress or the case of thermal expansion;
3. $v^* = 0.0$, $f = 0.732$; which implies that there is no transverse deformation.

Therefore, if an FBGS is used as an embedded sensor, it is necessary to make a correction of the sensitivity factor with respect to the transverse principal strain. Otherwise, only when the transverse principal strain of the optic fibre is not sensitive to the host's strain field can the sensitivity factor be regarded as a constant.

10.3.2.2 Temperature measurement alone

If a stand-alone FBGS is subjected to a temperature change, then:

$$\varepsilon_1 = \varepsilon_2 = \alpha \Delta T$$

where α is the isothermal expansion coefficient of the optic fibre. Substituting the above equation into Eq. (10.9), we can derive the following:

$$\frac{\Delta \lambda}{\lambda} = \left\{ \alpha \left[1 - \frac{n^2}{2}(2P_{12} + P_{11}) \right] + \xi^* \right\} \Delta T = (\alpha + \xi) \Delta T$$

Usually ξ is over ten times greater than α (for silica, $\alpha = 0.55 \times 10^{-6}$, $\xi = 8.3 \times 10^{-6}$), therefore the effect of thermal expansion on the measurement result may be negligible for such a stand-alone FBGS.

10.3.3 Temperature and strain coupling

As an embedded strain sensor, ideally, the measured strain ε_1 in Eq. (10.9) should represent the host strain in the optic fibre direction. The temperature compensation can be made simply by:

$$\varepsilon_1 = (\Delta \lambda / \lambda - \xi^* \Delta T)/f \quad [10.13]$$

For the germanium-doped silica core,[8] the thermal-optic coefficient ξ is approximately equal to 8.3×10^{-6}, then the revised constant ξ^* is 8.96×10^{-6}. If the measured strain is greater than 0.001 and the variation of the temperature is smaller than 10 °C, comparatively the term $\xi^* \Delta T$ in Eq. (10.13) is one order of magnitude smaller than that of the strain, and the temperature compensation would be unnecessary in some cases.

10.4 Simultaneous measurements of strain and temperature

10.4.1 Simultaneous measurement of axial strain and temperature

If the internal temperature of the host is unknown and its contribution to the shift of the wavelength is compatible to that of the strain, it is impossible to determine strain and temperature from Eq. (10.9) only. One of the most significant limitations of FBG sensors is their dual sensitivity to temperature and strain. This leads to difficulty in the independent measurements for these two measurands. The main approach is to locate two sensor elements which have different responses to strain and temperature. The sensor schemes are based on the combination of FBGs with different grating types, such as FBGs with different diameter, different Bragg wavelength, different codope, hybrid

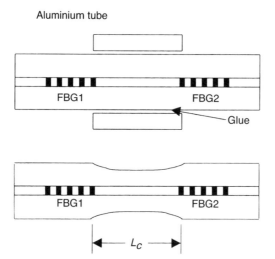

10.4 Structures of FBG cavity sensors: (a) FBG Fabry–Perot cavity, (b) FBG slightly tapered cavity.

FBGs and long period fibre grating, Fabry–Perot cavity, stimulated Brillouin scattering or fibre polarization rocking filter. The observables may be Bragg wavelength, intensity, Brillouin frequency or polarization rocking resonant wavelength.[12–19] This subsection introduces simultaneous axial strain and temperature measurements by using an FBG Fabry–Perot cavity[20,21] or superstructured FBGs.[22]

The structure of a FBG Fabry–Perot cavity sensor is shown in Fig. 10.4(a), which consists of two identical FBGs separated by a short cavity with a length of L_c. If the reflectivity of the two FBGs, $R_g(\lambda)$, is small, the reflection spectrum of the FBG Fabry–Perot cavity sensor, $R_{gc}(\lambda)$, is approximately given by:

$$\left. \begin{array}{l} R_{gc}(\lambda) = a_c R_g(\lambda) F(\lambda) \\ F(\lambda) = 1 + \cos[\phi(\lambda)] = 1 + \cos(4\pi n_c L_c/\lambda) \end{array} \right\} \quad [10.14]$$

where a_c is a constant, $F(\lambda)$ is an interference of the cavity, $\phi(\lambda)$ is the phase difference between the light reflected by the two FBGs, and n_c is the effective refractive index of the section. The reflection spectrum of the FBG Fabry–Perot cavity sensor is modulated by the cavity phase change. As a result, strain and temperature are encoded into the Bragg wavelength shift and the variation in the cavity's phase difference or optical path. The change in the phase difference can be decoded by measuring the change in the total reflected power or the reflection spectrum profile of light from the sensor.

When a minimum of $F(\lambda)$ occurs within the grating's main reflection band, the FBG Fabry–Perot cavity reflection spectrum is split into two peaks, one on each side of the Bragg wavelength λ_B. If this minimum λ_{min} coincides with λ_B the intensities of the two peaks become equal. With applied strain or temperature, both the spectrum and the interference function as a result of change in the phase difference variation. The respective shifts of λ_B and λ_{min} can be expressed by:

$$\begin{pmatrix} \Delta\lambda_B/\lambda_B \\ \Delta\lambda_{min}/\lambda_{min} \end{pmatrix} = \begin{pmatrix} K_{1T} & K_{1\xi} \\ K_{2T} & K_{2\xi} \end{pmatrix} \begin{pmatrix} \Delta T \\ \varepsilon \end{pmatrix} \qquad [10.15]$$

where K_{iT} and $K_{i\xi}$ are the temperature and strain coefficients of the FBG ($i = 1$) and the cavity ($i = 2$) sections, respectively. If the two sections have equal coefficients such as $K_{1\xi} = K_{2\xi}$ and $K_{1T} = K_{2T}$, the relative position between λ_{min} and λ_B will not change with applied strain or temperature. In this case the reflection spectrum of the FBG Fabry–Perot cavity sensor shifts, but its profile remains unchanged. Thus temperature and strain cannot be determined separately. However, if the FBG and cavity sections have different strain and temperature coefficients, that is $K_{1\xi} > K_{2\xi}$ or $K_{1T} > K_{2T}$, λ_B will move at a faster rate than λ_{min}. This results in a reduction in the intensity of peak 1 and an increase in the intensity of peak 2 when strain or temperature increases. When λ_B increases to coincide at the next maximum of $F(\lambda)$ at λ_{max}, peak 2 reaches a maximum value and peak 1 vanishes. Further increase in strain or temperature will result in two peaks in the FBG Fabry–Perot cavity reflection spectrum, but in this case the intensity of peak 2 (corresponding to the longer wavelength) will decrease, whereas the intensity of peak 1 (corresponding to the shorter wavelength) increases. Therefore, the intensity of the two peaks changes periodically with strain and temperature.

The intensity fluctuation of the light source can be eliminated by introducing a normalized parameter $M = (I_{P1} - I_{P2})/(I_{P1} + I_{P2})$, where I_{P1} and I_{P2} are the respective intensities of peaks 1 and 2. If the relationship between $\Delta\lambda_p$, M, ε and ΔT is assumed to be linear, then the two measurands can be determined simultaneously by measuring the changes in M and the wavelength shift $\Delta\lambda_p$ of either peak 1 or peak 2 with respect to strain and temperature:

$$\begin{pmatrix} \Delta\lambda_p \\ \Delta M \end{pmatrix} = \begin{pmatrix} A_{1T} & A_{1\xi} \\ A_{2T} & A_{2\xi} \end{pmatrix} \begin{pmatrix} \Delta T \\ \varepsilon \end{pmatrix} \qquad [10.16]$$

In order that the strain and temperature coefficients of the cavity section are different from those of the grating sections, a short (1 mm long) and thin aluminium tube (with an inside diameter of 0.3 mm and wall thickness of 0.15 mm) was glued onto the cavity section. This section is more difficult to stretch than the grating sections, and thus its strain coefficient is correspondingly smaller, so that $K_{2\xi} < K_{1\xi}$. On the other hand, its temperature coefficient becomes larger ($K_{2T} > K_{1T}$). This is because the thermal expansion coefficient of aluminium ($23.5 \times 10^{-6}/°C$) is much larger than that of the silica glass fibre ($0.55 \times 10^{-6}/°C$), and expansion of the aluminium tube due to temperature rise induces additional strain to the cavity section.

Figure 10.4(b) shows another FBG Fabry–Perot cavity structure, that is an FBG tapered cavity sensor, in which the cavity section is tapered slightly. The maximum change of diameter at the cavity section is smaller than 15%, hence no obvious changes in transmission and mode-effective refractive index are induced. The tapered cavity section possesses the same strain and temperature coefficients as those of the grating section. However, the average strain suffered by the tapered cavity (ξ_c) becomes larger than that at the grating section (ξ), which is given by $\xi_c = \eta\xi$, where η is an average ratio of cross-sectional areas between the grating and cavity sections. The relative movement between $F(\lambda)$ and $R_g(\lambda)$ remains zero when the temperature changes, therefore the spectral profile is only sensitive to the strain applied along the sensor:

$$\begin{pmatrix} \Delta\lambda_B/\lambda_B \\ \Delta\lambda_{min}/\lambda_{min} \end{pmatrix} = \begin{pmatrix} \alpha_1 & \beta_1 \\ \alpha_1 & \eta\beta_1 \end{pmatrix} \begin{pmatrix} \Delta T \\ \varepsilon \end{pmatrix} \qquad [10.17]$$

where α_1 and β_1 are the temperature and strain coefficients of the grating section, respectively.

Superstructured FBG sensors were developed for this purpose,[22] which have advantages of easier manufacturing and no need to alter the mechanical properties and geometry of fibre sensor. FBG sensors generally consist of a single FBG or a combination of FBGs written in low-birefringent optical fibre. In the former case, the Bragg wavelength shift can be used to measure the axial component of strain or a change in temperature. In the latter case, axial strain and temperature can be simultaneously determined according to the Bragg wavelength shifts, or the Bragg wavelength shifts and the intensities.

10.4.2 Simultaneous measurement of multi-axial strain and temperature

10.4.2.1 Polarization-maintaining (PM) FBG

The transverse strain-induced wavelength sensitivity of FBGs in silica optical fibres is low. For instance, in lateral compression, the changes in fibre birefringence are smaller than 10^{-6}. This level of birefringence corresponds to wavelength separations much smaller than the typical bandwidth of an FBG. However, we have shown in Section 10.3.1 that the sensitivity factor is a function of the two principal transverse strains and axial strain, which are normally unknown. The needs are apparent for multi-axial measurements of strain and temperature.

FBGs were written in a polarization-maintaining fibre for the measurement of lateral strains.[23-26] A polarization-maintaining fibre may consist of a circular core and inner cladding surrounded by an elliptical stress-applying region. Birefringence is induced by thermal stresses generated during the cool-down from the drawing temperature due to the geometric asymmetry of the stress-applying region. This stress-induced birefringence leads to different propagation constants for the two orthogonal polarization modes in the fibre. The x and y axes are parallel to the fast and slow axes of the principal polarization axes of the fibre, respectively, which correspond to the minor and major axes of the elliptical stress-applying region. Polarization-maintaining fibre has an initial birefringence that is sufficient to split the grating completely into two separate spectra. Because of the difference in effective refractive indices of the two orthogonal polarization modes, two effective FBGs result in one along the polarization axes by writing one FBG. For the case of low-birefringent FBG, if the temperature sensitivity of the FBG remains constant, the relative Bragg wavelength shifts can be written as:

$$\left. \begin{array}{l} \Delta\lambda_{ax}/\lambda_{ax0} = \varepsilon_3 - 0.5 \times n_{effx}^2[p_{11}\varepsilon_1 + p_{12}(\varepsilon_2 + \varepsilon_3)] \\ \Delta\lambda_{ay}/\lambda_{ay0} = \varepsilon_3 - 0.5 \times n_{effy}^2[p_{11}\varepsilon_2 + p_{12}(\varepsilon_1 + \varepsilon_3)] \end{array} \right\} \quad [10.18]$$

where λ_{ax} and λ_{ay} are the Bragg wavelength for the two orthogonal polarization modes, respectively, λ_{ax0} and λ_{ay0} are the initial unstrained Bragg wavelength for the two orthogonal polarization modes, respectively, and n_{effx} and n_{effy} are the effective refractive indices of the two orthogonal polarization modes, respectively. Supposing that the axial strain is known, the relative Bragg wavelength shifts can be written in matrix:

$$\begin{pmatrix} \Delta\lambda_{ax}/\lambda_{ax0} \\ \Delta\lambda_{ay}/\lambda_{ay0} \end{pmatrix} = \begin{pmatrix} K_{11} & K_{12} \\ K_{21} & K_{22} \end{pmatrix} \begin{pmatrix} \varepsilon_1 \\ \varepsilon_2 \end{pmatrix} \quad [10.19]$$

where the coefficient matrix contains $p_{11}, p_{12}, \varepsilon_3, n_{effx}$ and n_{effy}. For the case of high-birefringent FBG, such as polarization-maintaining FBG (PM-FBG), the sensor must be calibrated to the finite-element predictions by performing a least-squares fit to determine the coefficient matrix from the measured wavelength data. According to Eq. (10.19), the lateral components of the strain can be determined. In many structures, one would like an FBG sensor that could measure both lateral axes of strain, axial strain and temperature. An approach to solving this problem is to use dual overlaid FBGs at different Bragg wavelengths written onto a PM fibre.[27] If two FBGs of different Bragg wavelengths, such as 1300 nm and 1550 nm, were written at a single location in a PM fibre, four effective FBGs result in one along the corresponding polarization axis and at corresponding Bragg wavelength. The relative Bragg wavelength shifts can be written as:

$$\begin{pmatrix} \Delta\lambda_{ax}/\lambda_{ax0} \\ \Delta\lambda_{ay}/\lambda_{ay0} \\ \Delta\lambda_{bx}/\lambda_{bx0} \\ \Delta\lambda_{by}/\lambda_{by0} \end{pmatrix} = \begin{pmatrix} K_{11} & K_{12} & K_{13} & K_{14} \\ K_{21} & K_{22} & K_{23} & K_{24} \\ K_{31} & K_{32} & K_{33} & K_{34} \\ K_{41} & K_{42} & K_{43} & K_{44} \end{pmatrix} \begin{pmatrix} \varepsilon_1 \\ \varepsilon_2 \\ \varepsilon_3 \\ \Delta T \end{pmatrix} \qquad [10.20]$$

where λ_{bx} and λ_{by} are the Bragg wavelength for the two orthogonal polarization modes of the second FBG, respectively, and λ_{bx0} and λ_{by0} are the initial unstrained Bragg wavelength for the two orthogonal polarization modes of the second FBG, respectively. Assuming linearity in sensor response and that the element of the 4 × 4 coefficient matrix is independent of strain and temperature, and the matrix is not singular, the elements of the matrix can be determined by performing separate experimental and least-squares fitting calibrations of the response of the sensor to lateral strain, to axial strain and to temperature changes.

10.4.2.2 π-Phase-shifted FBG

Another approach is to use π-phase-shifted FBGs.[28,29] If a regular FBG is irradiated with UV light at a certain region in the middle of the FBG, the refractive index in the region is raised. Such processing produces two FBGs out of phase with each other, which act as a wavelength-selective Fabry–Perot resonator, allowing light at the resonance to penetrate the stop-band of the original FBG. The resonance wavelength depends on the size of the phase change. When the shifted phase is equal to π at a wavelength λ_0 in the stop-band of the original FBG, the strong reflections from the two FBG sections are out of phase, resulting in strong transmission at this wavelength. This post-processing FBG is called π-phase-shifted FBG (π-FBG). The transmission window of π-FBG can be made very narrow and is split in two when the FBG is birefringent. This sharpness permits very high accuracy measurement of the FBG birefringence. Furthermore, the birefringence required for separating the peak is much smaller than for regular FBGs and can be provided by the intrinsic birefringence of an FBG written in non-PM fibre.

The fibre has birefringence in the absence of an external load, of which several factors, such as geometric, UV-induced and stress-induced, may be at the origin. However, for mathematical convenience, it is assumed that the initial birefringence in the FBG is due to a residual strain state in the fibre core that is described by the principal strains ε_{10} and ε_{20} ($\varepsilon_{10} > \varepsilon_{20}$). These principal strains are in directions perpendicular to each other and to the fibre axis, and direction 1 makes an angle ϕ with the x axis. According to Eq. (10.18), the wavelength separation can be expressed as:

$$\Delta\lambda_0 = \Delta\lambda_{10} - \Delta\lambda_{20} \approx 0.5\lambda_0[(n_{\text{eff}1} + n_{\text{eff}2})/2]^2 \\ (p_{12} - p_{11})(\varepsilon_{10} - \varepsilon_{20}) \qquad [10.21]$$

Smart textile composites integrated with fibre optic sensors

The new wavelength separation can be obtained by a function of ϕ[28,29]

$$\Delta\lambda(\phi) = a + b\cos(2\phi) \qquad [10.22]$$

where a and b are positive values which are independent of ϕ. Thus, the larger the angle ϕ is, the lower the sensitivity to lateral strain.

10.5 Measurement effectiveness

For the FBGS to be an ideal embedded strain sensor in a textile composite, the following conditions will have to be met:[30] (1) the integrated optic fibre has little effect on the host strain field; (2) the axial strain of optic fibre ε_1 can represent the nearby host strain in the optic fibre direction; and (3) the effective Poisson's ratio v^* is constant during the measurement period. The embedded-optic-fibre–host system can be illustrated[31] as a composite comprising a cylindrical fibre and two concentric shell layers, with fibre, coating and host from the centre to the outer surface (Fig. 10.5). The height of the composite cylinder is $2H$, the radius of the fibre is R_1, which is also the interior radius of coating, R_2 is the outer radius of the coating and the inner radius of the host, and R_3 is the outer radius of the host. The following basic assumptions have been made so that the case can be simplified as an axial-symmetric problem: (1) The optical fibre, fibre coating and host are linear elastic. (2) The thermal expansion coefficients of fibre, coating and host are constants. (3) There is no discontinuity in displacement at the interfaces of fibre and coating, coating and host under loading. (4) Thermal load is uniform in the whole composite cylinder. (5) The two ends $(z = \pm H)$ are assumed to be free from any external force and there is no constraint on displacement.

10.5.1 General views of the normalized strain distribution

Figure 10.6(a), (b) and (c) illustrate the normalized strain distribution in the fibre, host and coating, respectively, along the fibre axial direction with various values of radius r. Figure 10.6(a) shows that the normalized strain of

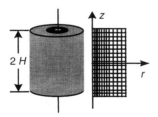

10.5 Three-layer composite model of host, coating, and fibre.

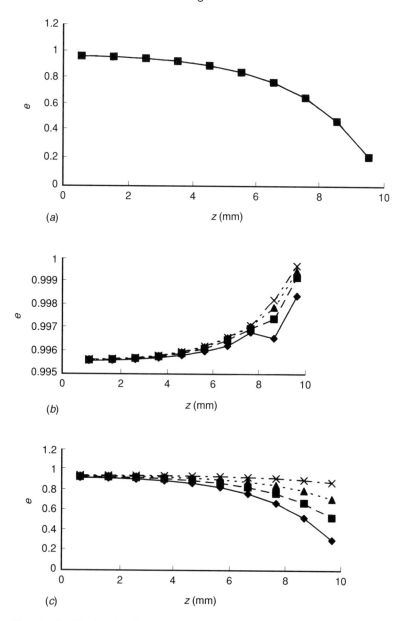

10.6 Strain distribution in the embedded optic fibre–host sytem: (a) distribution of fibre's axial strain at various radial positions, —♦— 0.007875; —■— 0.023625; —▲— 0.039375; —×— 0.055125; (b) distribution of host's axial strain at various radial positions, —♦— 0.248; —■— 0.498; —▲— 0.748; —×— 0.998; (c) distribution of coating's axial strain at various radial positions, —♦— 0.0705; —■— 0.0855; —▲— 0.1005; —×— 0.1155.

fibre declines with increasing length z. Because the thermal expansion coefficient of the optic fibre is more than one order of magnitude smaller than that of the host, the strain level of the optic fibre depends largely on the restriction of the host. As the middle part of the optic fibre is restricted more than the parts near the two ends, the strain of the middle optic fibre is closer to that of the host without the embedded fibre sensor ($e = 1$). The strain distribution curves are identical, regardless of the radial position in the fibre, thus the strain of the fibre can be regarded as a function of z only.

Figure 10.6(b) shows the reverse trends of strain distribution in the host. The strain level of the middle host is lower than that of the host near the boundary, which is because the middle host is restricted more by the fibre and the host near the boundary can expand more freely under a thermal load. The normalized strain values of the host are very close to 1 ($0.995 < e < 1$), which implies that the host with the embedded fibre sensor has a strain field very close to that without the embedded optic fibre.

Because of its lower elastic modulus, the strain distribution of the coating is significantly affected by both the fibre and the host, as shown in Fig.10.6(c). The strain distribution along z is similar with that of the fibre when r approaches R_1, and the strain near the outer surface of the coating varies little along z, like that of the host.

10.5.2 Effects of parameters on effectiveness coefficient

The term H_{95} has been introduced,[31] at which e has a value of 0.95. The physical meaning of H_{95} is that only when the z of the optic fibre grating is smaller than H_{95} is the measurement result of the fibre effective. The length of the zone represents the limits of effective measurement of host strain by an embedded FBGS. The longer the zone length, the more effective an FBGS in a host. Thus the relative length of the effective zone is defined as the effectiveness coefficient β by:

$$\beta = \frac{H_{95}}{H}$$

The effectiveness coefficient is influenced by a number of factors, the elastic modulus, Poisson's ratio of coating, tension stiffness ratio, the thickness of composite etc.[30,31] Figure 10.7 shows that increasing E_c leads to a sharp increment in β when E_c varies from 0.045 to 1 GPa. The curve then reaches a plateau and the effect on β becomes very small. Thus $E_c = 1$ GPa can be regarded as a threshold value for the specified conditions.

Figure 10.8 plots the effective Poisson's ratio of the optic fibre against the fibre length. Within a range from zero to 7 mm, the effective Poisson's ratio

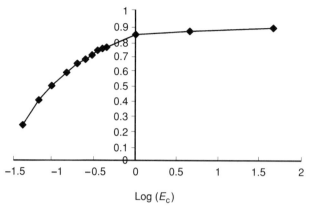

10.7 Effect of elastic modulus of coating on the effectiveness coefficient β of optic fibre.

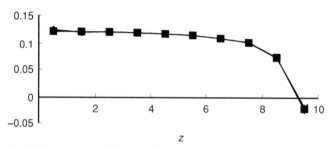

10.8 Distribution of fibre's effective Poisson's ratio along the fibre length at various radial positions.

declines slightly with the increment of z. Thus it can be regarded as a constant along the z axis except for the short portion near the boundary. The value of v^* near $z = 0$ is 0.13 which is not equal to the Poisson's ratio of the fibre (0.17). This indicates that the transverse strain of the optic fibre is not dominated by the host strain (otherwise it should be close to -1) but by the Poisson's ratio of the fibre.

In the particular case under our investigation, the elastic modulus of the coating should be equal to or greater than the threshold value $f = 0.78\,\text{GPa}$, if the measurement effectiveness is concerned (see Fig. 10.7). However, the increment of E_c will affect the value of the effective Poisson's ratio, v^*, then the sensitivity factor, f. Figure 10.9 plots v^* as a function of E_c, which exhibits the reversed trend compared with that of the effectiveness β. When the coating modulus is smaller than, or equal to, the threshold value of 1 GPa, the variation of v^* is rather small. By considering the effects of the coating elastic modulus on both measurement effectiveness and the effective Poisson's ratio,

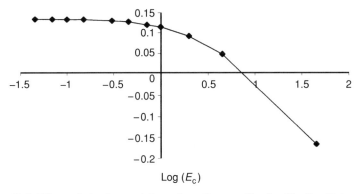

10.9 Effect of elastic modulus of coating on fibre's effective Poisson's ratio.

the optimal coating modulus should be chosen as the threshold value $E_c = 1$ GPa. This case illustrates the necessity of selecting optimal material properties of coating in order to make effective measurements.

10.6 Reliability of FBGs

A single-mode silica optic fibre has a typical cladding diameter of 125 microns and an outer diameter of 250 microns. It can be introduced into a textile preform in its manufacturing processes such as weaving, knitting and braiding. Alternatively, it can be introduced at the consolidation process of the textile composites. In both cases, caution should be exercised when integrating it into the textile structure. Apart from avoiding the damage to the optic fibre during manufacturing processes, the reliability of an embedded fibre Bragg grating sensor is influenced by a number of other factors as follows:[32]

1. Grating-making method: an optic fibre is normally decoated around the grating location before UV pulse laser exposure and recoated afterward. During the decoating and UV irradiation processes, the optic fibre may get damaged, as shown in Chapter 8. Further, the new coat may cause some variation in the fibre. The sensitivity to strain may be different for FBG sensors, even with the same central reflecting wavelength.
2. Location and direction of sensors in composite hosts: usually the FBG sensor is used to measure certain positional and directional normal strain in the host. However, in the procedure of embedding the FBG sensor into the composite host, some deviation of the optic fibre sensor may occur from the designed position and direction. To reduce the measurement error, determination of the real location of the grating of FBG sensors inside the composite may be a key issue, especially when the sensors are

embedded in a large gradient strain field such as near the tip of a crack. Furthermore, it is very difficult to precisely detect the real position and direction of the fibre grating sensor after it is embedded into a composite. The levels of errors induced by the deviation of position and direction of the sensors will be given in Section 10.7.

3   The interfaces of fibre/coating and coating/resin: stress concentration exists around the interfaces and may cause cracks, which may affect the measurement results. Researchers have introduced a debonding phenomenon into the interfaces between fibre and coating as well as between coating and host when fabricating the fibre Bragg optic strain sensors and integrating them into a laminated composite beam.[33] The study showed that the dual-ended sensors are reliable in cyclic bending deformation, while the single-ended sensors are not when their interface bonding fails. To reduce the measurement error, determination of the real location of the grating of FBG sensors inside the composite may be a key issue for future work, especially when the sensors are embedded in a large gradient strain field such as near the tip of a crack.

4   The environment and working conditions to which a textile composite is subject, and how long it is subject to them, will greatly influence the selection and integration of fibre optic sensors. The analysis of measurement effectiveness in the previous section is based on the composite model assuming perfect interfaces between the coating and fibre as well as between the coating and host. Another condition is that the strain measured by the optic fibre sensor is at a predetermined position and in a predetermined direction. Hence, the following section will investigate their effects when these assumed conditions are not met.

10.7 Error of strain measurement due to deviation of position and direction

10.7.1 Deviation of direction

Assume the principal strains in a structure are ε_1, ε_2 and ε_3. If an FBG sensor is embedded in the structure and the angles between the sensor and principal strains are α_1, α_2 and α_3, the strain induced along the FBG sensor is:

$$\varepsilon = [\cos\alpha_1 \quad \cos\alpha_2 \quad \cos\alpha_3] \begin{bmatrix} \varepsilon_1 & 0 & 0 \\ 0 & \varepsilon_2 & 0 \\ 0 & 0 & \varepsilon_3 \end{bmatrix} \begin{bmatrix} \cos\alpha_1 \\ \cos\alpha_2 \\ \cos\alpha_3 \end{bmatrix} \quad [10.23]$$

$$= \varepsilon_1 \cos^2\alpha_1 + \varepsilon_2 \cos^2\alpha_2 + \varepsilon_3 \cos^2\alpha_3$$

where $\cos^2\alpha_1 + \cos^2\alpha_2 + \cos^2\alpha_3 = 1$. The strain measured in the sensor is a function of the principal strains and the angles between them and the sensor. In practice, when a sensor is embedded in the structure, it is most likely to show deviation between the sensor and desired direction. Suppose the deviations of the sensor are $\Delta\alpha_1$, $\Delta\alpha_2$ and $\Delta\alpha_3$, then the strain in the sensor will be:

$$\varepsilon_{\Delta\alpha} = \varepsilon_1 \cos^2(\alpha_1 + \Delta\alpha_1) + \varepsilon_2 \cos^2(\alpha_2 + \Delta\alpha_2) + \varepsilon_3 \cos^2(\alpha_3 + \Delta\alpha_3) \quad [10.24]$$

According to Eq. (10.24), if one wants to measure one of the principal strains, the simplest method is to place sensors along the direction of the desired strain. For example, if ε_1 is desired, the sensor will be placed along this direction. Then we have $\alpha_1 = 0$ and $\alpha_2 = \alpha_3 = \pi/2$. The influence of other strains (ε_2 and ε_3) will be cancelled. If the deviations between the sensor and the desired direction are $\Delta\alpha_1$, $\Delta\alpha_2$ and $\Delta\alpha_3$, the strain along the sensor will be:

$$\begin{aligned}\varepsilon'_{\Delta\alpha} &= \varepsilon_1 \cos^2 \Delta\alpha_1 + \varepsilon_2 \cos^2(\pi/2 + \Delta\alpha_2) + \varepsilon_3 \cos^2(\pi/2 + \Delta\alpha_3) \\ &= \varepsilon_1 \cos^2 \Delta\alpha_1 + \varepsilon_2 \sin^2 \Delta\alpha_2 + \varepsilon_3 \sin^2 \Delta\alpha_3\end{aligned} \quad [10.25]$$

The relative error between the measured and real values of strain is:

$$error_1 = \left|\frac{\varepsilon'_{\Delta\alpha} - \varepsilon_1}{\varepsilon_1}\right| = \left|\frac{\varepsilon_1 \cos^2 \Delta\alpha_1 + \varepsilon_2 \sin^2 \Delta\alpha_2 + \varepsilon_3 \sin^2 \Delta\alpha_3 - \varepsilon_1}{\varepsilon_1}\right| \quad [10.26]$$

If $\Delta\alpha_1$, $\Delta\alpha_2$ and $\Delta\alpha_3$ are small quantities, Eq. (10.26) can be expressed as:

$$error_1 = \left|\frac{-\varepsilon_1\Delta\alpha_1^2 + \varepsilon_2\Delta\alpha_2^2 + \varepsilon_3\Delta\alpha_3^2}{\varepsilon_1}\right| \leq |\Delta\alpha_1^2| + \left|\frac{\varepsilon_2}{\varepsilon_1}\Delta\alpha_2^2\right| + \left|\frac{\varepsilon_3}{\varepsilon_1}\Delta\alpha_3^2\right| \quad [10.27]$$

In Eq. (10.27), the third and higher orders of deviations are omitted. The measurement error is quadratic to the deviations in the orientation angles.

10.7.2 Deviation of position

The strain field of a structure is related to the position and the deformation of the structure. Suppose the strain at point (x, y, z) can be expressed as $\varepsilon = \mathbf{f}(x, y, z, \delta)$, where ε is a tensor and δ represents the displacement vector of the structure. The variation of strain due to the deviations of the position $(\Delta x, \Delta y, \Delta z)$ can be expressed as:

$$\Delta\varepsilon = \frac{\partial \mathbf{f}(x, y, z, \delta)}{\partial x} \cdot \Delta x + \frac{\partial \mathbf{f}(x, y, z, \delta)}{\partial y} \cdot \Delta y$$

$$+ \frac{\partial \mathbf{f}(x, y, z, \delta)}{\partial z} \cdot \Delta z + E(\Delta x, \Delta y, \Delta z) \quad [10.28]$$

where $E(\Delta x, \Delta y, \Delta z)$ is the second and higher-order quantities of the position deviation $(\Delta x, \Delta y, \Delta z)$. When the deviations are small and $E(\Delta x, \Delta y, \Delta z)$ can be omitted, Eq. (10.28) can be simplified as:

$$\Delta \varepsilon \approx \frac{\partial f(x, y, z, \delta)}{\partial x} \cdot \Delta x + \frac{\partial f(x, y, z, \delta)}{\partial y} \cdot \Delta y + \frac{\partial f(x, y, z, \delta)}{\partial z} \cdot \Delta z \quad [10.29]$$

For a cantilever beam, if a concentrated load acts at its free end, the axial strain of an arbitrary point (x, y, z) in the beam due to the normal displacement δ of the free end is:

$$\varepsilon_1(x, y, z) = \frac{3\delta}{L^3}(L - x)z \quad [10.30]$$

where L is the length of the beam, x is the location of the point along the beam, y is the location of the point along the y-direction, and z is the distance between the point and the neutral axis of the beam.

The strain in the beam is independent of the position of y. That means the deviation of y will not influence the strain measurement. If there are any deviations of the test point $(\Delta x, \Delta y, \Delta z)$, according to Eq. (10.29) the variation of the strain will be:

$$\Delta \varepsilon_1 = -\frac{3\delta z}{L^3} \cdot \Delta x + \frac{3\delta(L - x)}{L^3} \cdot \Delta z \quad [10.31]$$

The relative error of strain deduced from the position deviations is:

$$error_2 = \left| \frac{\Delta \varepsilon}{\varepsilon(x, y, z)} \right| \leq \left| \frac{\Delta x}{L - x} \right| + \left| \frac{\Delta z}{z} \right| \quad [10.32]$$

If the sensors are placed near the fixed end ($x = x_{min} \approx 0$) and far from the neutral axis ($z = z_{max} \approx h/2$, h is the height of the cantilever beam) of the beam, the measurement error will be the minimum.

10.7.3 Deviation of both direction and position

If deviations of direction and position occur simultaneously, then the total relative error is:

$$error_{total} = error_1 + error_2 \leq |\Delta \alpha_1^2| + \left| \frac{\varepsilon_2}{\varepsilon_1} \Delta \alpha_2^2 \right| + \left| \frac{\varepsilon_3}{\varepsilon_1} \Delta \alpha_3^2 \right|$$

$$+ \left| \frac{\Delta x}{L - x} \right| + \left| \frac{\Delta z}{z} \right| \quad [10.33]$$

From Eq. (10.33) we know that, in order to obtain an accurate measurement result, the deviations of direction and position should be as small as possible.

This derivation is more suitable for the surface-mounted FBGs. If embedded in a composite, the localized heterogeneity will normally lead to a complicated strain field. If the strain is not uniformly distributed along FBG, grating apodization occurs. The strain may induce birefringence which may cause peak split. Hence, the analysis of measurement error must consider at least the above-mentioned factors.

10.8 Distributed measurement systems

By the means of measuring the shift of the peak wavelength of the reflective spectrum, the average strain over the whole grating length is determined, rather than the distribution of the strain. In many applications, the distribution of strain and temperature is the major concern, thus a number of methods can be implemented to serve this purpose. The grating length of FBGs can be made very short (e.g. 2 mm), thus the strain measured by the shift wavelength of the reflective spectrum can be regarded as a localized strain. The FGB arrays render themselves as ideal candidates for multi-point or quasi-distributive measurements by the multiplexing techniques used for several other optical fibre sensors, such as wavelength-division-multiplexing (WDM),[34] time-division-multiplexing (TDM) and spatial-division-multiplexing (SDM). Detailed reviews have been given by Rao,[35] Kersey et al.[36] and Othonos and Kalli.[37] Figure 10.10 shows a schematic diagram of the measurement systems, consisting of a broad-band LED, a coupler or optic switch, an optical analyser, and an embedded multiplexing FBG array by which we carried out our pseudo-distributed measurement of strain-temperature in textile composites.

In simple cases, if it is monotonic, the strain distribution can be determined along the grating by analysing the reflection spectrum. Volanthen and his colleagues developed a measurement system of distributed grating sensors to measure the wavelength and reflectivity of gratings as functions of time delay.[38–40] By using low-coherence reflectometry, this technique was applied to three-point bending experiments of a textile composite beam embedded with FBGs.[30]

10.9 Conclusions

In conclusion, fibre optic sensors are ideal candidates to be embedded in textile structural composites for monitoring manufacturing processes and internal health conditions. The sensors provide an effective means by which the distributions of a number of physical parameters, such as temperature, stress/strain, thermal expansion, pressure, etc., can be quantitatively determined. In integrating the sensors into textile composites, apart from the properties of the sensors themselves, the reliability of the sensors and sensing scheme as well

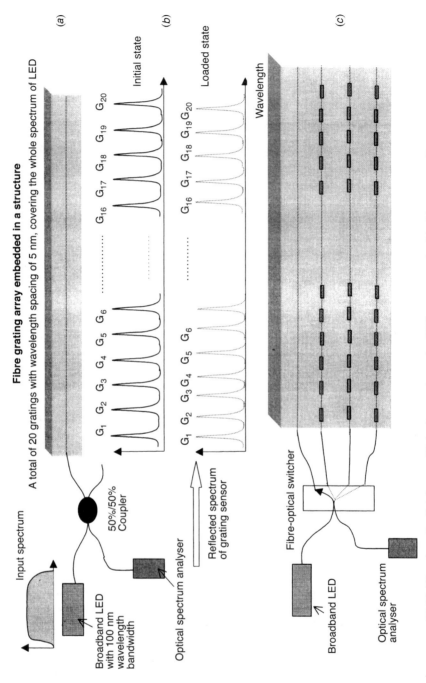

10.10 Wavelength-division-multiplexing systems by FBG arrays embedded in composites.

as the interaction between sensors and composites are very important issues to consider.

Acknowledgements

The author wishes to acknowledge the Research Grants Council of the Hong Kong SAR government for funding support (Project No. PolyU123/96E and PolyU5512/98E). The author also thanks her colleagues, Dr Du W.C., Dr Tang L.Q. and Dr Tian X.G. for their contributions in the development of theoretical analysis and experiments.

References

1. Du C W, Tao X M, Tam Y L and Choy C L, 'Fundamentals and applications of optical fiber Bragg grating sensors to textile composites', *J. Composite Struct.*, 1998, **42**(3), 217–30.
2. Matthews L K (ed.), *Smart Structures and Materials 1995, Smart Systems for Bridges, Structures, and Highways*, SPIE, 1995, **2446**.
3. Merzacher C I, Kersey A D and Friebele E J, 'Fibre optic sensors in concrete structures: a review', *Smart Mater. Struct.*, 1996, **5**, 196–208.
4. Hayes S A, Brooks D, Liu T Y, Vicker S and Fernando G F, 'In-situ self-sensing fibre reinforced composites', *Smart Structures and Materials 1996, SPIE*, **2718**, 376–84.
5. Hong C S, Kim C G, Kwon I and Park J W, 'Simultaneous strain and failure sensing of composite beam using an embedded fibre optic extrinsic Fabry–Perot sensor', *Smart Structures and Materials 1996*, SPIE, 1996, **2718**, 122–33.
6. Badcock R A and Fernando G F, 'Intensity-based optical fibre sensor for fatigue damage detection in advanced fibre-reinforced composites', *Smart Mater. Struct.*, 1995, **4**(4), 223–30.
7. Udd E, Fiber optic sensor overview. In *Fiber Optic Smart Structures*, ed. E Udd, John Wiley, 1995.
8. Measures R M, Fiber optic strain sensing. In *Fiber Optic Smart Structures*, ed. E Udd, John Wiley, 1995.
9. Butter C D and Hocker G P, 'Fiber optics strain gauge', *Appl. Opt.*, 1992, **1370**, 189–96.
10. Sirkis J S and Haslach H W Jr, 'Full phase-strain relation for structurally embedded interferometric optic fiber sensors', *Proc. SPIE*, 1990, **1370**, 248–59.
11. Jackson D A and Jones J D C, 'Fibre optic sensors', *Pot. Acta*, 1986, **33**, 1469–503.
12. Takahashi S and Shibata S, 'Thermal variation of attenuation for optical fibers', *J. Non-crystal. Solids*, 1979, **30**, 359–70.
13. Akhavan Leilabady P, Jones J D C and Jackson D A, *Optics Lett.*, 1985, **10**, 576.
14. Friebele E J, Putnam M A, Patrick H J et al., 'Ultrahigh-sensitivity fiber-optic strain and temperature sensor', *Optics Lett.*, 1998, **23**, 222–4.
15. Chan C C, Jin W, Rad A B et al., 'Simultaneous measurement of temperature and strain: an artificial neural network approach', *IEEE Photon. Technol. Lett.*, 1998, **10**, 854–6.

16 James S W, Dockney M L and Tatam R P, 'Simultaneous independent temperature and strain measurement using in-fiber Bragg grating sensors', *Electron. Lett.*, 1996, **32**, 1133–4.
17 Xu M G, Archambault J L, Reekie L et al., 'Discrimination between strain and temperature effects using dual-wavelength fiber grating sensors', *Electron. Lett.*, 1994, **30**, 1085–7.
18 Cavaleiro P M, Araújo F M, Ferreira L A et al., 'Simultaneous measurement of strain and temperature using Bragg gratings written in germanosilicate and boron-codoped germanosilicate fibers', *IEEE Photon. Technol. Lett.*, 1999, **11**, 1635–7.
19 Patrick H J, Williams G M, Kersey A D et al., 'Hybrid fiber Bragg grating/long period fiber grating sensor for strain/temperature discrimination', *IEEE Photon. Technol. Lett.*, 1996, **8**, 1223–5.
20 Du W C, Tao X M and Tam H Y, 'Fiber Bragg grating cavity sensor for simultaneous measurement of temperature and strain', *IEEE Photon. Technol. Lett.*, 1999, **11**, 105–7.
21 Du W C, Tao X M and Tam H W, 'Temperature independent strain measurement with a fiber grating tapered cavity sensor', *IEEE Photon. Technol. Lett.*, 1999, **11**(5), 596–8.
22 Guan B O, Tam H Y, Tao X M and Dong X Y, 'Simultaneous strain and temperature measurement using a superstructure fiber Bragg grating', *IEEE Photon. Technol. Lett.*, 2000, **12**(6), 1–3.
23 Kawase L R, Valente L C G, Margulis W et al., 'Force measurement using induced birefringence on Bragg grating', in *Proc. 1997 SBMO/MTT-S International Microwave and Optoelectronics Conference*, **1**, 394–6.
24 Lawrence C M, Nelson D V, Makino A et al., 'Modeling of the multi-parameter Bragg grating sensor', in *Proc. SPIE – Third Pacific Northwest Fiber Optic Sensor Workshop*, **3180**, 42–9.
25 Udd E, Nelson D and Lawrence C, 'Multiple axis strain sensing using fiber gratings written onto birefringent single mode optical fiber', in *Proc. 12th International Conference on Optical Fiber Sensors*, Williamsburg, USA, 1997, 48–51.
26 Udd E, Lawrence C M and Nelson D V, 'Development of a three axis strain and temperature fiber optical grating sensor', in *Proc. SPIE – Smart Structures and Materials 1997: Smart Sensing, Processing, and Instrumentation*, **3042**, 229–36.
27 Udd E, Schulz W L, Seim J M et al., 'Transverse fiber grating strain sensors based on dual overlaid fiber gratings on polarization preserving fibers', in *Proc. SPIE – Smart Structures and Materials 1998: Sensory Phenomena and Measurement Instrumentation for Smart Structures and Materials*, **3330**, 253–63.
28 LeBlanc M, Vohra S T, Tsai T E et al., 'Transverse load sensing by use of pi-phase-shifted fiber Bragg gratings', *Optics Lett.*, 1999, **24**, 1091–3.
29 Canning J and Sceats M G, 'π-phase-shifted periodic distributed structures in optical fibers by UV post-processing', *Electron. Lett.*, 1994, **30**, 1344–5.
30 Tao X M, Tang L Q, Du W C and Choy C L, 'Internal strain measurement by fiber Bragg grating sensors in textile composites', *J. Composite Sci. Technol.*, 2000, **60**(5), 657–69.
31 Tang L Q, Tao X M and Choy C L, 'Measurement effectiveness and optimisation of fibre Bragg grating sensor as embedded strain sensor', *Smart Mater. Struct.*, 1999, **8**, 154–60.

32 Tao X M, 'Integration of fibre optic sensors in smart textile composites – design and fabrication', *J. Text. Inst.*, 2000, **91**, Part 1 No.2.
33 Tang L Q, Tao X M, Du W C and Choy C L, 'Reliability of fibre Bragg grating sensors in textile composites', *J. Composite Interfaces*, 1998, **5**(5), 421–35.
34 Du C W, Tao X M, Tam Y L and Choy C L, 'Fundamentals and applications of optical fiber Bragg grating sensors to textile composites', *J. Composite Struct.*, 1998, **42**(3), 217–30.
35 Rao Y J, 'In-fibre Bragg grating sensors – Measurement and science technology', *Measurement Sci. Technol.*, 1997, **8**, 355–75.
36 Keysey A D, Davis M A, Patrick H J et al., 'Fiber grating sensors', *J. Lightwave Technol.*, 1997, **15**(8), 1442–63.
37 Othonos A and Kalli K, *Fibre Bragg Gratings: Fundamentals and Applications in Telecommunications and Sensing*, Artech House, 1999.
38 Volathen M, Geiger H, Cole M J, Laming R L and Dakin J P, 'Low-coherence technique to characterize reflectivity and time delay as a function of wavelength within a long fibre grating', *Electron Lett.*, 1996, **23**(8), 847–9.
39 Volathen M, Geiger H, Cole M J and Dakin J P, 'Measurement of arbitary strain profiles within fibre gratings', *Electron Lett.*, 1996, **23**(11), 1028–9.
40 Volathen M, Geiger H, Cole M J and Dakin J P, 'Distributed grating sensors using low-coherence reflectometry', *J. Lightwave Technol.*, 1997, **15**(11), 2076–82.

11
Hollow fibre membranes for gas separation

PHILIP J. BROWN

11.1 Historical overview of membranes for gas separation

During the last century and a half, the development of membranes has resulted in a voluminous amount of literature on the subject.

One review of membranes by Lonsdale[1] still contained over 400 references, even though the author claimed to have covered only the most relevant references in the article. This review also does not therefore intend to be a comprehensive review of the development of membrane technology; rather it is a short account of the interesting developments that have taken place along with the milestones in 'membranology' that have led to the development of hollow fibres for industrial gas separations.

The development of membrane technology is relatively recent, though membranes have been studied for over 200 years. Some of the early papers were quite outstanding and laid the foundations for today's understanding. In 1829, Thomas Graham[2] inflated water wet bladders containing air by inserting them into a jar filled with CO_2. He explained this effect as being due to the dissolution of the CO_2 in the water in the 'capillary canals' of the membrane, its diffusion through the membrane followed by the release of the CO_2 in the bladder.

J. K. Mitchell in 1831 experimented with the so-called 'penetrativeness' of fluids[3] and investigated the penetration of gases. Mitchell noticed that different gases escaped at different rates from natural rubber balloons and found that there was a 100-fold difference between the rates of escape between carbon monoxide and ammonia. Thomas Graham[4-6] investigated the diffusion of liquids and gases and recognized the potential for separation of gases by mechanical means. The important principles of molecular transport through porous and non-porous membranes were intuitively grasped in these early papers. Thomas Graham's experiments on the molecular mobility of gases were beautifully simple. Graham's diffusiometer as first constructed used a glass tube 10 inches (25.4 cm) in length and less than 1 inch (2.54 cm) in diameter. The tube was closed at one end with porous plaster of Paris (the plaster of Paris was later replaced by graphite), the tube was filled with hydrogen gas over a mercurial trough, and the graphite plate was covered with gutta percha. On removing the gutta percha, gaseous diffusion took place through the pores in the graphite. Graham had great insight into the mechanism of the diffusional processes, and said that: 'It seems that molecules only can pass; and they may be supposed to pass wholly unimpeded by friction, for the smallest pores in the graphite must be tunnels in magnitude to the ultimate atoms of a gaseous body.' Graham found that the passage times of different gases – H_2, O_2 and carbonic acid – showed a close relation to the reciprocal of the square of the densities of the respective gases. Graham used his diffusiometer not only to study single gases but also to partially separate mixed gases (Graham termed this atmolysis). Graham noted that each gas made its way through the graphite plate independently, each following its own rate of diffusion. Though Graham could be said to be the father of gas separation membranes, it was Fick[7] in 1855, using liquids, who described quantitatively the diffusional process. Fick realized that diffusion was analogous to heat conduction (Fourier's law), and with this hypothesis developed the laws of diffusion. The flux was then defined as shown in Eq. (11.1):

$$J = -AD\frac{\delta c}{\delta x} \qquad [11.1]$$

where c is the concentration, x is the distance, D is the diffusion coefficient, A is the area and J is the flux (one-dimensional). Fick thereafter developed his equational analogy further and obtained Eq. (11.2):

$$\frac{\delta c}{\delta x} = D\left[\frac{\delta^2 c}{\delta x^2} + \frac{1}{A}\frac{\delta A}{\delta x}\frac{\delta c}{\delta x}\right] \qquad [11.2]$$

In order to prove his hypothesis, Fick used two different diffusional-shaped cells containing water and placed a crystalline salt at the bottom of the cells. In case A, the cell was a cylindrical shape and in case B, the cell was a conical shape narrowing at the top.

A steady-state concentration gradient was achieved by changing the water at the top of the vessels. Fick found that, in case A, a linear relationship was observed between the specific gravity of the salt solution and distance down the tube, the gradient giving a measure of the diffusion coefficient. In case B, where the area for diffusion changes with time, he found that the results obtained by experiment compared well with those calculated, thus proving that the laws of diffusion were analogous to those of heat transfer. The calculation for the funnel was done using Eq. (11.2).

11.2 Development of membranes for industrial gas separation

It was, however, a number of decades before the principles laid down by Graham and Fick were to be put into use on a large scale. During the 1940s, uranium was needed for military atomic research and led to a large-scale gas separation membrane system. These membranes were porous and metallic, and were used to separate UF_6 isotopes. Isotope separation had already been achieved by Graham[8] in 1846. Polymers which could withstand the presence of HF were not available in the 1940s. The difference in molecular weight between $^{238}UF_6$ and $^{235}UF_6$ is small and yields a molecular weight ratio of 1.008. The gas mixture was therefore not easy to separate, and thousands of separation stages were needed.

A great breakthrough by Loeb and Sourirajan[9] (reverse osmosis membranes) came about in the early 1960s via the formation of ultrathin membranes of cellulose acetate (CA). These membranes were formed by the casting of a viscous solution of CA and then immersing the cast solution in water. The resulting precipitation gives rise to what is now known as an asymmetric membrane, which consists of a thin dense skin layer which is microporous, supported by a more highly porous sublayer. These thin membranes resulted in high flux and good salt rejection properties. The economic desalination of water was now realizable, and this sparked off a number of studies. Electron microscope[10] studies were carried out in order to determine the thickness of the dense skin layer and mechanisms, and explanations were sought for the formation of these skinned membranes.[11] Calculations as to the effective thickness of the CA membranes were done by Merton, Riley and Lonsdale,[1] and correlated well with those seen from electron microscopy at approximately 0.2 µm.

The adaptation of the principles used in the development of the membranes for the desalination of water via reverse osmosis was all that was needed to develop membranes for gas separation. It was realized that gas separation through dense polymer films was not practicable (due to the low gas flux) unless ultrathin dense films could be obtained. A thin film ensures a high flux even if the intrinsic permeability of the polymer is low. In 1970, using Loeb Sourirajan CA membranes which were dried, Merten and Gantzel[12] obtained a separation factor of approximately 40 for He/N_2, with high fluxes equivalent to those of a silicone rubber film 10 µm thick, but which would have a corresponding separation factor of only 1.5.

But undoubtedly the biggest breakthrough in the last 20 years or so was brought about by the Monsanto Company. Polysulfone, a polymer which at that time was considered to have a high intrinsic selectivity for gases but a low permeability, was utilized in the form of hollow fibres. The Monsanto

PRISM™ system, which used these fibres, was similar to the reverse osmosis system developed by Du Pont, the Permasep RO system. The hollow fibres were asymmetric, with thin skins on the outside supported by a porous layer. The skin in this case was microporous, and the breakthrough was to coat the fibres with a highly permeable polymer which blocked the pores. The separating layer was then the thin dense polysulfone skin layer and not the coating. The resulting fibres have a high flux and high gas separation performance. Henis and Tripodi (the inventors), along with the Monsanto company, issued a series of patents[13–16] in the early 1980s. Following the work of Henis and Tripodi, and with it the reality of commercial membrane systems for gas separation, the 1980s saw a time in which a plethora of work was done in developing the performance of membranes for gas separations. However, the work of Henis and Tripodi is worth describing in some detail since it was undoubtedly their work which caused many researchers to become involved in hollow fibre membranes for gas separations.

The United States Patent 4 230 463 (1980) by Henis and Tripodi describes multicomponent membranes for gas separations. This patent makes 78 claims and contains 7 figures. Sixty-three examples are given in all to demonstrate the flexibility of the invention, its limitations and successful use. A summarized description of the invention is given below.

The multicomponent membrane described in the invention exhibits a separation factor significantly greater than the intrinsic separation factor of the coating material which is used to 'occlude' the pores of a microporous polymer substrate. The intrinsic separation factor of a material is defined as the separation factor of a dry compact membrane of the material. The term 'significantly greater' was defined as being anywhere from 5 to 10% greater than the coating material and preferably 50% higher. The 'occluding contact' refers to the fact that, after coating the membrane, the amount of gas passing through the material of the membrane is increased relative to the amount of gas now passing through the pores. It is claimed that the porous separation membrane could be anisotropic and characterized by having a dense region in a barrier relationship to that of the porous separation membrane. The coating of porous membranes which have a relatively dense skin region provides enhanced flux through the multicomponent membrane. The coatings are applied in a liquid phase (usually a polysiloxane solution in isopentane) and, for advantageous coating, the liquid will adhere to the porous separation membrane to be utilized. It is preferable that the coating material polymerizes after the liquid has been applied to the porous separation material. Suitable porous separation substrates, e.g. polysulfone, cellulose acetate and polyimides, are some of those given. The porous separation membrane is preferably at least partially self-supporting and, with this in mind, hollow fibres were selected. Descriptions of polymer concentrations, the solvents for the spinning process and the spinning conditions are given in some of the examples in the patent. This patent in itself contains vast amounts of data, as well as a mathematical model (the resistance model), and it paved the way for others containing more detailed examples of spinning conditions and examples of the use of the resistance model. The resistance model was elaborated on by Henis and Tripodi in another paper,[17] and since this model is the basis of Monsanto's work, and since it has been used as a predictor by many other membrane researchers, it is shown below.

11.2.1 The resistance model approach to gas permeation

In the patents and papers which describe the resistance model, it is shown that permeation through composite hollow fibre membranes is analogous to electrical resistance. The symbols used to denote flux or permeation rate are the ones used by Henis and Tripodi. The permeation rate or flux Q_i is given by Eq. (11.3):

$$Q_i = \frac{P_i A \Delta c_i}{l} \qquad [11.3]$$

where P_i is the intrinsic permeability of the polymeric membrane to gas component i, A is the cross-sectional surface area of the membrane to component i, l is the thickness of the membrane area through which the component permeates, and Δc_i is the partial pressure difference across the membrane. The electrical analogy is that the permeation through a polymer membrane is equivalent mathematically to Ohm's law for flow through a resistor, Eq. (11.4):

$$I = E/R \qquad [11.4]$$

The current is analogous to permeation rate, the driving force or voltage is analogous to the concentration gradient or pressure differential, and the electrical resistance is then the analogue of resistance to permeate flow. The resistance to permeate flow can then be said to be R_i:

$$R_i = 1/P_i A \qquad [11.5]$$

Combining Eqs. (11.3) and (11.5), Henis and Tripodi derive Eq. (11.6):

$$Q_i = \Delta c_i / R_i \qquad [11.6]$$

The permeation behaviour of porous hollow fibre membranes is analogous then to the flow of electricity through an array of series–parallel resistors, and the membranes are thus called resistance model or 'RM' composite membranes. A cross-sectional representation of a membrane is shown in Fig. 11.1, along with its electrical analogue. Regions are defined as a skin of thickness l_2 (analogue R_2) defects or pores (analogue R_3), a highly porous substrate

Hollow fibre membranes for gas separation

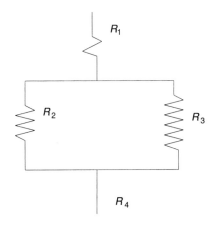

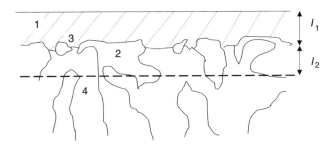

11.1 Representation of a porous asymmetric membrane and its electrical analogue. (From Henis & Tripodi.)

(analogue R_4), and the coating of thickness l_1 (analogue R_1). The total resistance to the flow R_t (see Eq. (11.7)) is a function of the resistance to flow in each of the regions defined within the membrane:

$$R_t = R_1 + \frac{R_2 R_3}{R_2 + R_3} + R_4 \qquad [11.7]$$

The porous substrate resistance R_4 is very small and can be disregarded. The coating R_1 substantially increases the resistance to the passage of gas. The blocked pores resist gas flow one million times more than an open pore. Assuming that the coating layer fills pores to a depth equal to that of the skin layer, the resistance in the pores is given by Eq. (1.8) and the resistance to the coating material is also given:

$$R_{3,i} = \frac{l_2}{P_{1,i} A_3} \quad \text{and for the coating} \quad R_{1,i} = \frac{l_1}{P_{1,i} A_1} \qquad [11.8]$$

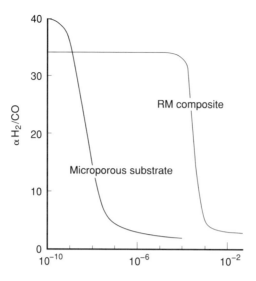

11.2 Effect of surface porosity on separation factor.

The skin resistance is given by Eq. (11.9):

$$R_{2,i} = \frac{l_2}{P_{2,i} A_2} \qquad [11.9]$$

The total flux for the composite membrane can be arranged to find the gas flux per unit area and per unit pressure drop, which is equivalent to the thickness corrected permeability or permeation rate (P/l) or P':

$$\left[\frac{P}{l}\right] = \left[\frac{l_1}{P_{1,i}} + \frac{l_2}{P_{2,1} + P_{1,i}(A_3/A_4)}\right]^{-1} \qquad [11.10]$$

The surface porosity is low in hollow fibre composite membranes; A_3 is much less than A_2. For simplicity, it can also be assumed then that $A_1 = A_2$.

The separation factor for two gases i and j is the ratio of the (P/l) values (see Eq. (11.11)):

$$\alpha_j^i = \frac{(P/l)_i}{(P/l)_j} \qquad [11.11]$$

The effect of the surface porosity on separation factors for H_2/CO for porous substrates and RM composite membranes is shown in Fig. 11.2.

The advantage of the RM composite membrane over a porous substrate membrane is that a far greater variability in the surface porosity of the fibre can be obtained without seriously affecting the flux and separation factor.

Work done by Peinemann et al.[18] uses the resistance model for highly microporous membranes. They showed that the supporting layer in thin film composite membranes can be at least as important as the selective layer. In an extreme case, where N_2 and C_2Cl_4 were separated using a polysulfone–silicone rubber composite, they found that, although the silicone coating was defect free and 87 μm thick, the intrinsic C_2Cl_4/N_2 selectivity of 50 was never obtained. This was due to the high permeability of tetrachloroethylene vapour through the silicone. The non-selective microporous support now resisted the C_2Cl_4 vapour to the same order of magnitude as the silicone layer itself. The Peinemann paper highlights the fact that a non-selective support layer can restrict the flow of a highly permeable gas and reduce separation factors to below those expected.

The Monsanto invention led to the commercial PRISM™ separator.[19] Since the work of the Monsanto team, an international effort has been put into membrane research.

11.2.2 Polymer development

Due to the successful implementation of polysulfone in hollow fibre membranes, this system has become the one to outperform. The intrinsic selectivity of polysulfone is high, with a CO_2/CH_4 separation factor of about 29, an H_2/CH_4 separation factor of about 54, and an H_2 permeability of about 10 Barrer. A substantial amount of work has been done in trying to develop new polymers which are more permeable and more selective than polysulfone, while at the same time still tractable. The greatest success appears to have come from the polyimide class of polymers, though others are worthy of note. The work of Koros et al.[20] showed the relationship between the chemical structures in a series of polyimides and their gas separation properties. The polyimides studied are shown in Fig. 11.3.

One of the Koros polymers in particular (6FDA–ODA) exhibited some quite remarkable permeability/separation characteristics. The general rule for a group of polymers such as poly(paraphenylene oxide), polysulfone, polycarbonate and cellulose acetate is that, as the gas selectivity goes up, the permeability goes down. The 6FDA–ODA polyimide does not conform to this general trend. The permeability for 6FDA–ODA to CO_2 at 35 °C at 10 atm was 23 Barrer with a CO_2/CH_4 selectivity of 60.5. This compares to a permeability of 5.7 Barrer and a CO_2/CH_4 selectivity of 27 for polysulfone. The superior performance of the polyimides had shown that, through controlling both the intrasegmental mobility and intrasegmental packing (preferably reducing both), new CO_2/CH_4 could be designed with high permeability and selectivity.

Similar work since by Stern et al. has been done in order to further the

11.3 Polyimides studied by Koros et al.

11.4 Poly(1,3phenyl1,4phenyl) 4 phenyl1,3,4triazole.

understanding of the structure property relationships in similar polyimides.[21] Stern used the (6FDA) monomer with a series of diamines to show that relatively small structural modifications of a polyimide can result in large changes in gas permeability. The results from Stern's study are intriguing and, within the polyimide series, the selectivity decreases with increasing permeability, but to a smaller extent than with standard commercial polymers. Stern in discussion points out that anomalies can still occur. For instance, within one polyimide series, the polymer determined to have the smallest interchain spacing and the highest density did not exhibit the lowest permeability and highest selectivity as expected. On the whole, it seems likely that in order to understand the structure/permeability relationships more fully, information on the free volume distribution within the polymer will be required.[22] Smolders et al.[23] worked on developing tailor-made polymers for gas separation. As well as looking at polyimides, they have another polymer (a polytriazole) which was shown to have good permeability and selectivity characteristics. The polytriazole polymer (see Fig. 11.4) has a CO_2 permeability of 10–20 Barrer and a CO_2/CH_4 selectivity of 60.

The polytriazoles are thermally and chemically resistant and are therefore attractive materials for membrane manufacture. The polytriazoles are soluble in formic acid and the polymer can be formed into membranes via a phase inversion technique. One of the problems with the polytriazole is the limited number of solvents available, which in effect reduces the processing flexibility. A slightly different type of membrane but one worthy of note is the ion-exchange type membrane. Here, selective transport of H_2, CO_2 and H_2S has been achieved using facilitated transport gel–ion exchange membranes.[24] The best commercial polymers available achieve an H_2/CO_2 separation factor of around seven to ten, whereas gel–ion exchange membranes can give a much better performance. The ion-exchange membranes in this case are made from polyperfluorosulfonic acid (PFSA) and are modified to form a gel for use as the support for the solvent and carrier. These membranes have hydrophilic regions into which a solvent which has the desired complexing agent is embedded. Typically, H_2O is used as the solvent, with ethylene diamine as the carrier. Selectivities have been achieved of over 400 for CO_2/H_2, with permeabilities of 11 000 Barrer for CO_2. The high intrinsic permeability is in itself not enough to obtain high productivity for commercial use, as the membrane must also be thin. In one assessment for the economic case for using an immobilized liquid membrane, the film thickness becomes an important economic consideration. Reducing the membrane thickness from 50 μm to 30 μm and increasing the CO_2 permeability from 800 to 2000 Barrer results in a calculated reduction in the required capital investment from $172 MM to $12 MM. The difficulty with this type of membrane is to create a mechanically stable gel within a module and to match the properties available from the thick films in the thin ones. If the technical problems can be overcome, facilitated transport membranes will become an economically viable gas separation system.

During the late 1980s, the Japanese C1 chemistry programme put a substantial effort into developing efficient membranes for gas separations. The Japanese budget was approximately £7.5 million on membrane development over 7 years.

The interest of the Japanese lay mainly in the separation of O_2 from air and H_2 from CO. Companies which were interested in hydrogen separation were Ube Industries Ltd., Toyobo Co. Ltd. and Sumitomo Electric Industries Co., whilst those interested in oxygen research were Toyobo, Nitto Electric, Torray, Kururay, Asahi Glass, Teijin and Matsushita Electric. An investigation of porous organic, inorganic and non-porous organic membranes was carried out. The membranes developed during the project exceeded the targets set at the beginning. Materials were evaluated in flat film form and then hollow fibre form. The performance of the various membranes was summed up in a DTI report.[25] Ube developed the polyimide membrane, Sumitomo focused on plasma polymerized membranes, with permeability ratios recorded as high as 300 for H_2/CO, and Toyobo developed a polysulfoneamide membrane and found that increasing the amide content increased the polymer's selectivity. The polyimides made by Ube had better properties than conventional polymers, with a CO_2/CH_4 selectivity of 46, but not as high as those developed by Koros et al., referred to earlier. Reiss et al. achieved a remarkable selectivity using polyaniline polymers.[26] Tailoring the gas selectivity for polyaniline membranes involves doping polymer films with counter ions of an appropriate size. The selectivity reported for H_2/N_2, O_2/N_2 and CO_2/CH_4 surpasses the highest previously reported values of other permeable polymers. Polyaniline $(C_6H_4NH)_n$ was made into a mechanically robust film from a 5% w/v solution in N-methylpyrrolidinone. The cast films were dried in a vacuum for 24 hours and then tested in the 'as cast' form. Samples of the 'as cast' film were doped in 4.0 M acid solution for 15 hours. The films were undoped using 1.0 M NH_4OH for 24 hours. Samples were then redoped by immersion for 12 hours in a 0.0175 M solution of the same acid used in the initial doping, which resulted in a partially doped film. The effects of the doping and redoping on polymer permeability and selectivity were, to say the least, remarkable. One example of a redoped film gave an incredible separation factor for H_2/N_2 of 3590, with a hydrogen permeability of 10 Barrer. In another example, the dopant cycling procedure achieved a CO_2/CH_4 selectivity of 640, well over 100 times that possible with polysulfone.

However, despite the promise of the polyaniline polymer, its use to manufacture hollow fibre membranes that can compete commercially have so far eluded researchers in the field. This may be due to the difficulties encountered in hollow fibre spinning processes.

11.3 Theories of permeation processes

The permeation of gases through polymer membranes depends upon whether the membrane is porous or dense. If the membrane is porous, the gas flow is predominantly controlled by the mean free path of the gas molecules and the pore size. The mechanism of flow through porous membranes has been discussed elsewhere.[27,28] In short, the gas flow regime is contributed to by Poiseuille flow and Knudsen flow, the amount of each contribution being defined by pore size, pressure, viscosity and the molecular weight of the gas involved. However, microporous membranes exhibit low gas selectivity, as shown earlier in the resistance model approach.

For more efficient separations, dense polymer membranes are used, although they will usually have a high microporous substructure. In dense polymer membranes, gas flow across the membrane is via a solution diffusion mechanism. In simple terms, the gas is sorbed onto the polymer, then diffuses down a concentration gradient via Fick's law (Eq. (11.1)) and finally desorbs from the low pressure side of the membrane. A large number of publications have been written on the subject of the mechanism of gas permeation, and some key arguments have been described elsewhere.[29-40]

11.4 Phase inversion and hollow fibre membrane formation

The term phase inversion refers to the process by which a polymer solution inverts into a three-dimensional network. Initially, the solvent system is the continuous phase, and after phase inversion the polymer is the continuous phase.[41] Four phase inversion processes exist: (1) the dry process, in which a volatile solvent is lost and phase inversion occurs, (2) the wet process, in which solvent is exchanged for non-solvent and precipitation occurs, (3) the thermal process, where a latent solvent (a substance which is only a solvent at elevated temperatures) is used, involving the cooling of the polymer solution which

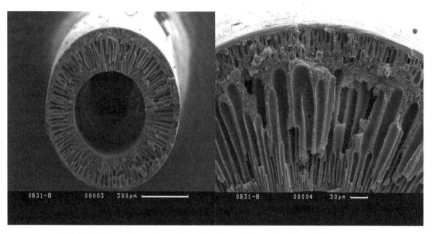

11.5 Finger-type pores in polyetherketone hollow fibres.

then gels, (4) the polymer-assisted phase inversion or PAPI process, where the system contains a solvent and two compatible polymers from which a dense film is cast. The dense film morphology is known as an interpenetrating network, IPN. After complete or partial solvent evaporation from the IPN, the film is usually immersed in water or a solvent which is a solvent for one polymer but not the other.

The hollow fibre membranes reviewed in the present work here are formed using the wet phase inversion process. The wet phase inversion process has been very well described by Strathmann.[42,43] In his work, Strathmann casts various polymer solutions onto glass plates and then immerses the cast film into a bath of precipitation fluid. This kind of precipitation is fast and a skinned membrane structure forms. The skin formation is explained by Strathmann on the basis of the concentration profiles of the polymer, solvent and the precipitant which occurred during the phase inversion process.

In skin-type membranes, like hollow fibre membranes, two characteristic structures can be formed. One structure is sponge-like below the skin, and one has finger-shaped pores below the skin. These finger-shaped pores can be clearly seen in the polyetherketone hollow fibre (see Fig. 11.5).[43,44]

These structures are different when different precipitation conditions are used.[44] The finger-like structures which form cannot be entirely explained by the thermodynamic and kinetic approach. The skin formation is the same as that for sponge-type membranes, but, when the shrinkage stress in the skin cannot be relieved by creep relaxation of the polymer skin, the skin ruptures. The points at which the skin splits form the initiation points for fingers. The exchange of solvent with precipitant in the finger is much faster than through the unfractured skin region. The fingers form a layer of precipitated polymer around them; the rest of the polymer solution is thus protected from the precipitant and, between the pores, a sponge-like structure is formed. In the case of polysulfone hollow fibre membranes, it was found by Senn[45] that the reduction of these finger-like pores or macrovoids improved the gas separation performance of the fibres. Elimination of macrovoids was achieved by making the appropriate selection of polymer, solvent and precipitant, and then obtaining the correct spinning conditions.

Many investigations have been made to examine the effects of phase inversion on membrane morphology, and the related effects of spinning conditions on hollow fibre membrane morphology and the ensuing gas separation properties. Kesting and Fritzche[46] argue that considering the disadvantages of synthesizing new polymers, namely, increased cost of materials and problems of poorer processing and mechanical properties, better use should be made of commercially available polymers. The work by Monsanto[47] that developed polysulfone second generation hollow fibre membranes justifies this point of view. Thus, the spinning conditions can themselves impact molecular packing, polymer density and interchain displacement to enhance gas separations. The second-generation polysulfone hollow fibre membranes show a fourfold increase in oxygen permeability, with no loss in oxygen/nitrogen selectivity[47] when compared to the original fibres.

The factors affecting membrane structure and performance are many. For example, the better the compatibility between polymer and solvent, i.e. the smaller the solubility parameter disparity between polymer and solvent, the longer it takes to remove solvent from the polymer, therefore the slower the precipitation rate. The greater the precipitant polymer solubility parameter disparity, the faster the precipitation will occur. Increases in the compatibility of polymer and precipitant tend to give membranes with a sponge-type structure (optimum for gas separation), whilst decreasing compatibilities lead to finger-like structures. The solvent precipitant interaction is also important, and a measure of this is the heat of mixing. In systems such as polyetherketone, where the only suitable solvent is 98% sulfuric acid, the heat of mixing solvent and non-solvent (water) is so great that it is difficult to obtain sponge-like membrane structures.[44] However, with polysulfone and polyimides, many types of hollow fibre morphologies can be obtained due to the large numbers of solvents and non-solvent combinations available. Perhaps this is why polysulfone, polyethersulfone and polyimide polymers are still being investigated for their phase separation characteristics.[46-53]

Additives can be added to the precipitant or polymer solution, and these change the activity coefficients for the polymer, solvent or precipitant. In the case of the precipitant, additives such as salts reduce the rate of precipitation and a sponge structure is favoured. In the case of the polymer spinning solution, the same additives generally increase the rate of precipitation and a finger structure is therefore favoured. Increasing the polymer concentration tends to result in sponge-type structures, whilst low polymer concentrations result in finger-type structures.

The higher the polymer concentration, the higher the strength of the surface precipitated layer, since a higher polymer concentration exists at the point of precipitation and this tends to prevent the initiation of finger pores. Increasing the viscosity of the polymer solution has the same effect. It was shown by Senn[45] that a sponge-type structure was preferable to a finger-like structure, both in terms of permeation rate and separation factor for polysulfone hollow fibre membranes; though some of the early photographs of Monsanto hollow fibres showed the fibres to have finger-type structures, it is unlikely that these fibres are the ones actually used in the PRISM™ system.

Moreover, with hollow fibre systems the formation shape and size of the hollow fibre lumen is dependent upon the injection rate and the composition of the internal coagulant, as well as the air-gap in the dry jet wet spinning system. An essential function of the injection fluid is to produce a circular lumen of sufficient diameter to allow the unimpeded flow of the faster permeant gas.

The effects of spinning conditions on fibre morphology and performance have been discussed recently for polyimides,[52,53] polyetherketone[44] and polysulfone. Many papers have been published, examples of which can be found in the following references.[45-51]

11.5 Future hollow fibre membranes and industrial gas separation

Membranes have to compete with the existing gas separation technologies of pressure swing adsorption, cryogenics and absorption. The advantages and disadvantages of each separation technology have been well established.[54]

Within this context, the use of hollow fibre membranes for industrial separation processes has developed, and they can now compete effectively in some cases with conventional processes in terms of energy and capital costs. Membranes for gas separation have developed significantly in the last 20 years; however, there is still a need for high temperature and chemically resistant hollow fibre membranes that exhibit good selectivity and gas permeability. In spite of the developments in gas separation membranes, there are still only a few types of hollow fibre membranes that are commercially available. Polymers currently in use include polysulfone (Monsanto), polycarbonate (Dow Chemical Co.) and polyimide (Medal Co.). Ceramic membranes have found broad application for chemical and temperature resistance, but these are far more expensive and more difficult to make than polymeric membranes.[55]

With this in mind, new better polymeric hollow fibre membranes are still needed that have unique combinations of mechanical toughness, high modulus, thermo-oxidative and hydrolytic stability, resistance to organic solvents and retention of physical properties at very high temperatures up to 250 °C. The challenge continues.

References

1. Lonsdale H K, 'The growth of membrane technology', *J. Memb. Sci.*, 1982, **10**, 81.
2. Hwang S T and Kammermeyer K, '*Membranes In Separations*', John Wiley, New York, 1975 (Prologue by Karl Sollner).
3. Mitchell J K, 'On the penetrativeness of fluids', *Roy. Inst. J.*, 1831, **2**, 101, 307.
4. Graham T, 'A short account of experimental researches on the diffusion of gases through each other, and their separation by mechanical means', *Q. J. Sci.*, 1829, **2**, 74.
5. Graham T, 'On osmotic force', *Phil. Trans.* (The Bakerian lecture), 1854, 177.
6. Graham T, 'On the molecular mobility of gases', *Phil. Trans.*, 1863, 385.
7. Fick A E, 'Ueber diffusion', *Pogg Ann.*, 1855, **94**, 59.
8. Graham T, 'On the motion of gases', *Phil. Trans.*, 1846, 573.
9. Loeb S and Sourirajan S, 'Sea water demineralisation by means of a semipermeable membrane', UCLA Water Resources Centre Report WRCC-34, 1960.
10. Riley R L, Merten U and Gardiner J O, 'Replication electron microscopy of cellullose acetate membranes, *Desalination*, 1966, **1**, 30.
11. Strathmann H and Kock K, 'The formation mechanism of asymmetric membranes', *Desalination*, 1977, **21**, 241.
12. Gantzel P K and Merten U, 'Gas separations with high-flux cellulose acetate membranes', *Ind. Engr. Chem. Process Des. Devel.*, 1970, **9**, 331.
13. Monsanto Co., US Pat 4 230 463, 1980.
14. Monsanto Co., UK Pat 2 100 181 A, 1980.
15. Monsanto Co., US Pat 4 364 759, 1982.
16. Monsanto Co., UK Pat 2 047 162 A, 1980.
17. Henis J M S and Tripodi M K, *J. Memb. Sci.*, 1981, **8**, 233.
18. Peinemann K V, Pinnau I, Wijmans J G, Blume I and Kuroda T, 'Gas separation through composite membranes', *J. Memb. Sci.*, 1988, **37**, 81.
19. Backhouse I, 'Recovery and purification of industrial gases using prism separators' in *Membranes in Gas Separations*, 4 BOC Priestly Conference, Leeds, 1986, 265.
20. Koros W J, Kim T H, Husk G R and O'Brien K C, 'Relationship between gas separation properties and chemical structures in a series of aromatic polyimides, *J. Memb. Sci.*, 1988, **37**, 45.
21. Stern S A, Mi Y and Yamamoto H, 'Structure/permeability relationships of polyimide membranes II, *J. Polym. Sci.*, Part B, 1990, **28**, 2291.
22. Private Communication, Dechamps A and Smolders C A.
23. Smolders C A, Mulder M H U and Gebben B, 'Gas separation properties of a thermally stable and chemically resistant polytriazole membrane, *J. Memb. Sci.*, 1989, **46**, 29.
24. Pellegrino J J, Ko M, Nassimbene R and Einert M, 'Gas separation technology',

Proceedings of the International Symposium on Gas Separation Technology, Antwerp, Belgium, 445, 1989.
25 'Modern Japanese membranes', ed. King R, 'Reports from a DTI Sponsored Mission to Japan', 1988.
26 Reiss H, Kaner R B, Mattes B R and Anderson M R, 'Conjugated polymer films for gas separations', *Science,* 1991, **252**, 1412.
27 Hwang S T and Kammermeyer K, *Membranes in Separations,* John Wiley, New York, 1975.
28 Brown P J, 'The modification of polysulphone hollow fibre membranes for gas separation', PhD Thesis, University of Leeds, 1991.
29 Paul D R and Erb A J, *J. Memb. Sci.,* 1981, **8**, 11.
30 Vieth W R, Howell J M and Hsieh J H, 'Dual sorption theory', *J. Memb. Sci.,* 1976, **1**, 177.
31 Petropoulos J H, 'On the dual mode gas transport model for glassy polymers', *J. Polym. Sci. Part B,* 1988, **26**, 1009.
32 Raucher D and Sefcik M D, 'Industrial gas separations', *ACS Symp. Ser. 233,* Amer. Chem. Soc., Washington DC, Chapters 5 and 6, 1983.
33 Datyner A and Pace R J, 'Statistical mechanical model for diffusion of simple penetrants in polymers', *J. Polym. Sci., Part B,* 1979, **17**, 437.
34 Chern R T, Koros W J, Sanders E S, Chen H S and Hopfenburg H B, 'Industrial gas separations', *ACS Symp. Ser. 233,* Amer. Chem. Soc., Washington DC, Chapter 3, 1983.
35 Sefcik M D, 'Dilation of polycarbonate by carbon dioxide', *J. Polym. Sci., Part B,* 1986, **24**, 935.
36 Koros W J and Fleming G K, Comments on measurement of gas-induced polymer dilation by different optical methods, *J. Polym. Sci. Part B,* 1987, **25**, 2033.
37 Meares P, 'The diffusion of gases through polyvinyl acetate', *J. Amer. Chem. Soc.,* 1954, **76**, 3415.
38 Crank J and Park G S, *Diffusion in Polymers,* Academic Press, New York, 1968.
39 Sweeting O J and Bixler H J, *Science and Technology of Polymer Films,* John Wiley, New York, 2 Chapter 1, 1971.
40 Paul D R and Maeda Y, 'Effect of antiplasticization on gas sorption and transport. I. Polysulphone, *J. Polym. Sci. Part B,* 1987, **25**, 957.
41 Kesting R E, *Materials Science of Synthetic Membranes,* ed. Lloyd, Chapter 7, 131, 1984.
42 Strathmann H and Kock K, 'The formation mechanism of phase inversion membranes', *Desalination,* 1977, **21**, 241.
43 Strathmann H, *Materials Science of Synthetic Membranes,* ed. Lloyd, Chapter 8, 165, 1984.
44 Brown P J, Rogers-Gentile V and Ying S, 'Preparation and characterisation of polyetherketone hollow fibre membranes for gas separation', *J. Textile Inst.,* 1999, December 90 Part 3, 30.
45 Senn S C, 'The preparation and characterisation of hollow fibre membranes for gas separation', PhD Thesis, University of Leeds, 1987.
46 Kesting R E and Fritzche A K (eds), *Polymer Gas Separation Membranes,* 2nd edn, John Wiley, New York, 77, 1993.
47 Kesting R E, Fritzche A K, Murphy M K, Cruse C A, Handermann A C and Moore

M D, 'The second-generation polysulfone gas-separation membrane I', *J. Appl. Polym. Sci.*, 1990, **40**, 1557.

48 Brown P J, 'The effect of residual solvent on the gas transport properties of polysulphone hollow fibre membranes', Textile Institute Fibre Science Group Conference, Paper 11, 1989.

49 Brown P J, East G C and McIntyre J E, 'Effect of residual solvent on the gas transport properties of hollow fibre membranes', *Polymer Commun.*, 1989, **31**, 156.

50 Chung T and Hu X, 'Effect of air gap distance on the morphology and thermal properties of polyethersulphone hollow fibres', *J. Appl. Polym. Sci.*, 1997, **66**, 1067.

51 Chung T, Teoh S K, Lau W W and Srinivasan M P, 'Effect of shear stress within the spinneret on hollow fiber membrane morphology and separation performance, *Ind. Eng. Chem. Res.*, 1998, **37**, 3930.

52 Chung T, Xu Z and Huang Y, 'Effect of polyvinylpyrrolidone molecular weights on morphology, oil/water separation, mechanical and thermal properties of poletherimide/polyvinylpyrrolidone hollow fibre membranes, *J. Appl. Polym Sci.*, 1999, **74**, 2220.

53 Wang Dongliang, Li K and Teo K, 'Phase separation in polyetherimide/solvent/nonsolvent systems and membrane formation', *J. Appl. Polym Sci.*, 1999, **71**, 1789.

54 Proceedings of the 4th BOC Priestly Conference, Special Publication 62, Royal Society of Chemistry, 1986.

55 Bhave Ramesh R, *Inorganic Membrane Synthesis, Characteristics, and Applications*, van Nostrand Reinhold, New York, 1991.

12
Embroidery and smart textiles

BÄRBEL SELM, BERNHARD BISCHOFF AND
ROLAND SEIDL

12.1 Introduction

At first glance, embroidery technology seems to have little in common with intelligent textiles and high-tech clothing, since embroidery is generally defined as follows: 'The decoration of woven or knitted textile fabrics or other surfaces, e.g. leather, through the application of threads or other decorative objects (beads, cords, applications), by sewing them in or on in an arrangement designed to achieve a pattern on the ground fabric'.[1]

The 1990s, however, taught us otherwise. Embroidery is the only textile technology in which threads can be arranged in (almost) any direction. The boom in technical textiles has, all of a sudden, created an interest in embroidery technology for mechanical engineering. Arranging high performance fibres, such as Kevlar, carbon fibres, PPS, PBO, etc. in components, according to the lines of force acting upon them, results in elements combining high performance with small mass.

What was until recently unthinkable has now come about: medicine has discovered embroidery. This technology allows wound dressing to be constructed in such a way that embroidered, three-dimensional structures become functional in the healing of wounds.

A combination of aesthetics and innovation is represented in the embroidered stamp, which was introduced as a world novelty in the year 2000 in St. Gallen, Switzerland. Embroidery technology – until very recently used 'just' for decorative purposes – is now slowly penetrating the field of technical textiles. This technology offers unique possibilities and a potential for 'smart textiles'.

12.2 Basics of embroidery technology

Almost all peoples of all regions can look back on a tradition of hand embroidery many centuries old. This was used then, and it still is, for home decorations, clothing, and textiles for cult or religious purposes. The variety of stitches, materials and patterns, which are characteristic features of particular

Embroidery and smart textiles 219

eras and countries, is huge and beyond visualization. The ground fabric is held in a round frame, the embroiderer pierces the ground fabric by hand with a needle from the top, and the other hand, beneath the embroidery frame, receives the needle, carefully pulls tight the thread and sends the needle to the top again close to the previous stitch. The stitches used include outline embroidery stitches, such as rope, coral, cable, chain or split stitches; flat-mass embroidery stitches, such as couching, satin stitch, darning, cross stitch or appliqué; and shading stitches, such as the feather stitch.

12.2.1 Combined embroidery techniques

Combined embroidery techniques mean that, apart from the ground material and basic embroidery, other additional materials and techniques are used. These products are still categorized as embroidery, even though other techniques such as padding, shearing, etc. are also involved. This is explained by the fact that these are primarily embroidery products. All other techniques are subordinate and merely serve the purpose of the embroidery process, exercising no major influence on the end result.

Combined embroidery techniques without cutwork mean that there are no openings in the ground material. The use of additional materials is frequent with this technique and, less often, other techniques are applied, such as shearing of applications. These combined techniques are divided up into braid embroidery, cord and ribbon embroidery (soutache), appliqués and quilting, and pearl and tinsel embroidery (sequins). The combined embroidery techniques with cutwork comprise thread tensioning and thread pull-out.

12.2.2 Embroidery machines

The fundamental principle of hand embroidery machines is to increase performance by simultaneous operation of a number of needles. Needles with points on either end and eyes in the middle are used. The frame, which holds the ground material, moves to produce a pattern. The needles move in and out of the material along their longitudinal axis. The pattern control is by hand, with a pantograph following an embroidery drawing.

The shuttle embroidery machine is essentially based on the same design as the hand embroidery machine. The most important components of such machines are:

- frame to stretch and move the ground material
  pattern control by hand: pantograph machines
  pattern control by a drive: automatic embroiderer
  electronic pattern control: Positronic

- needle bar with yarn guide
- shuttle bar with drive
- central drive unit.

12.3 Embroidery for technical applications – tailored fibre placement

With the aid of embroidery, through fibre orientation, fibre-composite components can be created to withstand complex forces. This is called tailored fibre placement, TFP.[2]

The process applied is based on embroidery technology, modified to suit the specific technical requirements. A 'cord' of reinforcing fibres is sewn onto the ground material. The arrangement (guiding) of the cord is CNC controlled, and the 'embroidery pattern' is produced with the aid of a computer program. The reproducibility of technical embroidery is good and a change of pattern can be rapidly effected.

The cord can be made of glass, carbon or aramid fibres. Hybrid yarns, consisting of reinforcement and matrix fibres, can also be applied.

A variety of materials can be used as the ground material on which the cord is embroidered, such as woven and non-woven fabrics or foils. In the process, these materials become part of the composite and have to be coordinated with the matrix material. This applies equally to the binding ends, used for the production of the cord. In this way, for instance, glass wovens and polyester ends can be used for a glass/epoxy resin system. For a carbon/PEEK system, PEEK foil and PEEK ends are used. The main advantage, when compared to standard textile technologies, is that reinforcement threads can be arranged in any direction, i.e. at any angle between $0°$ and $360°$. Thus embroidery allows an accumulation of reinforcement material by repeated embroidering across the same place.

To turn the embroidered goods into composites, conventional technologies can be used, such as layout by hand, resin injection processes or pressing. The composite can consist of embroidered products only, or of a combination of standard reinforcement structures.

Examples of typical construction components are:

- rotating machine elements, subjected to centrifugal forces, must be reinforced by radial and tangential structures
- compression-tension bars reinforced by loops and by radial and tangential structures
- levers, subjected to pressure and inflection loads, reinforced by cords, which are arranged along the main stress lines.

12.4 Embroidery technology used for medical textiles

The field of medical textiles is vast. Clothing for hospital staff, surgery blankets, textiles for hygienic and orthopaedic purposes, dressings, implants and much more are included in the wider sense. For future development, products for the healing and construction of human tissue are of particular interest, and this is also the main focus of this chapter. Up to now, products manufactured with embroidery technology have not been available in the medical field. Even though the technique of embroidering has been known for centuries, this field of work has never been touched. However, the first medical products have now been manufactured and are being launched on the market.

For many years, textile structures have been successfully used for wound treatment, operations and implants. The popular range of these products include plasters, wound gauze, dressing material, surgical stitch material, vessel prosthesis and hernia nets. There is a differentiation between external and internal (implant) application. Most products are woven or knitted textile materials, which, for the various applications, are made up according to individual requirements. Figure 12.1 illustrates an embroidered stent.

Today, the use of textile structures in medicine is a growing market. There are constantly new discoveries concerning cellular components, and the reactions of the human body to new and existing substances. New ideas and improvements are needed in order to meet the requirements of this new field.

This is an opportunity for embroidery technology. The process of embroidering allows threads to be arranged on a flat substrate in any direction, thus enabling

12.1 Embroidery stent, as used for the repair of abdominal aortic aneurysms.

the production of any desirable textile form. This is particularly advantageous with relatively small motifs. As opposed to weaving, where threads are arranged at rigid angles, embroidery also enables rounded patterns. Additionally, made-up embroidery goods are dimensionally stable – unlike knitted fabrics. A further particularity of embroidery is that threads can be placed on top of one another, achieving three-dimensional characteristics. Modern embroidery machines are equipped with colour change devices, enabling switching from one material component to another during the course of production.

Refined and highly sophisticated software enables such flexible production methods. This combines graphics input, of entering images and patterns, with machine language in the embroidery works. The draft of an embroidery pattern is entered into a CAD (computer-aided design) programme, after which it is immediately available for production in machine-readable code. This enables rapid prototyping, so that prototypes to customer requirements can be produced within one day. Ideas from doctors and hospitals can be turned into prototypes equally fast. These rapidly produced embroidered patterns provide a helpful basis for discussion and communication with people from other specialized fields, thus accelerating the development processes.

For medical products, special thread materials are embroidered onto a substrate according to the form required. If a substrate is chosen that is made of polyvinyl alcohol (PVA) or cellulose acetate (CA), it can be dissolved in water or acetone respectively after the embroidery process. The design must ensure that the thread structures hold together securely. Local reinforcements and functional elements can easily be incorporated into the embroidered goods. In an initial phase, the research team with J. G. Ellis[3] has developed embroidered textile implants. They designed hernia patches, implants for intervertebral disc repair and a stent for the repair of abdominal aortic aneurysms (Fig. 12.1). They even used embroidery technology to place and fix rings made of Nitinol shape memory.

In the development of medical textiles, polyester is frequently used. If specialized biocompatible materials are used, they are brought in at a later, more advanced stage of development, due to their high cost (up to 4000 US$/kg). Various synthetic or natural polymer fibres feature very specific structural and mechanical properties, favoured by tissue engineering as bone, cartilage or skin replacements. Karamuk and Mayer[4] carried out tests with embroidered materials that decompose inside the body. These were threads made of polyglycolic acid (PGA). In vitro tests (Fig. 12.2) showed that forces in embroidery goods can be controlled by embroidery technology; in this way, the mechanical properties of the textile can be adapted to those of the body tissue, and inflammatory reactions can be avoided.

The research team, including E. Karamuk,[5] developed a textile wound dressing for the treatment of chronic slow-healing wounds. These embroidered

Embroidery and smart textiles 223

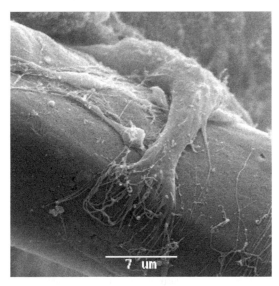

12.2 Cellular growth on a single fibril.

12.3 Electron microscopic view of the embroidered layer of a Tissupor wound dressing.

goods have a three-dimensional structure (Fig. 12.3) and achieve the desired reconstruction of skin tissue. Great attention was paid to the selection of the yarns used: monofilament and multifilament were chosen. In their interstices, multifilaments provide small pores for the body tissue to grow. The spacing between the yarns has larger pores and allows blood vessels to grow into the variously sized holes provided. Apart from the structural configuration of the embroidery goods, stiff elements are incorporated to effect local stimulation of the body tissue. Clinical studies were carried out successfully with patients who suffered from deep tissue defects due to bedsores, and with patients who suffered from surface wounds due to ulcers caused by bad circulation in their legs.

The growing field of tissue engineering will require more textile structures to support human cells, to reconstruct tissue or entire organs. Possible alternatives, such as transplants or genetic engineering, still leave too many questions unanswered. That is why the means of embroidery technology will yield further products and solutions. Further development work will result in the provision of tailor-made substrates for medical textiles for the healing and regeneration of various wounds and organs.

12.5 Embroidered stamp – gag or innovation?

The world's first embroidered stamp, presented to the public in June 2000, provides further evidence to prove that embroidery technology can be used outside the field of fashion. The agreement on an appropriate size for the forming of proper textile structures was an essential factor. Furthermore, a self-adhesive coating had to be chosen, which would meet the requirements of the post office as well as stamp collectors. Although this project was very successful, it cannot be assumed that embroidery technology will be used as standard technology to produce stamps, since the cost is very high. The limited time available to translate this innovative idea into practice hardly allowed for pure research on the subject. This example demonstrates however that, with flexibility existing technologies can be combined and applied to new areas for

12.4 Production of embroidered stamps on a large embroidery machine.

12.5 A square of four stamps on the embroidery machine.

which they were not originally intended. The production of the embroidered stamps on a large embroidery machine is shown in Fig. 12.4 and 12.5.

12.6 Summary

The combination of embroidery technology and intelligent textiles was, until very recently, a contradiction – now they are one and the same. The examples of 'unconventional' uses of embroidery, as presented in this chapter, indicate the potential still to be exploited. However, we are at the very beginning of exploring this field, and extensive research and development work will be required for the eventual manufacture of commercial products. Tailor-made and/or three-dimensional textiles capable of withstanding forces will be with us and clothe us – the smart textiles of the new millennium.

References

1 Seidl R, 'Manuscript of lecture on embroidery', Swiss Textile College, 2000.
2 Seidl R, 'Manuscript of lecture on technical textiles', Swiss Federal Institute of Technology, Zurich, 2000.
3 Ellis J G, 'Embroidery for engineering and surgery', *Textile Institute World Conference*, Manchester (UK), 2000.
4 Karamuk E and Mayer J, 'Embroidery technology for medical textiles and tissue engineering', *Tech. Text. Internat.*, July/August 2000, 9–12.
5 Karamuk E, 'Embroidery technology for medical textiles', *Medical Textiles*, 1999, Bolton, UK.

13
Adaptive and responsive textile structures (ARTS)

SUNGMEE PARK AND
SUNDARESAN JAYARAMAN[*]

13.1 Introduction

The field of textiles has been instrumental in bringing about one of the most significant technological advancements known to human beings, i.e. the birth of the computer, which spawned the information/knowledge revolution being witnessed today. It is only appropriate that this field take the next evolutionary step towards integrating textiles and computing, by designing and producing intelligent textiles that can adapt and respond to the wearer's needs and the environment. In this chapter, we discuss the need for the new generation of adaptive and responsive textile structures (ARTS) and present the design and development of the Georgia Tech Wearable Motherboard™ (GTWM), the first generation of ARTS. We will discuss the universal characteristics of the interface pioneered by the GTWM, or the smart shirt, and explore the potential applications of the technology in areas ranging from medical monitoring to wearable information processing systems. We will conclude the chapter with a discussion of the research avenues that this new paradigm spawns for a multidisciplinary effort involving the convergence of several technical areas – sensor technologies, textiles, materials, optics and communication – that can not only lead to a rich body of new knowledge but, in doing so, enhance the quality of human life.

13.2 Textiles and computing: the symbiotic relationship

John Kay's invention of the flying shuttle in 1733 sparked off the first Industrial Revolution, which led to the transformation of industry and subsequently of civilization itself. Yet another invention in the field of textiles – the Jacquard head by Joseph Marie Jacquard (*ca.* 1801) – was the first binary information processor. Ada Lovelace, the benefactor for Charles Babbage who worked on the analytical engine (the predecessor to the modern day

---

[*] To whom correspondence should be addressed.

computer), is said to have remarked, 'The analytical engine weaves algebraic patterns just as the Jacquard loom weaves flowers and leaves.' Thus, the Jacquard mechanism that inspired Babbage and spawned the Hollerith punched card has been instrumental in bringing about one of the most profound technological advancements known to humans, viz. the second Industrial Revolution, also known as the Information Processing Revolution or the Computer Revolution.[1]

13.2.1 The three dimensions of clothing and wearable information infrastructure

Humans are used to wearing clothes from the day they are born and, in general, no special 'training' is required to wear them, i.e. to use the interface. It is probably the most universal of human–computer interfaces and is one that humans need, use and are very familiar with, and that can be enjoyed and customized.[2] Moreover, humans enjoy clothing, and this universal interface of clothing can be 'tailored' to fit the individual's preferences, needs and tastes, including body dimensions, budgets, occasions and moods. It can also be designed to accommodate the constraints imposed by the ambient environment in which the user interacts, i.e. different climates. In addition to these two dimensions of functionality and aesthetics, if 'intelligence' can be embedded or integrated into clothing as a third dimension, it would lead to the realization of clothing as a personalized wearable information infrastructure.

13.2.2 Textiles and information processing

A well-designed information processing system should facilitate the access of information Anytime, Anyplace, by Anyone – the three As. The 'ultimate' information processing system should not only provide for large bandwidths, but also have the ability to see, feel, think and act. In other words, the system should be totally 'customizable' and be 'in-sync' with the human. Clothing is probably the only element that is 'always there' and in complete harmony with the individual (at least in a civilized society!). And textiles provide the ultimate flexibility in system design by virtue of the broad range of fibres, yarns, fabrics and manufacturing techniques that can be deployed to create products for desired end-use applications. Therefore, there is a need for research in textiles that would result in a piece of clothing that can serve as a true information processing device, with the ability to sense, feel, think and act based on the wearer's stimuli and the operational environment. Such an endeavour can facilitate personalized mobile information processing (PMIP) and give new meaning to the term man–machine symbiosis. The first step in creating such an intelligent or adaptive and responsive textile structure (ARTS) has been taken with the design and development of the Georgia Tech Wearable Motherboard™, also known as the smart shirt.[3]

13.3 The Georgia Tech Wearable Motherboard™

In the spring of 1996, the US Navy Department put out a broad agency announcement inviting white (concept) papers to create a system for the soldier that was capable of alerting the medical triage unit (stationed near the battlefield) when a soldier was shot, along with some information on the soldier's condition characterizing the extent of injury. As such, this announcement was very broad in the definition of the requirements, and specified the following two key broad objectives of the so-called sensate liner:

- Detect the penetration of a projectile, e.g. bullets and shrapnel
- Monitor the soldier's vital signs.

The vital signs would be transmitted to the triage unit by interfacing the sensate liner with a personal status monitor developed by the US Defense Advanced Projects Research Agency (DARPA).

13.3.1 The name 'Wearable Motherboard'

As the research progressed, new vistas emerged for the deployment of the resulting technology, including civilian medical applications and the new paradigm of personalized information processing using the flexible information infrastructure. Therefore, we coined the name Wearable Motherboard™ to better reflect the breadth and depth of the conceptual advancement resulting from the research. Just as chips and other devices can be plugged into a computer motherboard, sensors and other information processing devices can be plugged into the sensate liners produced during the course of the research. Therefore, the name Wearable Motherboard is apt for the flexible, wearable and comfortable sensate liners. The name also represents (1) a natural evolution of the earlier names sensate liner and woven motherboard, (2) the expansion of the initial scope and capability of a sensate liner targeted for combat casualty care to a much broader concept and spectrum of applications and capabilities – much like a platform to build upon, and (3) the symbiotic relationship between textiles and computing that began with the Jacquard weaving machine.

13.3.2 Detailed analysis of the key performance requirements

Quality function deployment (QFD) is a structured process that uses a visual language and a set of inter-linked engineering and management charts to transform customer requirements into design, production and manufacturing process characteristics.[4] The result is a systems engineering process that prioritizes and links the product development process to the design so that it assures product quality as defined by the customer. Additional power is derived when QFD is used within a concurrent engineering environment. Therefore, the QFD approach was adopted for the research that also served as an example of integrated product/process design (IP/PD) or concurrent engineering.

The first step in this QFD process is to clearly identify the various characteristics required by the customer (the US Navy, in this research) in the product being designed.[5,6] Therefore, using this information on the two key performance requirements, an extensive analysis was carried out. A detailed and more specific set of performance requirements was defined with the result shown in Fig. 13.1. These requirements are functionality, usability in combat, wearability, durability, manufacturability, maintainability, connectability and affordability. The next step was to examine these requirements in depth and to identify the key factors associated with each of them. These are also shown in the figure. For example, functionality implies that the GTWM must be able to detect the penetration of a projectile and should also monitor body vital signs – these are the two requirements identified in the broad agency announcement from the Navy.

Likewise, as shown in the figure, wearability implies that the GTWM should be lightweight, breathable, comfortable (form-fitting), easy to wear and take off, and provide easy access to wounds. These are critical requirements in combat conditions, so that the protective garment does not hamper the soldier's performance. The durability of the GTWM is represented in terms of a wear life of 120 combat days and its ability to withstand repeated flexure and abrasion – both of which are characteristic of combat conditions. Manufacturability is another key requirement, since the design (garment) should eventually be produced in large quantities over the size range for the soldiers; moreover, it should be compatible with standard issue clothing and equipment. The maintainability of the GTWM is an important requirement for the hygiene of the soldiers in combat conditions; it should withstand field laundering, should dry easily and be easily repairable (for minor damages). The developed GTWM should be easily connectable to sensors and the personal status monitor (PSM) on the soldier. Finally, the affordability of the proposed GTWM is another major requirement, so that the garment can be made widely available to all combat soldiers to help ensure their personal survival, thereby directly contributing to the military mission as force enhancers.

Thus, in the first step of the conceptual design process, the broad performance requirements were translated into a larger set of clearly defined functions along with the associated factors (Fig. 13.1).

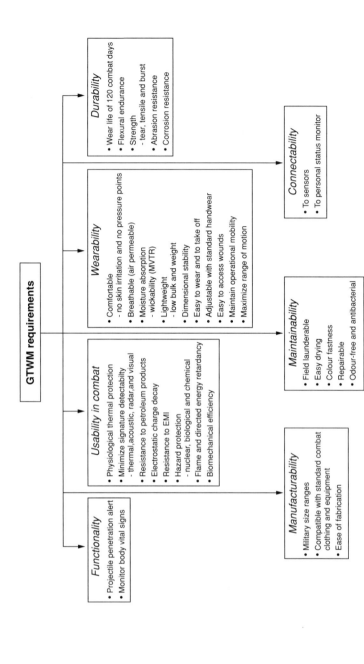

13.1 GTWM: performance requirements.

Adaptive and responsive textile structures (ARTS) 231

13.3.3 The GTWM design and development framework

Once the detailed performance requirements were defined, the need for an overall design and development framework became obvious. However, no comprehensive framework was found in the literature; therefore, one was developed. Figure 13.2 shows the resulting overall GTWM design and development framework and encapsulates the modified QFD-type (quality function deployment) methodology developed for achieving the project goals. The requirements are then translated into the appropriate properties of GTWM: sets of sensing and comfort properties. The properties lead to the specific design of the GTWM, with a dual structure meeting the twin requirements of 'sensing' and 'comfort'. These properties of the proposed design are achieved through the appropriate choice of materials and fabrication technologies by applying the corresponding design parameters as shown in the figure. These major facets in the proposed framework are linked together as shown by the arrows between the dotted boxes in Fig. 13.2. The detailed analysis of the performance requirements, the methodology and the proposed design and development framework can be found elsewhere.[7] This generic framework can be easily modified to suit the specific end-use requirements associated with the garment. For instance, when creating a version of GTWM for the prevention of sudden infant death syndrome (SIDS), the requirement of 'usability in combat' would not apply.

13.3.4 The design and structure of GTWM

This structured and analytical process eventually led to the design of the structure. The Wearable Motherboard consists of the following building blocks or modules that are totally integrated to create a garment (with intelligence) that feels and wears like any typical undershirt. The modules are:

- a comfort component to provide the basic comfort properties that any typical undergarment would provide to the user,
- a penetration sensing component to detect the penetration of a projectile,
- an electrical sensing component to serve as a data bus to carry the information to/from the sensors mounted on the user or integrated into the structure,
- a form-fitting component to ensure the right fit for the user, and
- a static dissipating component to minimize static build-up when the garment is worn.

The elegance of this design lies in the fact that these building blocks (like LEGO™ blocks) can be put together in any desired combination to produce structures to meet specific end-use requirements. For example, in creating a

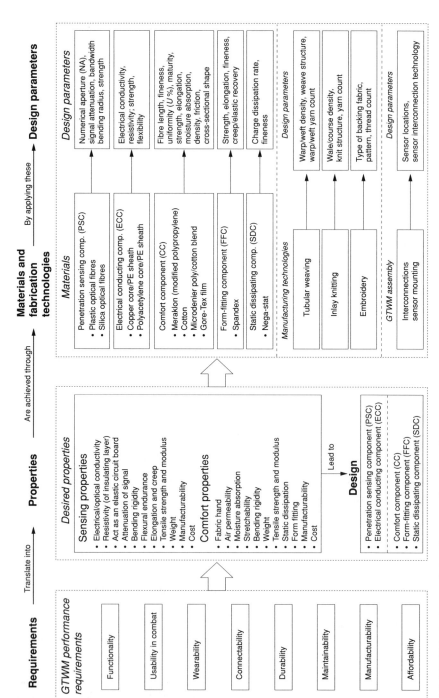

13.2 GTWM: design and development framework.

GTWM for healthcare applications, e.g. patient monitoring, the penetration sensing component will not be included. The actual integration of the desired building blocks will occur during the production process through the inclusion of the appropriate fibres and yarns that provide the specific functionality associated with the building block. In and of itself, the design and development framework resulting from this research represents a significant contribution to systematizing the process of designing structures and systems for a multitude of applications.

13.3.5 Production of GTWM

The resulting design was woven into a single-piece garment (an undershirt) on a weaving machine to fit a 38–40 in chest. The various building blocks were integrated into the garment at the appropriate positions in the garment. Based on the in-depth analysis of the properties of the different fibres and materials, and their ability to meet the performance requirements, the following materials were chosen for the building blocks in the initial version of the smart shirt:[7]

- Meraklon (polypropylene fibre) for the comfort component
- plastic optical fibres for the penetration sensing component
- copper core with polyethylene sheath and doped nylon fibres with inorganic particles for the electrical conducting component
- Spandex for the form-fitting component
- Nega-Stat™ for the static dissipating component.

The plastic optical fibre (POF) is spirally integrated into the structure during the fabric production process without any discontinuities at the armhole or the seams using a novel modification in the weaving process. With this innovative design, there is no need for the 'cut and sew' operations to produce a garment from a two-dimensional fabric. This pioneering contribution represents a significant breakthrough in textile engineering because, for the first time, a full-fashioned garment has been woven on a weaving machine.

An interconnection technology was developed to transmit information from (and to) sensors mounted at any location on the body, thus creating a flexible bus structure. T-Connectors – similar to button clips used in clothing – are attached to the yarns that serve as the data bus to carry the information from the sensors, e.g. EKG (electrocardiogram) sensors on the body. The sensors plug into these connectors and, at the other end, similar T-connectors are used to transmit the information to monitoring equipment or DARPA's personal status monitor. By making the sensors detachable from the garment, the versatility of the GTWM has been significantly enhanced. Since the shapes and sizes of humans will be different, sensors can be positioned on the right

234 Smart fibres, fabrics and clothing

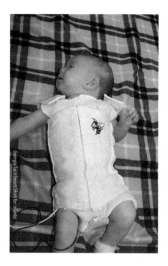

13.3 Wearable Motherboard™ on subjects. The optical fibres are lit, indicating the shirt on the left is 'armed' for detecting penetration.

locations for all users and without any constraints imposed by the GTWM. In essence, the GTWM can be truly 'customized'. Moreover, it can be laundered without any damage to the sensors themselves. In addition to the fibre optic and specialty fibres that serve as sensors, and the data bus to carry sensory information from the wearer to the monitoring devices, sensors for monitoring the respiration rate, e.g. RespiTrace™ sensors, have been integrated into the structure, thus clearly demonstrating the capability to directly incorporate sensors into the garment.

Several generations of the woven and knitted versions of the Wearable Motherboard have been produced (see Fig. 13.3). The lighted optical fibre in the figure illustrates that the GTWM is 'armed' and ready to detect projectile penetration. The interconnection technology has been used to integrate sensors for monitoring the following vital signs: temperature, heart rate and respiration rate. In addition, a microphone has been attached to transmit the wearer's voice data to recording devices. Other sensors can be easily integrated into the structure. For instance, a sensor to detect oxygen levels or hazardous gases can be integrated into a variation of the GTWM that will be used by firefighters. This information, along with the vital signs, can be transmitted to the fire station, where personnel can continuously monitor the firefighter's condition and provide appropriate instructions, including ordering the individual to evacuate the scene if necessary. Thus, this research has led to a truly and fully customizable 'Wearable Motherboard' or intelligent garment.

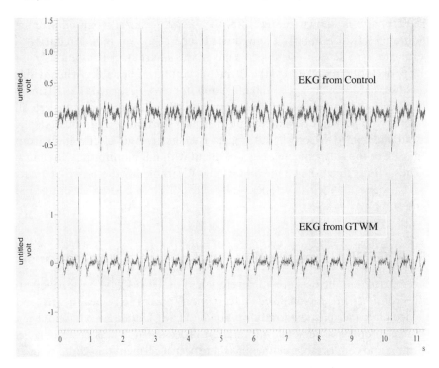

13.4 EKG trace from the GTWM.

13.3.6 Testing of the GTWM

The penetration sensing and vital signs monitoring capabilities of the GTWM were tested. A bench-top set-up for testing the penetration sensing capability was devised. A low-power laser was used at one end of the plastic optical fibre (POF) to send pulses that 'lit up' the structure indicating that the GTWM was armed and ready to detect any interruptions in the light flow that might be caused by a bullet or shrapnel penetrating the garment. At the other end of the POF, a photo-diode connected to a power-measuring device measured the power output from the POF. The penetration of the GTWM resulting in the breakage of the POF was simulated by cutting the POF with a pair of scissors; when this happened, the power output at the other end on the measuring device fell to zero. The location of the actual penetration in the POF could be determined by an optical time domain reflectometer, an instrument used by telephone companies to pinpoint breaks in fibre optic cables.

The vital signs monitoring capability was tested by a subject wearing the garment and measuring the heart rate and electrocardiogram (EKG) through the sensors and T-connectors. In Fig. 13.4, the EKG trace from the Wearable

Motherboard is shown along with the control chart produced from a traditional set-up. Similarly, the wearer's temperature was monitored using a thermistor-type sensor. A subject wearing the smart shirt continuously for long periods of time evaluated the garment's comfort. The subject's behaviour was observed to detect any discomfort and none was detected. The garment was also found to be easy to wear and take off. For monitoring acutely ill patients, who may not be able to put the smart shirt on over the head (like a typical undershirt), Velcro™ and zipper fasteners are used to attach the front and back of the garment, creating a garment with full monitoring capability.

Thus, a fully functional and comfortable Wearable Motherboard or smart shirt has been designed, developed and successfully tested for monitoring vital signs.[3]

13.4 GTWM: contributions and potential applications

This research on the design and development of the GTWM has opened up new frontiers in personalized information processing, healthcare and telemedicine, and space exploration, to name a few.[8] Until now, it has not been possible to create a personal information processor that was customizable, wearable and comfortable; neither has there been a garment that could be used for unobtrusive monitoring of the vital signs of humans on earth or space, such as temperature, heart rate, etc. Moreover, with its universal interface of clothing, the GTWM pioneers the paradigm of an integrated approach to the creation and deployment of wearable information infrastructures that, in fact, subsume the current class of wearable computers.

13.4.1 Potential applications of the GTWM

The broad range of applications of the GTWM in a variety of segments is summarized in Table 13.1. The table also shows the application type and the target population that can utilize the technology. A brief overview of the various applications follows.

13.4.1.1 Combat casualty care

The GTWM can serve as a monitoring system for soldiers that is capable of alerting the medical triage unit (stationed near the battlefield) when a soldier is shot, along with transmitting information on the soldier's condition, characterizing the extent of injury and the soldier's vital signs. This was the original intent behind the research that led to the development of the GTWM.

Adaptive and responsive textile structures (ARTS)

Table 13.1 Potential applications of the Wearable Motherboard

Segment	Application type	Target customer base
Military	Combat casualty care	Soldiers and support personnel in battlefield
Civilian	Medical monitoring	Patients: surgical recovery, psychiatric care
		Senior citizens: geriatric care, nursing homes
		Infants: SIDS prevention
		Teaching hospitals and medical research institutions
	Sports/performance monitoring	Athletes, individuals Scuba diving, mountaineering, hiking
Space	Space experiments	Astronauts
Specialized	Mission critical/hazardous applications	Mining, mass transportation
Public safety	Fire-fighting	Firefighters
	Law enforcement	Police
Universal	Wearable mobile information infrastructure	All information processing applications

13.4.1.2 Healthcare and telemedicine

The healthcare applications of the GTWM are enormous; it greatly facilitates the practice of telemedicine, thus enhancing access to healthcare for patients in a variety of situations. These include patients recovering from surgery at home, e.g. after heart surgery, geriatric patients (especially those in remote areas where the doctor/patient ratio is very small compared to urban areas), potential applications for patients with psychiatric conditions (depression/anxiety), infants susceptible to SIDS (sudden infant death syndrome) and individuals prone to allergic reactions, e.g. anaphylaxis reaction from bee stings.

13.4.1.3 Sports and athletics

The GTWM can be used for the continuous monitoring of the vital signs of athletes to help them track and enhance their performance. In team sports, the coach can track the vital signs and the performance of the player on the field and make desired changes in the players on the field depending on the condition of the player.

13.4.1.4 Space experiments

The GTWM can be used for the monitoring of astronauts in space in an unobtrusive manner. The knowledge to be gained from medical experiments in space will lead to new discoveries and the advancement of the understanding of space.

13.4.1.5 Mission critical/hazardous applications

Monitoring the vital signs of those engaged in mission critical or hazardous activities such as pilots, miners, sailors, nuclear engineers, among others. Special-purpose sensors that can detect the presence of hazardous materials can be integrated into the GTWM and enhance the occupational safety of the individuals.

13.4.1.6 Public safety

Combining the smart shirt with a GPS (global positioning system) and monitoring the well-being of public safety officials (firefighters, police officers, etc.), their location and vital signs at all times, thereby increasing the safety and ability of these personnel to operate in remote and challenging conditions.

13.4.1.7 Personalized information processing

A revolutionary new way to customize information processing devices to 'fit' the wearer by selecting and plugging in chips/sensors into the Wearable Motherboard (garment).

A detailed analysis of the characteristics of the GTWM and its medical applications can be found elsewhere.[2]

13.4.2 Impact of the technology

The Wearable Motherboard™ technology has the potential to make a significant impact on healthcare while enhancing the quality of life. For instance, patients could wear the GTWM at home and be monitored by a monitoring station (similar to home security monitoring companies), thereby avoiding hospital stay costs and reducing the overall cost of healthcare. At the same time, a home setting can contribute to faster recovery. As another example, when a baby version of the GTWM is used for monitoring infants prone to SIDS (sudden infant death syndrome), it can shift the focus from the treatment of infants who have suffered brain damage due to apnea to the prevention of the damage in the first place. Because the GTWM can be tailored, it can be used across the entire population spectrum, from infants to senior citizens of both genders.

Adaptive and responsive textile structures (ARTS)

13.4.2.1 Product versatility

The 'plug and play' feature in the GTWM greatly broadens its application areas. For instance, athletes can choose to have one set of sensors to monitor their performance on the field, while firefighters could have a different set of sensors, e.g. heart rate, temperature and hazardous gases for their application. Thus, the GTWM is a versatile platform and serves as a true motherboard.

13.4.2.2 Product appeal

The GTWM is similar to any undershirt and is comfortable, and easy to wear and use. By separating the sensors from the garment, the maintenance of the garment has been enhanced. The current versions of the garment can be machine-washed. Initial tests have demonstrated the reliability of the system to continuously monitor the various vital signs. In terms of affordability, the anticipated cost of producing the smart shirt is in the $35 range. The costs associated with the required sensors and monitoring would vary depending on the individual application. Thus, conceptually, the smart shirt can be

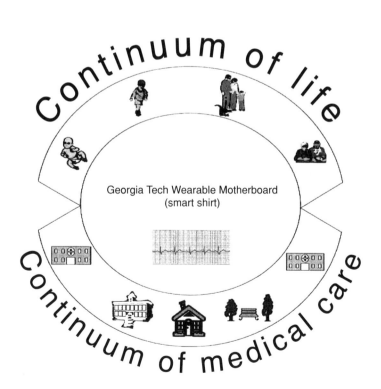

13.5 The twin continua of life and medical care.

likened to a home alarm system. Just as the overall cost of the home monitoring system will depend on the number of points monitored, the types of sensors used and the desired response, the final cost of the smart shirt will also vary. This ability to customize greatly enhances the appeal of the GTWM to a wide cross-section of the population.

Thus, the smart shirt will have a significant impact on the practice of medicine, since it fulfils the critical need for a technology that can enhance the quality of life while reducing healthcare costs across the continuum of life, i.e. from newborns to senior citizens and across the continuum of medical care, i.e. from homes to hospitals and everywhere in-between as shown in Fig. 13.5. The potential impact of this technology on medicine was further reinforced in a special issue of *Life Magazine – Medical Miracles for the Next Millennium*, autumn 1998 – in which the smart shirt was featured as one of the '21 breakthroughs that could change your life in the 21st century'.[9]

13.5 Emergence of a new paradigm: harnessing the opportunity

One of the unique facets of the GTWM is that there are no seams or 'breaks' in the plastic optical fibre, which circumnavigates the garment from top to bottom. This pioneering contribution represents a significant breakthrough in textile engineering because, for the first time, a full-fashioned garment has been *woven* on a weaving machine. With this innovative process, there is no need for the conventional cut and sew operations to produce a garment from a two-dimensional fabric.

Although it started off as a 'textile engineering' endeavour, the research has led to an even more groundbreaking contribution with enormous implications: the creation of a wearable integrated information infrastructure that has opened up entirely new frontiers in personalized information processing, healthcare, space exploration, etc. Therefore, there is an exciting and unique opportunity to explore this new paradigm on two major fronts, i.e. mobile wearable information processing systems (MWIPS) and vital signs monitoring systems (VSMS) that can not only lead to a rich body of new knowledge but in doing so, enhance the quality of human life. The two fronts should eventually converge and give rise to a generation of personalized mobile information processing systems (PMIPS) with embedded intelligence that can sense, adapt and respond to the needs of the wearer and the environment. Thus, the Wearable Motherboard, with its truly universal human interface of a garment, can serve as the integration framework for the realization of affective and invisible computing.

Today, when a microwave oven is used, the individual is totally hidden from or unaware of the microprocessor built into the oven. Likewise, research in

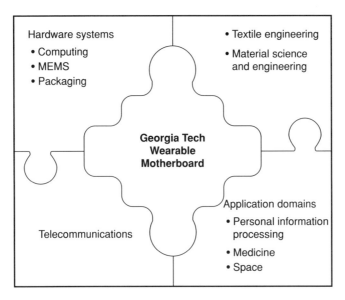

13.6 An interdisciplinary research approach to personalized mobile information processing.

PMIPS should lead to smart clothing where the 'intelligence' is embedded into the clothing, and the user can harness the required information processing capabilities without being an expert in the domain of computer hardware or software. Thus, research must explore in depth the promise of the Wearable Motherboard technology, while simultaneously engineering the transformation from an innovative concept whose feasibility has been conclusively demonstrated to a robust system with a multitude of real-world applications including mobile personal information processing, multimedia-rich computing, medicine and space exploration.

13.5.1 Need for interdisciplinary research

Research in this area must expand on the symbiotic relationship between art, science and engineering pioneered in the development of the GTWM, where the fine art of weaving has been skilfully blended with scientific principles and engineering design to create a unique and innovative structure/system that is of practical significance to humankind. Furthermore, the design and development process involved a convergence of several technical areas – sensor technologies, textiles, materials, optics and communication – making it an ideal example of a multidisciplinary approach to problem-solving. This successful paradigm must be applied during the research for the realization of PMIPS. As illustrated in Fig. 13.6, it must bring together researchers from the complementary disciplines essential for exploring the new paradigm:

- textile engineering and materials science and engineering
- computing hardware, electronics packaging and microelectromechanical systems (MEMS)
- telecommunications
- application domain: medicine, space, information processing, networking, etc.

Such an interdisciplinary approach will not only lead to effective solutions but also to significant breakthroughs in the field.

13.5.2 A research roadmap

We will now identify key areas of research in ARTS, exploring the new paradigm that will lead to the realization of PMIPS.

13.5.2.1 Sensor development – design and development of wearable, interconnectable sensors/microchips

In the current generation of the GTWM, sensors for heart rate monitoring are affixed to the body and connected to the T-connectors on the garment. This could be uncomfortable/difficult for certain users such as invalid persons, burn victims and infants with sensitive skin. There is a need for a sensor that can be integrated into the structure and held against the user's body using the form-fitting component of the GTWM. New sensors that can be integrated into the GTWM must be identified, designed and developed. The sensors must be rugged enough to withstand laundering and 'field use', and/or should also be easily detachable from the garment. The sensors must also be interconnectable, compact and have a high degree of reliability. This research calls for collaboration between the fields of textile engineering, materials science and engineering, microelectronics, MEMS and biomedical engineering.

13.5.2.2 Interconnection technology development

The interconnection technology used in the present version of the GTWM is manual and represents the first attempt at realizing such interconnections in flexible textile materials. There is a need for an interconnection technology that can provide precise, rugged and flexible interconnections using an automated process suitable for mass production. An interconnection technology for mounting microchips on the GTWM and technology for coupling optical fibres to sensors (as opposed to interconnecting them) should also be developed. This research calls for collaboration between the fields of textile engineering, microelectronics, electronic packaging and material science.

13.5.2.3 Wireless communications technology development

Wireless transmission capability is critical for the effective use of the GTWM in the various applications (see Table 13.1) so that the data can be received, decoded and analysed, and appropriate action taken. For example, if the patient recovering at home from heart surgery is wearing the smart shirt, the EKG (electrocardiogram) needs to be transmitted to the hospital on a regular basis. This monitoring will help the patient feel 'secure' and will facilitate recuperation while simultaneously reducing the costs and time associated with recovery. Moreover, in the event of an emergency, the doctor can be notified instantaneously, leading to prompt and effective treatment. Since the current version of the GTWM does not have wireless transmission and reception capability, wireless technologies such as those used in cellular phones, radio modems and baby monitors must be integrated into the GTWM. Important factors that would determine the selection of this technology are cost, the ability to interface with existing computer hardware, bandwidth, speed and size. The technology must be capable of providing bidirectional, multimodal multimedia communications between the sensors, cameras and microphones mounted on the GTWM and remote monitoring stations. This research calls for collaboration between the fields of microelectronics, telecommunications and textile engineering (to ensure seamless integration of the communications device with the garment).

13.5.2.4 Development of decoding/image compression software

There is a need for software that can decode the transmitted data, e.g. vital signs, and display it in a meaningful format. The software should also have the ability to record the data for later analysis. The key considerations in the selection of this software are cost, the ability to run on multiple platforms and a user-friendly interface. The software system must also analyse the decoded data and automatically raise an alarm if the vital signs are beyond certain previously defined thresholds. The important factor that would be considered is the ability to automatically alert emergency medical personnel using telephones, pagers or the Web. This research calls for collaboration between the fields of telecommunications, electrical engineering and medicine.

13.5.2.5 Development of new materials

The required set of properties for the materials used in the GTWM (comfort fibres, conducting fibres, optical fibres, etc.) will depend on the application, the sensor suite, etc. For example, chemical resistance will be necessary for fibres used in fire-fighting applications; however, chemical resistance is not very important when it comes to using these fibres in baby clothes. Likewise, to build or embed 'memory' into clothing – that is analogous to the computer's memory – there is a need for research in fibres and materials that can provide such characteristics eventually leading to 'on-off' capabilities being embedded in the fibres themselves. As yet another example, there is a need for fibres that can integrate properties of stretch, comfort and conductivity in the same single fibre. Research in this area calls for collaboration between the fields of textile engineering and materials science and engineering.

13.5.2.6 Development of next generation ARTS using MEMS devices

ARTS represent the new class of textile garments that can sense the vital signs of the individual wearing the garment (say, the GTWM), analyse the data using built-in intelligence and provide a suitable response based on the analysis. For example, some individuals are susceptible to anaphylaxis reaction (an allergic reaction) when stung by a bee, and need a shot of epinephrine (adrenaline) immediately to prevent serious illness or even death. Therefore, there is a critical need for research that could lead to the development and incorporation of (1) appropriate sensors on the GTWM to detect the anaphylaxis reaction (or a diabetes shock), (2) 'adaptive' mechanisms that can monitor the wearer's vital signs and create a response, and (3) a built-in feedback mechanism, e.g. a MEMS device, that can effect the responsive action (administer an injection). This research calls for collaboration between the fields of MEMS, medicine and textile engineering.

Thus the research roadmap presented in this section can be utilized to build on the Wearable Motherboard concept and technology to create adaptive and responsive textile structures that will pave the way for PMIPS.

13.6 Conclusion

The Jacquard weaving machine was the precursor to today's powerful computers. Similarly, the Wearable Motherboard is a versatile and mobile information infrastructure that can be tailored to the individual's requirements to take advantage of the advancements in telemedicine and information processing and thereby significantly enhance the quality of life. Moreover, the Wearable Motherboard technology provides a strong foundation or platform for further advancements in the area of personalized mobile wearable information processing systems that will lead to pervasive computing. In short, clothing can indeed have the third dimension of 'intelligence' embedded into it and spawn the growth of individual networks or personal networks where each garment has its own IN (individual network) address much like today's IP (internet protocol) address for information processing devices. When such IN garments become the in thing, personalized mobile information processing will have become a reality for all of us!

Acknowledgements

The authors would like to thank Dr Eric Lind of the US Department of Navy, Mr Don O'Brien of the US Defense Logistics Agency and Dr Rick Satava of DARPA for identifying the need for a soldier protection system and for providing the funds to carry out this research under Contract N66001-96-C-8639. They also thank the Georgia Tech Research Corporation for providing matching funds for the research. Thanks are due to Ms Sharon Abdel-Khalik, RN, at Crawford Long Hospital, Atlanta, Georgia, and Dr Robert Gunn, MD, and his colleagues at the Department of Physiology, Emory University School of Medicine, Atlanta, Georgia, for their help in testing the heart rate monitoring capabilities of the smart shirt. Thanks are also due to Dr Gary Freed, Director of the Emory Egleston Apnea Center, for his insight and input on the problem of SIDS. They also thank Dr Bob Graybill of DARPA for providing funds under Contract F30602-00-Z-0564 to carry out research on the PMIPS paradigm presented in this chapter. Finally, the authors would like to acknowledge the contributions of Dr Chandramohan Gopalsamy and Dr Rangaswamy Rajamanickam to the initial development of the Wearable Motherboard™.

References

1 Jayaraman S, 'Designing a textile curriculum for the '90s: a rewarding challenge', *J. Textile Inst.*, 1990, **81**(2), 185–94.
2 Gopalsamy C, Park S, Rajamanickam R and Jayaraman S, 'The Wearable Motherboard™: the first generation of adaptive and responsive textile structures (ARTS) for medical applications', *J. Virtual Reality*, 1999, **4**, 152–68.
3 The Georgia Tech Wearable Motherboard™: the intelligent garment for the 21st century, http://vishwa.tfe.gatech.edu/gtwm/gtwm.html.
4 Akao Y, *Quality Function Deployment*, Productivity Press, Cambridge, MA, 1990.
5 Anon, *Proceedings of the DLA/ARPA/NRaD Sensate Liner Workshop*, Columbia, South Carolina, USA, April 11, 1996.
6 Anon, *Proceedings of the Pre-Proposal Sensate Liner Workshop*, Phoenix, Arizona, USA, June 27, 1996.
7 Rajamanickam R, Park S and Jayaraman S, 'A structured methodology for the design and development of textile structures in a concurrent engineering environment', *J. Textile Inst.*, 1998, **89**(3), 44–62.
8 Park S, Gopalsamy C, Rajamanickam R and Jayaraman S, 'The Wearable Motherboard™: An information infrastructure or sensate liner for medical applications', *Studies in Health Technology and Informatics*, IOS Press, 1999, **62**, 252–8.
9 Giri P, 'The smart shirt', Special Issue of *LIFE Magazine, Medical Miracles for the Next Millennium*, October 1998, p. 65, New York, NY.

14
Wearable technology for snow clothing

HEIKKI MATTILA

14.1 Introduction

Not so long ago, terms like 'intelligent textiles' and 'smart clothing' started to appear in seminar presentations and academic discussions. The aim seemed to be to develop textile materials which have temperature regulating, electroconductive and other advanced properties. Wearable computer research was focusing on how to link computer hardware to clothing, glasses and other accessories worn by the user. The armed forces both in the USA and Europe were carrying out their own projects in this field.

In 1998, representatives from two Finnish universities and some industrial companies discussed future research projects and recognized smart textiles to be both challenging and interesting. The problem was how to convince private and public sponsors to allocate enough funds for such a new and rather sci-fi research area. Therefore, an extraordinary approach was selected. Let's first make a prototype. By combining electronics and other unusual devices of existing technology to a garment, we could prove that such cross-scientific research and development is possible and worth funding in the future.

For the cross-scientific approach, a network of four university departments and four industrial companies was formed. The Institute of Electronics and the Institute of Textiles from Tampere University of Technology, the Institute of Industrial Arts and the Institute of Textile Design from University of Lapland, the snowmobile suit manufacturer Reima-Tutta Oy, compass and navigating systems producer Suunto Oyj, heart rate monitor producer Polar-Electro Oy, and DuPont Advanced Fibre Systems were the participants. In addition, Siemens and Nokia Mobile Phones assisted with GSM (global system for mobile communication) communications. Altogether, more than ten researchers worked on the project.

The snowmobile suit was selected as the prototype. A snowmobile is a vehicle used for work and leisure in harsh arctic winter conditions. The driver may be alone or in a group in remote areas. His health and even his life may depend on the vehicle and his clothing. In the event of an accident, he might need help fast. Reima-Tutta Oy is one of the world's leading snowmobile suit manufacturers. Thus, the team had expert knowledge of conventional snowmobile clothing and the problems associated with it.

Work started in the autumn of 1998 with a brainstorming phase which ended in the spring of 1999. A product design phase followed and the prototype was tested in Northern Lapland during the winter of 1999–2000. Maximum publicity was one of the objectives of the project. A well-planned press release and a show were organized for spring 2000, and the show was filmed by more than ten international TV stations. Finally, the experiences from the prototype and the smart clothing philosophy were displayed in the Finnish pavilion at the Expo 2000 World Fair in Hanover.

The budget for the project was approximately one million US dollars, half of which came from TEKES, the National Technology Agency, which is the main financing organization for applied and industrial R&D in Finland. The other half was paid by Reima-Tutta Oy, the snowmobile suit manufacturer. For a firm with an annual turnover of 40 million dollars, this was its biggest ever investment into a single R&D project. A management board was formed for the project, with representatives from each party. The board closely followed the work of the research team and offered advice and comments when needed.

14.2 Key issues and performance requirements

The targeted user for the suit was defined as an experienced snowmobile driver who knows how to move around in an arctic environment and who has basic first-aid skills. There were several problems that the suit should solve in the case of an emergency. Firstly, in case the person gets lost he must be able to locate himself and to know which way to go in order to reach a road or village. Also, the rescue units must be able to locate him if he needs help. He must have access to local weather forecasts as well as know the times of sunrise and sunset. In case his snowmobile breaks down, he must be able to walk away in deep snow or survive long enough to be rescued. In case he falls through ice into water, he must be able to get out of the water and to survive in sub-zero temperatures long enough to be rescued. If he has an accident and falls unconscious, the suit should send an automatic distress signal, together with enough information regarding his condition for the rescue units to know how quickly he must be found. Furthermore, the suit should provide shelter and make life as comfortable as possible in the arctic climate.

The objective was to design and build as many different features into the prototype as possible. The electronics and other devices should be part of the garment, rather than being inserted into pockets and picked out when needed. All wires and other devices must be hidden and the garment must look, inside and out, like a clothing product. The product must also be comfortable to wear. Besides the suit, underwear also had to be designed, in order to facilitate heart rate monitoring.

The maximum weight for the suit was agreed to be 4.5 kg. All the devices must work in $-20\,°C$ for 24 hours. Furthermore, everything should be operational if the person perspires or even if he falls into water. The user interfaces must be designed so that they can be used without removing heavy gloves or if the fingers are numb from cold. The product must be washable in a normal washing machine.

Budget and time limitations meant that no basic research could be done and the inventing of totally new devices should be kept to a minimum. This forced the research team to use existing technology, but applications could be new.

14.3 The prototype

The purpose of the suit itself is to keep the user warm and dry. A snowmobile suit is usually a two-piece garment consisting of high waistline pants and a jacket with hood. The suit must be warm and weathertight, breathable but watertight. All seams are usually taped for this purpose. Durable and watertight material is normally used at the seat and knee areas. Special shock-absorbing padding may be used in the knee, elbow and shoulder areas. The team selected a breathable material, which lets the humidity out but prevents water from getting inside. For comfort and safety, the suit should maintain thermal equilibrium. The possibility of using phase change materials (PCMs) was analysed. PCMs are materials containing, either in the fibre or in the coating, microcapsules with a substance which changes phases from solid to liquid and back, according to a certain temperature range. In this process, heat is either released or absorbed, and a more equal temperature is maintained within the wearer's personal microclimate. Materials like Outlast and Gore-Tex were selected for this purpose. In case of emergency, once the distress signal has been sent, the rest of the energy could be used for the signal light and for heating critical areas of the body. Heating, however, requires a lot of energy and in a practical environment it may not be feasible to use the portable energy sources for such a purpose.

Several sensors were used for monitoring the body functions and movements of the user. Sensors monitoring heart rate were attached to the underwear. The rest of the sensors were attached to the suit itself. There are six sensors for measuring the outside temperature and four for measuring the inside temperature. The temperature differences are used for analysing where and in which position the user is, i.e. is he on his back in the snow or is he just lying down by a fire? All the electronic devices were attached to the suit itself, including the processor and communications network, with interfaces in different parts of the garment for connecting the necessary components. The user interface and interface for recharging the batteries would also be part of the suit. This modular solution makes it possible to have different versions, thus giving the consumer the option to buy only the devices he needs. Also, only the user interface, the interface for power recharging, wiring and other interfaces must be washable, while the rest of the components can be removed. Even so, they must be resistant to shock, water and bending.

Sensors monitoring the heart rate had to be placed close to the skin. They were sewn permanently to the underwear. By monitoring the heart rate, it is possible to tell, for example, how much pain the user is suffering. Body temperature was measured in three different places. Other sensors were used for monitoring temperatures outside and inside the suit. By measuring the acceleration forces, it was possible to determine whether the user had been in collision with something, whether he is still moving or what position he is lying in. This information was gathered by an accelerometer developed by Polar Electro Oy. The CPU unit containing the programs needed for operating the whole suit and analysing various features was placed inside the jacket at the back of the user.

The global positioning system (GPS) uses several orbiting satellites for radio navigation, providing the user with the ability to pinpoint his exact position anywhere in the world. By receiving signals from a minimum of four satellites simultaneously, it is possible to determine the latitude, longitude and altitude of the user very accurately. The GPS system was installed into the central processing unit. The user could not only locate himself, but he could also be found by others. A compass with a GOTO arrow was included, with direction and distance to the desired destination also displayed. However, special development work had to be carried out in order to make the GOTO arrow operational, even if the person is standing or lying still. In a normal GPS, the GOTO arrow works only if the user is moving.

GSM was selected for data transfer. Instead of using a mobile telephone, the Siemens Cellular Engine M20 was installed as part of the central processing unit. This device is capable of sending and receiving short messages.

The power supply was the tricky part. After checking all the possibilities, from solar panels to utilizing static electricity, conventional batteries had to be selected. The smallest battery to guarantee power for 24 hours would weigh 600 to 700 g. The battery is rechargeable either at home or from the snowmobile during the drive.

The objective was to design a single user interface for all the electronics in the garment. The interface had to be part of the suit and usable with the left or right hand without removing gloves. This eliminated any solid or embroidered keyboards. In fact, nothing existed that fulfilled the requirements, and therefore it had to be designed for this project. The research team came up with

14.1 The user interface.

a brilliant idea, a winding palm-size display unit attached with a cable to the chest of the garment. When the unit is pulled from the chest, the menu in the display changes accordingly. There are three steps, each about 5 cm apart. At the first step with the unit nearest the wearer, the display shows the time, heart rate, power level, the level of GSM signal and coordinates. The next step shows the second menu, and so on. To operate the system, it is only necessary to squeeze the unit, as shown in Fig. 14.1.

In addition to electronics, other survival equipment was developed. These features are displayed in Fig. 14.2. It is quite usual to drive over frozen lakes and rivers with a snowmobile, and sometimes the ice gives way. Two awls attached to the sleeves will help the user to get out of the water and back on to the ice. At the back of the garment there is an airtight sack, and by getting into

Wearable technology for snow clothing

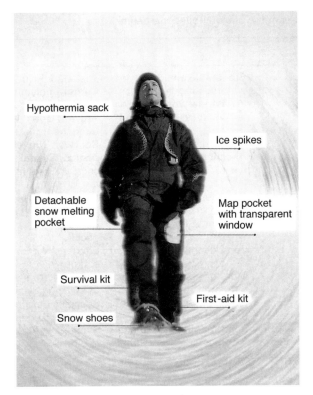

14.2 Survival features in the suit.

the sack the user can prevent hypothermia for as long as 4 hours. One of the thigh pockets is detachable and fireproof, and can be used for melting snow over a fire. In such situations, to avoid hypothermia, it is important to drink liquid. Waterproof matches, collapsible snow shoes, and a first-aid kit are also provided. A summary of all features is presented in Table 14.1.

In case of an emergency, the wearer of the suit can send a distress signal by squeezing the user interface for 30 seconds. The rescue team can confirm the signal back to him and also read his coordinates. Through various sensors, the suit monitors the condition of the user, for example, heart rate, g-forces, temperature, pain, hypothermia, and so on. If a critical situation is diagnosed by the software, a distress signal will be sent automatically. The interface, however, warns the user first and he has 30 seconds to cancel the signal if necessary. In addition to just the signal and coordinates, the rescue team will be able to see whether the person moves around, and in what position he is lying, whether he is in the water, his body temperature, the outside temperature, etc.

Table 14.1 Features of the snowmobile smart suit

Problem	Solutions	Technology
Cold and wet environment	Breathable and waterproof outer fabrics PCM underwar for temperature regulation	Taslan, polyamide, Cordura polyamide fabrics Aracon, Gorix, Outlast underwear materials
Disorientation	Location by GPS Direction and distance to desired destination	Rockwell-Jupiter GPS system, Suunto electronic compass, real time GOTO arrow
Fall through ice into water, hypothermia	Get into hypothermia sack to maintain heat Melt snow for drinking water Send distress signal	Hypothermia sack, warming underwear Detackable fireproof pocket Siemens M20 GSM engine
Accident, health problem, unconsciousness	Monitor heart rate, body temperature, outside temperature, g-forces and body movement Send manual or automatic distress signal	Polar Electro heart rate monitoring system Specially designed Hitachi CPU unit Siemens M20 GSM engine
Snowmobile breaks down	Collapsible snow shoes	
Local weather conditions	Weather Service Finland report	Siemens M20 GSM
How to operate the features with one hand and gloves on	Winding, palm size display unit attached by a cable to the chest of the garment with changing menu according to distance. Operated by squeezing the unit	PIC processor controls power management Electronic compass Patented Yo-Yo user interface

14.4 Conclusions

The prototype snowmobile suit developed within this project is a smart garment. Cross-scientifically, electronic and textile innovations were combined, and the objectives were achieved. The suit works as planned. It is not likely that the prototype costing 1 million dollars to create will be commercially exploited in its exact shape and form. However, much was learned. Several innovations, including the user interface, were patented and will be commercially applied in the future. The garment manufacturer Reima-Tutta Oy launched snowboard clothing with electronic devices in the autumn of 2000.

Some of the objectives proved to be difficult to achieve at the prototype stage. For example, to make the garment fully washable could not be done. Only the wiring and interfaces could be treated so that it is possible to wash the product in a normal washing machine. All the other devices, including CPU, GSM engine, user interface, etc., had to be removed. What was, however, achieved was to guarantee that the suit and all electronics would remain fully operational even if the user falls into water. Gorix is a fabric that can be heated electrically. The idea of using extra power for heating critical body areas would have worked well except that the batteries used for operating the electronics could maintain heating levels for only a very short period. Therefore, this idea had to be abandoned.

The project received more publicity than expected. Magazine articles and programmes by ten European TV stations, including *BBC World*, spread the word worldwide. This has convinced the participating companies and universities that there is a demand for smart clothing. Similar ventures by Philips, StarLab's I-Wear and the haute couture fashion house Oliver Lapidus confirm this. As a result, Reima-Tutta Oy has set up a wearable intelligence laboratory called Clothing + with ten researchers. The lab will work as part of a network connecting the other firms and universities that participated in the original project. The objective will be to create new innovations for adding intelligent features to apparel products.

15
Bioprocessing for smart textiles and clothing

ELISABETH HEINE AND HARTWIG HOECKER

15.1 Introduction

The application of enzymes in food processing, in the paper and leather industries, as additives in washing powders, and in the desizing process of cotton is well established. However, biocatalysis has also entered textile processing. Enzymes, biocatalysts with specific and selective activity, are today produced by biotechnological processes in great amounts and constant quality, and are therefore applicable to large-scale processes.

In view of the new applications resulting from the design of enzymes for specific processes, there is a demand for extensive collaboration between bio- and textile chemists.

Besides natural protein fibres, like wool and silk, and natural cellulose fibres, like cotton, flax and hemp, synthetic fibres are also targets for biocatalysed processes. In cotton finishing, chemical processes are widely substituted by enzyme-catalysed processes. In addition to the well-known biostoning and biofinishing, attributes like 'used look' and 'modified handle' are realized by enzymatic finishing. Moreover, there is potential for replacing the alkaline scouring in cotton pretreatment by the use of enzymes like, for example, pectinases. Catalases are added to destroy the residual peroxide in bleaching baths, to make easy reuse of the liquor possible, leading to an environmentally friendly and cost-effective process. In wool finishing, enzymes, mainly proteases, are used to achieve shrinkproofing. The properties of wool textiles like handle, whiteness and lustre are modified by enzyme-catalysed reactions as well. In early wool processing stages, like raw wool scouring and carbonization, the prospects of enzyme application are assessed. Furthermore, bioprocesses are described leading to pilling reduction and dyeability improvement. The degumming of silk, traditionally performed by the aid of soap, alkali or acid, is achieved by proteases. To improve the quality and consistency of flax, microbial or dew-retting is replaced by a specific enzymatic retting. Moreover, the dyeability of flax is improved by the enzyme-catalysed degradation of pectic substances without damaging the cellulose components. Hemp is enzymatically modified with regard to crystallinity, accessibility and 'pore structure'. Via the controlled enzyme-catalysed fibrillation of lyocell fibres, the so-called 'peach-skin' effect is produced. There is a broad range of applications and a multitude of prospects for the use of enzymes in textile processing, leading to a positive impact on the environment. This chapter surveys recent developments in the field of enzymatic textile processing and discusses the advantages and limitations of these finishing processes.[1]

The use of enzymes in food processing, in the paper and leather industry, as additives in washing powders, and in the desizing process of cotton is well established. In the meantime, enzymatic processes have been developed, which aim at modifying the appearance and handle of textiles made of wool and cotton.

Enzymes are biocatalysts with selective and specific activity, accelerating distinct reactions and remaining unchanged after the reaction. From an ecological and economical point of view, the moderate reaction parameters of enzyme-catalysed processes and the possibility of recycling enzymes are particularly attractive. Today, enzymes are produced by biotechnological processes in great amounts and constant quality, thus allowing the use of enzymes in large-scale processes. Advances in the field of genetic engineering enable enzyme manufacturers to design an enzyme for a special process, e.g. with regard to temperature stability or pH optimum. The design of an enzyme for a special purpose requires an understanding of the catalytic action of the enzyme on a specific substrate, i.e. in the case of natural fibre substrates, the designer of an enzyme process needs a certain knowledge of wool and/or cotton morphology, of the effect of a special enzyme on the fibre components, and consequently on the properties of the fibre material as a whole. Furthermore, to evaluate the enzymatic process, the results of the enzymatic treatment are to be compared to the results of usual chemical processing.

The first enzymatic process in textile finishing was the desizing process using amylases. Many areas of textile finishing have been conquered since then. Today, prospects lie in the field of developing new durable press finishes for cotton, e.g. by cross-linking, in the field of effluent colour removal, and in the field of composting synthetic fibres.[2] Due to the proteinic nature of the enzymes, safety in handling enzymes is often questioned because repeated inhalation of protein material can cause allergic reactions in some individuals. It is important to notice that there is no evidence to indicate that enzyme allergies are developed through skin contact.[3] Enzymes can be safely handled by using safe product design, by engineering controls, by safe working practices, and by appropriate personal protective equipment. In product design, enzyme aerosols and powdered formulations should be avoided, whereas granular (with low dusting capabilities) and liquid formulations (with low mechanical action or in closed vessels) can be recommended. The market potential of enzymes is considerable. A market study says that there should be growth from US$ 350 million in 1992 to US$ 588 million in 2000, and that the biggest potential concerns new applications in the paper, chemical and pharmaceutical industries and in waste treatment.[4]

15.2 Treatment of wool with enzymes

15.2.1 Morphology of wool

Wool as a complex natural fibre being mainly composed of proteins (97%) and lipids (1%) is an ideal substrate for several enzyme classes (e.g. proteases, lipases). The wool fibre consists of two major morphological parts: cuticle and cortex. The cuticle is composed of overlapping cells surrounding the inner part of the fibre, the cortex. The latter is built up of spindle-shaped cortex cells which are separated from each other by the cell membrane complex.[5] The cuticle is subdivided into two main layers: exo- (with a- and b-layer) and endocuticle, and an outermost membrane called the epicuticle, causing the Allwoerden reaction of wool fibres treated with chlorine water.[6] One important component of the cuticle is 18-methylicosanoic acid.[7] In a model of the epicuticle drawn up by Negri et al.,[8] this fatty acid is bound to a protein matrix to build up a layer. According to Negri et al.,[9] this layer, referred to as the F-layer,[10] can be removed by treating wool with alcoholic alkaline or chlorine solutions, thus enhancing wettability. Another important characteristic is the cross-linking of the exocuticle, e.g. the a-layer contains 35% cystine residues. In addition to the normal peptide bonds, the cuticle is cross-linked by isodipeptide bonds, ε-(γ-glutamyl) lysine.

The hydrophobic character of the a-layer, in particular, caused by the large amount of disulfide cross-links and the bound lipid material, is the origin of the diffusion barrier, e.g. for dye molecules. Therefore, the composition and morphology of the wool surface is primarily modified in fibre pretreatment processes.

15.2.2 Heterogeneous reactions – the catalytic action of enzymes on wool and cotton

Using wool or cotton as substrates for enzyme-catalysed reactions, a special type of enzyme kinetics is followed. In the heterogeneous system of the soluble enzyme and the solid substrate enzyme, diffusion plays a more decisive role than in a homogeneous system where both enzyme and substrate are soluble. In the heterogeneous reaction, the kinetics depend not only on the concentration of the reaction partners, and on the temperature and the pH value of the liquor; as additional parameters, the diffusion of the enzyme to and into the

Bioprocessing for smart textiles and clothing 257

solid phase of the substrate and the diffusion of the reaction products out of the solid phase into the liquor have to be considered.[11]

The reaction products, e.g. peptides in the case of wool and oligosaccharides in the case of cotton, when diffusing out of the fibre, act as a substrate in the liquor. Thus, part of the enzymes is bound to the soluble substrate in the treatment bath.

The diffusion of the enzyme from the liquor into the fibre (wool) resembles the diffusion of a dyestuff. The following steps are to be considered:

1 Enzyme diffusion in the bath
2 Adsorption of the enzyme at the fibre surface
3 Diffusion from the surface into the fibre interior
4 Enzyme-catalysed reaction

The complex structure of natural fibres, especially of wool, complicates enzymatic fibre modification. Enzymes like proteases and lipases catalyse the degradation of different fibre components of a wool fibre, thus making reaction control difficult.

Once having diffused into the interior of the fibre, proteases hydrolyse parts of the endocuticle and proteins in the cell membrane complex, thus, if not controlled, leading to a complete damage of the wool fibre. Thus, at least for some applications, it is desirable to restrict the enzymatic action to the fibre surface, e.g. by immobilizing the enzyme.[11]

The results of treating wool with proteolytic enzymes will be unpredictable in the absence of detailed knowledge of (a) the processing history of the substrate and (b) how specific process conditions affect subsequent enzymatic treatments. To elucidate this point, the effects of adsorbed ionic species on enzyme/substrate adsorption and reaction kinetics were studied.[12]

For cotton, the restriction of the enzyme to the fibre surface is easily achieved, because cellulose, being a highly crystalline material and comprising only small amounts of amorphous parts, makes the diffusion of enzymes into the interior of the cotton fibre nearly impossible. Thus, by regulating enzyme dosage and choosing the type of enzyme, the catalytic action of the enzyme is confined to the surface of cotton and to the amorphous regions, leaving the fibres as a whole intact.

15.2.3 Shrinkproofing

One of the intrinsic properties of wool is its tendency to felting and shrinkage. There are different theories concerning the origin of wool felting. The hydrophobic character and the scaly structure of the wool surface are the main factors causing the differential frictional effect (DFE) that results in all fibres moving to their root end when mechanical action is applied in the wet state.[13]

Therefore, shrinkproofing processes aim at the modification of the fibre surface either by oxidative or reductive methods, and/or by the application of a polymer resin onto the surface. The most frequently used commercial process (the chlorine/Hercosett process) consists of a chlorination step followed by a dechlorination step and polymer application.[13] The chlorination results in the oxidation of cystine residues to cysteic acid residues in the surface of the fibre, and allows the cationic polymer to spread and adhere to the wool surface. Chlorination produces byproducts (AOX) which appear in the effluent and ultimately may generate toxicity in the whole food chain by being taken up by aquatic organisms. There is therefore an increasing demand for environmentally friendly alternatives.

Taking into account the problems related to the conventional antifelting process mentioned above, it becomes obvious that most of the enzymatic processes concern the development of an alternative method for shrinkproofing. The requirements for an enzymatic process were discussed by Haefely;[14] the antifelting effect should be achieved without application of a synthetic resin, only 'soft chemistry' should be applied and the whole process should be environmentally friendly, producing no harmful substances – a premise that is not yet fulfilled in all of the processes using enzymes as fibre-modifying agents. In some of the earliest enzyme finishing processes, wool was pretreated by gas chlorination (Chlorzym process[15]) or by H_2O_2 (Perzym process[16]) prior to incubating the fibres with papain, the plant derived protease, and bisulfite. These processes resulted in a complete removal of the cuticle cells. Because of the high prices of the enzymes used and the non-tolerable weight loss of the fibres, these early combined enzymatic processes never achieved an industrial scaling-up.

The major part of enzymatic processes published in the last few years also comprises combined processes. In 1983, a process was described for rendering wool shrink resistant, which completely descaled the fibres.[17] The treatment used potassium permanganate ($KMnO_4$) as a pre-oxidizing agent and a subsequent proteolytic treatment, and gave a certain pilling resistance for wool fabrics. Not only the early ones, like the Chlorzym process, but also current enzymatic processes still include the use of chlorinating agents. Inoue[18] described a three-step process: the first step comprises the application of a mixture of papain, monoethanolamin hydrosulfite and urea, the second step a treatment with dichloroisocyanuric acid, and the last step again an enzyme treatment resulting in a reduced area shrinkage of the fabric treated in that way. Connell et al.[19] combined a protease pretreatment either with a wet chlorination or with an oxidative treatment with sodium hypochlorite and potassium permanganate and additional polymer application to achieve a reduced area shrinkage of the fabric.

Not only chemical but also physical pretreatment processes are combined with the enzyme treatment of wool. In a process patented by Nakanishi and Iwasaki,[20] a low temperature plasma was applied to the fibres prior to a treatment with polymeric shrinkproofing agents. The enzyme was used to remove fibres protruding from the surface of the fabric, thus achieving a higher softness, too. Ciampi et al.[21] combined a protease treatment with a heat treatment in saturated steam, and Fornelli and Souren[22] described the use of high frequency (HF) radiation on enzyme-treated material.

In 1991, the Schoeller Superwash 2000 treatment of wool was reported by Schindler.[23] The treatment is a three-step process consisting of a so-called black-box pretreatment, an enzyme treatment and the application of an AOX-low polyamide resin. In the process reported by Aulbach[24] (also referred to as the Schoeller process) a 'certain minimum chemical pretreatment' was described as necessary for the enzyme process. In this process, a superwash-standard is achieved without the application of a resin. Fornelli[25,26] described the use of a resin as 'still imperative' to achieve a sufficient antifelting effect in the enzyme process reported ('BIO-LANA') because by the 'enzymatic filing' of the 'scaly epicuticle' of the wool fibre, only a certain, but insufficient, degree of antifelting was obtained. This was confirmed by Riva et al.,[27] who reported the role of an enzyme in reducing wool shrinkage. The *Streptomyces fradiae* protease used is described to be 'effective in reducing wool shrinkage' but not to 'confer the desirable shrink-resistant levels for severe machine-washed wool'. The shrinkage results were not improved by the additional use of sodium sulfite. Proteases were applied either after oxidative treatment using hydrogen peroxide or plasma treatment. In both cases, the felting resistance was improved significantly by the enzyme post-treatment.[28] The shrinkage of wool is reduced after protease treatment, but concerning the efficiency of a protease treatment alone it was stated that Basolan DC treatment, for example, was still superior.[29] In view of this, the question was raised how much shrinkproofing wool would need.[30] A practical approach was reported, showing that the degree of felting of enzyme-treated wool washed in household laundry machines is comparable to chlorinated wool.

The common characteristic of the so far reported processes is the application of proteolytic enzymes. In the following section, a survey of processes is given where other enzyme classes are applied.

The enzymatic treatment for shrinkproofing wool described by King and Brockway[31] is one of the few enzymatic processes working without the aid of a chemical or physical pre- or post-treatment. In this application, the enzyme PDI (protein disulfide isomerase), which rearranges disulfide bonds, is applied to washed material constituting wool. The PDI-treatment resulted in a non-shrinking material in contrast to the non-treated, misshaped one. Ogawa et al.[32] described the application of the enzyme transglutaminase for shrinkproofing wool. The enzyme introduces additional cross-links into the substrate, leading to an improved shrinkage resistance, anti-pilling and hydrophobicity 'without impairing the texture' of the material.

15.2.4 Handle modification

Softness plays an essential role in qualifying textile products. Consequently, a further goal beside the antifelting effect in the pretreatment and processing of wool fibres is the production of a 'cashmere-like' handle. It is the aim of the processes to modify wool fibres to reduce prickle and to enhance softness and lustre, thus improving the characteristics of wool as a basic material for high quality textiles. Prickle is caused by wool fibre endings that are stiff enough to stimulate nerve endings situated directly below the skin's surface. Therefore, a reduction of the bending modulus is aimed at. Besides that, fibre diameter and fabric construction play decisive roles concerning handle. Thus, most of the wool treatments to improve handle attempt to reduce the fibre diameter, e.g. by the complete descaling of the wool.[33] The descaling is performed by pretreating the fibres with $KMnO_4$ and ammonium sulphate, acetic acid and bisulfite, and a subsequent treatment with a proteolytic enzyme. Descaling was also achieved by the application of a heat-resistant neutral protease resulting in a 'cashmere-like' feel.[34] The combined use of the chlorinating agent dichloroisocyanurate and a proteolytic enzyme was also reported to improve handle properties.[35] In contrast to the descaling of wool, some investigations follow another attempt to improve handle. Benesch[36] described the complete removal of degraded or damaged portions of wool fibres, not of the cuticle alone, by a protease treatment, followed by a rinse with formic acid and by the application of a softener. A cashmere-like handle was also achieved by treating wool with sodium dichloroisocyanurate, neutralizing, incubating with papain and steaming at 100 °C.[37] Wool/cotton blends were treated by cellulases and proteases, resulting in a soft feel of the material.[38]

By carefully controlled treatments with proteolytic enzymes, it was possible to reduce the buckling load of wool yarns. It was shown that softness is improved and the subjectively perceived prickle of fabrics is reduced.[39] The new yarn-bundle compression test discriminates between wool samples that differ by about 1 μm in mean fibre diameter.

15.2.5 Whiteness and lustre

Another important factor in wool fibre pretreatment is the enhancement of whiteness and lustre. Bleaching of wool is necessary, especially when dyeing in pastel shades is desired. Using proteolytic enzymes alone[40] or in combination with H_2O_2,[41] the degree of whiteness and the hydrophilicity of the fibres were increased compared to the sole oxidative treatment.[42] Whiteness in wool bleaching is enhanced by the presence of a protease with both peroxide bleaching and using dithionite or bisulphite.[43] The treatment is accompanied by losses in weight and strength.

15.2.6 Dyeing behaviour

The suitability of enzymes for improving the dye uptake of wool was studied by integrating the enzymatic process in a pretreatment step prior to dyeing,[28] and by using them as auxiliary agents in wool dyeing.[44] It was shown that bulky dyestuffs are taken up more equally and in higher amounts by the enzyme-treated samples.[45] The fastness of dyed wool samples to artificial light is enhanced after enzyme treatment.[44]

15.2.7 Carbonization

Beside wool itself, natural soilings like vegetable matter (grass seeds, burrs) and skin flakes can be enzymatically modified. If not completely degraded and removed from the textile goods, vegetable matter and skin residues will lead to uneven dyeing and printing. The vegetable matter is normally removed by carbonizing, a process where wool is impregnated with sulfuric acid and then baked to char the cellulosic impurities. The residuals are then crushed and extracted from the wool as carbon dust by brushing and suction. Research work has been done to replace carbonizing by the use of enzymes like cellulases and ligninases, with the aim of reducing wool fibre damage and effluent load, and to save energy.

Some patents concern the enzymatic decomposition of plant constituents in wool. Sawicka-Zukowska and Zakrzewski[46] reported the removal of plant impurities from wool by the use of hydrolases, lyases and oxidoreductases. The amount of sulfuric acid used for carbonization was reduced by the action of cellulolytic and pectinolytic enzymes.[47] After having incubated wool with cellulases, burr removal became easier by weakening the cohesion between burr and wool.[48] No chemical or physical damage to the wool was observed after mechanical removal of the enzyme-treated burrs. Brahimi-Horn et al.[49] gave a survey of the use of enzymes on a range of model compounds for wool grease and on vegetable matter. The treatment of wool by a mixture of cellulases, pectinases and ligninases did not impair wool fibres. Liebeskind et al.[50] produced commercially unavailable lignin peroxidases with the aim of degrading especially the lignin of the burrs. The process of lignin degradation is a long-term process and in the time recorded (24 h) H_2O_2 was added several times. The wool was not attacked by this procedure, but neither were the burrs markedly modified. The 'Biocarbo' process was introduced, applying pectolytic and cellulytic enzymes by padding and washing in an acidic enzyme bath.

After drying, vegetable residues are removed by carding or beating; the method is recommended for wools containing less than 3% vegetable matter.[51] In contrast to this, the removal of vegetable matter from wool by using a trash tester (USTER MDTA 3) was not enhanced after treatment with hydrolytic enzymes (pectinases, hemicellulases, cellulases).[52] There are prospects for the use of oxidative enzymes in this field.

15.2.8 Silk

The degumming process of silk as another representative of natural proteinaceous fibres is normally performed by chemical treatment using soaps, alkali, acids or water. By this process, sericine, the amorphous protein glue, is removed from the silk thread liberating two highly crystalline fibroine fibres. Besides chemical degumming, the proteolytic removal of sericine is also successfully applied.[53] It was shown that the application of ultrasound accelerates the degumming process when using the proteolytic enzyme papain, whereas in the case of Alcalase™ (NOVO Nordisk A/S) and trypsin, the beneficial effect of the physical treatment is lower.[54]

15.2.9 Summary

Of all the processes described, only a few are 'pure' enzymatic processes, i.e. enzyme processes without any pre- or post-treatment. In two processes, other enzyme classes than proteases are used: transglutaminase[32] and protein disulfide isomerase.[31] Proteases are used to cut off the damaged fibres[36] or to achieve certain texturing effects.[38,55] The combined processes using proteases include the additional application of a resin[19,20,23,25,26] where the enzyme is used as a pre- or post-treatment. Oxidative[20–22,33,35,37,56] and physical pretreatments[20–22] are also combined with the application of enzymes. Then, emphasis is put on softening or improving the handle, followed by dyeing acceleration,[57–59a,c] bleaching[40,41] and pilling resistance.

In the combined processes, the cuticle surface is modified by the respective pretreatment removing lipids and cleaving disulfide bridges and, by the proteolytic post-treatment, enhancing the number of hydrophilic binding sites in the fibres. Both pre- and post-treatment lead to electrostatic repulsion of the fibres, to an enhanced degree of fibre swelling and to an improved dye uptake. If the fibre modification is successfully performed by the use of enzymes alone, either other enzyme classes than proteases are applied[11,31,32] or proteases are used not to modify the fibre itself but to completely degrade and remove fibres having been damaged in the course of earlier processing steps.

The combined enzymatic processes, including chemical pretreatment and especially chlorination steps, are not real alternative processes because AOX is still produced, even though in reduced amounts.

By the complete descaling of the fibres on the one hand both lustre and handle are enhanced and feltability is reduced. On the other hand, the fibre cortex is being exposed, thus leading to a weakening of the fibre. Regarding the fibre damage by enzymatic, especially proteolytic action, it can be stated that in order to achieve the desired effect, either the enzymatic action has to be controlled, e.g. diffusion control by enzyme immobilization,[59b] or the enzyme has to be specially 'designed', e.g. by genetic engineering. By the latter processes, a new enzyme is created or a usual enzyme is modified in such a way that only a distinct part of the target substrate is altered. Fornelli describes enzymes suitable for wool finishing as 'intelligent' or 'magic'.[25,26,42] Actually, the user of enzymes takes advantage of the enzyme's inherent characteristics of specificity and selectivity, the enzyme under use being either native or especially designed for a distinct process.

15.3 Treatment of cotton with enzymes

15.3.1 Morphology of cotton

Cotton is built up of cellulosic and non-cellulosic material. The outermost layer of the cotton fibre is the cuticle covered by waxes and pectins, followed inwards by a primary wall built up of cellulose, pectins, waxes and proteinic material.[60] The inner part of the cotton fibre consists of the secondary wall subdivided into several layers of parallel cellulose fibrils and the lumen. The smallest unit of the fibrils is the elementary fibril, consisting of densely packed bundles of cellulose chains.[61] In longitudinal direction, highly ordered (crystalline) regions alternate with less ordered (amorphous) regions. The non-cellulosic material is composed of pectins, waxes, proteinic material, other organic compounds and minerals. Additionally, sizes and soilings are added to cotton fibres. The whole material adhering to the fibre is up to 20% of the fibre weight. The non-cellulosic material is situated in or on the primary wall, the outer layer of a cotton fibre making up 1% of the fibre diameter. The secondary wall, which amounts to 90% of the fibre weight, is mainly composed of cellulose.

The enzymatic hydrolysis of cotton is performed by cellulase, being composed of at least three enzyme systems working together synergetically.[62,63] Endo-β(1,4)-glucanases (1) hydrolyse chains of native cellulose, thus degrading structures of low crystallinity and producing free chain endings. Cellobiohydrolases (CBHs) (2) degrade cellulose from the chain end, liberating cellobiose, which is hydrolysed by β-(1,4)-glucosidase (3) to glucose units. The most important cellulase-producing organisms are fungi of the genus *Trichoderma*, *Penicillium* and *Fusarium*.

Pectin is the generic term for polysaccharides like galacturonans, rhamnogalacturonans, arabinans, galactans and arabinogalactans. The hemicelluloses, consisting of xylans, glucomannans and xyloglucans, build up the other group of non-cellulosic material. In cotton, pectins mainly consist of α-(1,4)-bound polygalacturonic acid, the carboxylic groups being esterified with methanol. Pectinolytic enzymes are polygalacturonases cleaving the α-(1,4)-glycosidic bonds of pectin.

15.3.2 Enzymatic processing of cellulosic fibres

The enzymatic treatment of cotton can be subdivided into three major topics. The first one concerns the cellulosic part of the cotton and consequently the enzymes used are mainly cellulases. The effects achieved by these enzymes are pilling reduction, increase in softness, and amelioration of handle, surface structure and fabric appearance. These effects are summarized as 'biopolishing', a term created by the Danish company NOVO Nordisk. In the second application of enzymes in cotton finishing, the stone-wash process of denim material is replaced by the use of cellulases. By the enzyme treatment, the amount of stone material required can be reduced or even completely replaced and, unlike the stone-wash process, the enzyme treatment can be prolonged without damaging the textile material. The third area of application concerns the removal of natural fibre-adhering material (pectin) as a preparatory procedure for the subsequent processing steps. Pectinases, or a combination of pectinases and cellulases, are the enzymes used for degradation of the pectin material. The influence of interfering factors like surfactants, electrolytes and dyestuff materials on the quality of biofinish processes has been monitored via weight loss determination.[64] Cellulase treatment is not just for cotton but also for synthetic cellulosic fibres like lyocell/tencel and rayon, where it improves softness, drapeability, pilling tendencies and post-laundry appearance. On cellulose acetate, it is described as giving little benefit.[65] Using enzymes for the modification of lyocell in fibre blends, the enzyme product always has to be adjusted to the needs of the accompanying fibre.[66]

15.3.3 Biopolishing

The softening effects achieved by enzymatic treatment of cotton textiles were already patented 20 years ago.[67] The treatment did not lower the tensile strength while softening the textile. In 1981, a process was patented which comprises swelling of the cellulosic fibres with sulfuric acid, followed by a treatment with cellulases to give a modified fabric with soft handle.[68] The first real cellulase treatment of cellulosic fibres was published in 1988.[69] The process led to a weight reduction of the fibres, and a weight loss of 3–5% was evaluated as the optimum to obtain a soft handle and a better surface appearance. A combined process for cotton softening was patented in 1990.[70]

Bioprocessing for smart textiles and clothing 265

A low-temperature plasma was applied prior to the cellulase treatment. The effect was measured by the determination of the tear strength. By treating cotton fabrics with cellulases, e.g. from *Trichoderma reesei*, an improved softness and fabric appearance and a high dyeing yield are achieved, the weight loss of the fibres remaining low.[71] Seko et al.[72] described the use of cellulases produced by bacteria of the genus *Bacillus* on a cotton knit, enhancing softness and hygroscopicity and preserving excellent fibre tensile strength. In 1992, Pedersen et al.[73] reported the 'biopolishing' effect. The basis of biopolishing is a partial hydrolysis of the cellulosic fibres,[74] resulting in a certain loss of tensile strength. The optimum effect was achieved, with a loss in strength of 2 to 7%. The short fibre ends emerging from the fabric surface were enzymatically hydrolysed but an additional mechanical treatment was necessary to complete the process, i.e. to remove the fibres normally leading to pilling. The surface appearance of the fabrics treated in that way was improved. The effect was permanent, i.e. even after several machine washings, the textiles remained almost entirely without pills. In contrast to the softeners applied to the fibre surface, the water regain was not decreased by the enzymatic treatment.[75a,b,c] The process parameters and the commercial name of the enzyme used were disclosed by Bazin et al.[76] In 1993, a biopolishing process was patented by Videbaek and Andersen[77] With a surface application of the enzymes instead of batch processing, the strength loss of the fibres was minimized in the softening treatment of towelling.[78] The treatment of cotton/wool blends with cellulases and proteases resulted in fabric softening.[38] Cellulase-treated cotton fabrics were compared to alkali-treated polyester fabrics with 5% weight loss each; in both cases bending rigidity was decreased.[79] Clarke introduced an enzymatic process to reduce pilling, especially from garment dyed goods made of cotton or wool.[80] In the case of wool, a reduction reagent was added. The mechanism of cellulose degradation was described by Almeida and Cavaco-Paulo[81] The enzymatic hydrolysis of the cellulose mainly occurs in the amorphous regions. Consequently, a pretreatment decreasing the degree of crystallinity of the cellulose will enhance the enzymatic hydrolysis rate, e.g. mercerized cotton will be more prone to enzymatical modification than non-pretreated cotton. The authors define the softness of a fabric: the mechanical properties of a soft fabric are characterized by low bending and shearing stiffness and a high degree of elongation. A long-term enzymatic treatment exhibits a negative effect causing fibre damage and a decline of handle properties. Koo et al.[82] reported the use of cellulases on cotton dyed with different dye classes. A fabric dyed with direct or reactive dyes seemed to inhibit the cellulase action, documented by the fact that the higher the dye concentration on the fibres, the lower the weight loss of the fibres after enzyme treatment. In contrast, the catalytic action of cellulases applied to fabrics dyed with vat dyes was not inhibited (this is an important fact concerning the biostoning of cotton referred to later in this article). The authors also confirm that the fuzziness of the garments was reduced by degradation of the fragments of cotton fibrils, and as a consequence the treated fabrics showed less colour fading after laundering. Because of its lower degree of polymerization as compared with cotton, rayon is more sensitive to losses of strength when treated with cellulases.[83]

Generally it is accepted that short treatment time and surface-sensitive mechanical impact lead to a targeted surface modification of cotton, whereas continuous installations without liquor turbulence are unsuitable for this purpose.[84] Mechanical agitation during enzyme treatment affects not only tactile and aesthetic qualities but also thermal comfort performance.[85] Considering the variety of dyeing machines and methods that involve very different modes and levels of agitation, the effect of mechanical agitation on the quality of the cellulase-treated textile goods is of high practical interest. It was elucidated that endoglucanase activity increased at high agitation levels, thus leading to a higher risk of fabric strength loss.[86] Besides mechanical agitation, fabric processing history and fabric construction are as important parameters as the choice of enzyme composition and concentration.[87]

The components of the cellulase enzyme system were purified and analysed for their impact on cotton modification. Regarding cotton samples with same weight loss levels, either due to the catalytic action of cellobiohydrolase (CBH I) or endoglucanase, more strength loss was observed in the case of the endoglucanase-treated sample, but at the same time, positive effects on bending behaviour and pilling were achieved.[88] Compared to an endoglucanase-enriched preparation, the use of whole cellulase is more effective regarding surface fuzz removal, whereas the endo-enriched enzyme is less aggressive and causes less fibre damage, thus being more suitable for delicate knits.[89] Enzymatic hydrolysis of cotton fabrics dyed with direct dyes leads to weight losses equal to that of undyed fabrics, whereas on cotton dyed with reactive dyes, enzymatic degradation seems to be inhibited.[90]

15.3.4 Biostoning

Many casual garments are treated by a washing process to give them a 'worn appearance'. A well-known example is the stone-washing of denim jeans; the blue denim is faded by the abrasive action of pumice stones. In the 'biostoning' process, cellulases (neutral or acidic) are used to accelerate the abrasion by loosening the indigo dye on the denim.[91] The process is highly environmentally acceptable because of replacing or reducing the amount of stones, thus protecting the machines and avoiding the occurrence of pumice dust in the laundry environment. The process has already found broad acceptance in the denim-washing industry. Tyndall reported the use of cellulases in combination with a mechanical action for improving softness and surface appearance.[92] Enzymatic treatment led to a worn appearance of textile goods dyed with 'indigo, sulfur, pigments, vats or other surface dyes'. In this treatment, a certain degree of weight and strength loss has to be accepted. A broad survey of the denim finishing programme was given by Ehret,[93] describing the indigo dyeing process, the stone-washing process and its substitution by the cellulase treatment. To gain the desired 'old look' of denim material, it is important to minimize back staining, being highest at pH 5. Ehret stated that cellulases hydrolysing cellulosic material in the neutral pH range therefore, are preferably used for jeans finishing. Zeyer et al.[94] drew up an empirical model based on the observations made during the enzymatic decolourization of cellulose fabrics. The authors conclude that fibre surface friction plays an important role in the enzymatic decolourization of cotton. They state that mechanical action opens the outermost layers of the cellulosic crystal, thus increasing the enzyme-accessible part of the cellulose and allowing the enzymatic removal of the dye.

Indigo backstaining during biostoning was studied. It was concluded that backstaining increases with treatment time and enzyme concentration. Using different enzymes, the backstaining is low in case of a neutral cellulase (pH 6–8) and high in case of an acid cellulase (pH 4.5–5.5).[95]

15.3.5 Bleach cleanup

After bleaching cotton with H_2O_2, the bleaching liquor cannot be used for the next treatment step, e.g. dyeing, because of the oxidative effect of the residual H_2O_2. The degradation of this residual H_2O_2 in the bleaching bath by the enzyme catalase makes replacement of the treatment liquor or the washing of the goods unnecessary. Thus the same liquor can be used for the next processing step, leading to a saving of time, of waste water, and of energy.[96] The process of removing residual peroxide from cotton material was evaluated not only from an ecological but also from an economical point of view, with the result that, by the application of catalase the amount of fresh water used for rinsing, cooling and heating is reduced and waste water is saved, leading to total cost savings of 6 to 8% per year.[97]

15.3.6 Bioscouring

Effective dyeing and printing of textiles require uniform adsorbency. Furthermore, no disturbing amounts of dirt, sizes or adhering natural material should remain on the fibres. Sixty percent of all the problems occurring in the dyeing and finishing of cotton originate from a wrong or uneven pretreatment.[98]

Therefore, the pretreatment is the most important step in cotton finishing. Consequently, the adhering natural material and the sizes have to be removed. Surface waxes produce a water-repellent fibre surface. Seed husks, pectins, hemicelluloses and sizes fix part of the dyestuff and therefore cause uneven dyeing. Furthermore, when using sodium hypochlorite as a bleaching agent, this material, if not removed,[99] leads to the formation of AOX in the effluent. Therefore, prior to the bleaching process an alkaline boiling off is performed to reduce the amount of cotton-adhering material.

Regarding the enzymatic removal of natural material-adhering cotton, Bach and Schollmeyer examined the degradation of the cotton pectin by pectinolytic enzymes.[100] The pectin was either extracted from the cotton and used as a substrate in homogeneous solution, or degraded directly on the cotton surface in heterogeneous reaction. The degree of degradation was determined by measuring the amount of reducing groups being released into the solution. The consequences of this degradation were published by Bach and Schollmeyer in 1993.[101] The degree of whiteness after enzymatic degradation of the pectin was lower compared to alkaline boiling-off. However, combining the enzyme treatment (a simultaneous treatment of pectinase and cellulase) or the alkaline boiling-off with an alkaline H_2O_2 bleaching, the total degree of whiteness was higher in combination with the enzyme treatment. Roessner[102] confirmed that the degree of whiteness of a cotton sample only treated with enzymes was lower by 8–10% than the degree of whiteness of an alkaline boiled-off material. The effect of the enzymatic treatment was also determined by measuring the wettability of differently pretreated fabrics. The differences were documented by the TEGEWA drop test. Desized fabrics were either enzyme or alkali treated. Differences in wettability occurring in this treatment step were removed by an additional peroxide bleaching process. The combined process, including an enzyme treatment, delivered results comparable to the alkali treatment.

Factors influencing scouring are the nature of the substrate, the kind of enzyme used, the enzyme activity, the use of surfactants and mechanical impact.[103] It was observed that, during pectinase scouring, much less wax was removed compared with the alkaline scouring. If the treatment was combined with surfactant treatment, results equivalent to alkaline scouring could be achieved.[104] A water treatment at 100 °C is reported to increase the effectiveness of the subsequent scouring of cotton fabric with a combination of pectinase, protease and lipase, whereas the use of these enzymes alone showed very little effect.[105] A new pectinase (BioPrep™, NOVO Nordisk A/S) was screened that is stable under alkaline conditions, i.e. the optimum conditions for removal of cotton-adhering substances. Pectinase action does not create full wettability alone. Surfactants, Ca-binding salts, emulsifiers and high treatment temperatures complete the removal of the Ca-pectate/wax complex.[106]

The relative colour depths of caustic soda-scoured and environmentally friendly bioprepared fabrics do not show significant differences.[107]

The effect of scouring with enzymes was compared to conventional caustic soda treatment and solvent extraction. Whereas caustic soda treatment resulted in the highest deterioration on a molecular level, but led to a high level of whiteness, the solvent-extracted samples showed superior tensile strength and the bioscoured samples the best softening effect.[108]

15.3.7 Linen

Linen is the most sensitive fibre concerning treatment with cellulases. It has been worked out that mono-component cellulases are necessary to limit the enzyme treatment to the desired effects like enhanced handle or used look on pigment-dyed or printed textiles. To improve dyeing and wrinkle recovery, a major proportion of the pectic substances has to be eliminated without damaging the cellulosic part of the fibres. It is described that the control of this reaction is difficult in practice.[109] The quality of flax fibres can be enhanced via controlled enzyme retting, thus minimizing over-retting and reduced fibre strength.[110,111]

15.3.8 Hemp

Hemp fabrics were modified by using cellulase, hemicellulase and cellulase in combination and cellulase plus β-glucosidase. The hydrolysis rate and product properties like tensile strength, crystallinity and pore structure were investigated. It was shown that mechanical agitation has the greatest impact on the fabrics when applied during the enzyme treatment. The largest total porosity and the highest number of small pores was achieved after a treatment using cellulase alone.[112]

15.3.9 Lyocell

Lyocell is a synthetic cellulosic and a high-fibrillating fibre. The high fibrillation rate leads to interesting handle and look, although dyeing and finishing are more difficult due to the higher risk of local damage.[84] During processing, a primary fibrillation occurs due to hydro-mechanical action. After that stage the cellulase treatment is performed. By the catalytic action of these enzymes, fibrillated fibres are removed. In a subsequent secondary fibrillation step, short fibrils that cannot connect to form pills are produced, thus leading to a change in handle and appearance (peach skin). Endoglucanase-enriched enzyme preparations produce superior fabric handle and minimize seam damage.[113]

15.3.10 Summary

Most of the work done in the field of enzymatic cotton processing concerns the reduction of pilling and fuzzing of cotton fabrics by the catalytic action of cellulases. Furthermore, casual wear is produced by the catalytic action of cellulases in the biostoning process, replacing stone-washing.[93,94] In some of the processes, the enzyme treatment is combined with mechanical action to enhance the accessibility of the cotton substrate.[92] Another approach is the combination of a plasma treatment with the enzyme treatment.[70] A lot of research work has been done in the field of enzymatic scouring of cotton, with the aim of replacing the alkaline boiling-off.[101,102] The kinetics of enzymatic pectin degradation was investigated to study and optimize the process.[100] Bast fibres like flax and hemp, and synthetic cellulosic fibres like lyocell are also targets for enzyme-catalysed processes and have already been successfully modified.

15.4 Enzymatic modification of synthetic fibres

Polycaprolactone fibres were modified by the catalytic action of a lipase. The fibres were not stretched because the enzyme-catalysed degradation declines with rising degrees of stretch. After enzyme hydrolysis, fibres in unstretched form show irregular structures, whereas stretched fibres show longitudinally oriented structures.[114] Lipases improve wetting and absorbency of polyester fabrics. Compared to the reduced strength and mass from alkaline hydrolysis, the enzyme-treated fabrics showed full strength retention. The wettability effect proved to be due to hydrolytic action rather than protein adsorption.[115]

15.5 Spider silk

Spider dragline silk is stronger than steel and has a tensile strength approaching that of Kevlar. It is remarkable in its extremely high elasticity.[116] This unusual combination of high strength and stretch renders this material extremely attractive for researchers. Spider's dragline silk is mainly constituted of glycine (42%) and alanine (25%), and the remainder is predominantly built up of glutamine, serine and tyrosine.[117] Poly-alanine of 5 to 10 residues builds up a β-sheet and accounts for most of the crystalline fraction (30%). The crystalline domains are bonded via glycine-rich regions (β-turns) and are embedded in amorphous regions.[118] In humid surroundings the fibres supercontract, achieving 60% of their former length. It is the aim of research work to produce spider silk proteins biotechnologically in requested quantities and process them to fibres in industrial scale.[119] There are two main tasks within this field. First, the cloning and recombinant expression of spider silk proteins and second, to elucidate fundamentals and requirements for the processing of spider silk proteins to fibres. Spider silk analogues have been expressed in bacterial cells[120] but the overexpression of larger protein segments has not been achieved up to now.[119] Model substances like degummed natural silk from the mulberry silkworm *Bombyx mori* were used for solubility studies and for testing a laboratory scale spinning device to produce filaments from polymer solutions of small volume.[119] The future will show if appropriate technological parameters, possibly including biotechnical processing, will be feasible.

15.6 'Intelligent' fibres

Textiles contributing to thermal regulation via the incorporation of 'phase change' materials are related to wear comfort, especially in sports and leisure wear. Perspectives are opening up for these materials also in the field of medical textiles[121] and protective clothing with shape memory for fire brigades and racing and petrol pump attendants is under development.[122] The latter contains shape memory materials in a layer that at certain high temperatures contribute to the formation of an insulating layer by returning to their original shape. In this way, it protects the human from being overheated. The perspectives for biotechnical processing in the field of specialty fibres ranges from the enzyme-catalysed functionalization of fibres to the inclusion and thus immobilization of enzymes in or on those fibres.

15.7 Conclusions

The use of enzymes in cotton finishing has found much broader acceptance than the use of enzymes in wool finishing. Enzymatic processes are already well established in the cotton industry. The terms biofinishing and biopolishing are not only advertising slogans but also stand for ecologically acceptable processes. Compared to cotton fibres, wool is a more complex fibre material. The composite structure and the accessibility even of the bulk part of the fibres complicate the restriction of the enzyme treatment to the fibre surface. The complete degradation of fibres or fibre ends from textiles by enzymatic treatment leads to the desired effects of lustre and softness in cotton finishing. In wool finishing, single fibres have to be modified to achieve, for example, the antifelting effect, soft handle and lustre. Therefore, reaction control plays a more important role in wool finishing due to the possibility of enzymes diffusing into and damaging the wool fibres. Wool fibres are often pretreated by chemical or physical means prior to the enzyme treatment, to enhance the accessibility of the cuticle, to shorten the treatment time and to restrict the enzyme to the fibre surface.

Hence, the design of specialized enzymes reacting with only one specific component of the fibre and/or the production of diffusion-controlled enzymes might be a solution to develop future biotechnical processes for textile finishing. Thus, not only optimization of process parameters like pH value, temperature, ionic strength and knowledge on technological items like mechanical impact due to different machine devices, but rather completely engineered enzymes will lead us to 'tailor-made' smart products and processes. There is therefore a need for close cooperation between textile and bio-chemists that will lead to new enzymes and new fibres. As one example, there are vast perspectives in the field of 'intelligent' textile materials. In view of this, it should never be forgotten that enzymes should not be used for the sake of enzymes. However, it has been shown in many fields that ecology and economy profit from intelligent enzymatic processes.

The major advantage of enzymatic processes is the possibility of using conventional technology already existing in textile plants. Enzyme formulations should be applied in solution, to avoid dust formation and to reduce the known allergizing potential of protein material when inhaled. A heat treatment is sufficient to stop the enzymatic action irreversibly. Thus the transfer of an enzymatic process developed on laboratory scale into the textile industry should be possible without great delay.

Acknowledgements

We gratefully acknowledge the 'Society of Dyers and Colourists' for granting the permission to reproduce parts of the article cited as Ref. 1.

References

1. Heine E and Hoecker H, 'Enzyme treatments for wool and cotton', *Rev. Prog. Col.*, 1995, **25**, 57–63.
2. Etters J N and Annis P A, 'Textile enzyme use: a developing technology', *Am. Dyestuff Reporter*, 1998, **5**, 18–23.
3. Enzyme Technical Association, Washington, 'Safe handling of enzymes', *Text. Chem. Col. & Am. Dyest. Rep.*, 2000, **32**(1), 26–7.
4. Anon., Seifen, Oele, Fette, Wachse, 1994, **120**(9), 549–50.
5. Fraser R D B, Gillespie J M, MacRae T P and Marshall R C, Int. Rep. CSIRO, Div. Prot. Chem., Parkville, Australia (1981).
6. Allwoerden K, *Z. ang. Chem.*, 1919, **32**, 120.
7. Kalkbrenner U, Koerner A, Hoecker H and Rivett D E, 'Die internen Lipide der Cuticula von Wolle', *Schrift DWI*, 1990, **105**, 67.
8. Negri P, Cornell H J and Rivett D E, 'A model for the surface of keratin fibers', *Text. Res. J.*, 1993, **63**(2), 109.

9 Negri A P, Cornell H J and Rivett D E, 'The modification of the surface diffusion barrier of wool', *JSDC*, 1993, **109**, 296A.
10 Leeder J D and Rippon J A, 'Changes induced in the properties of wool by specific epicuticle modification', *JSDC*, 1985, **101**, 11.
11 Heine E, PhD Thesis, Aachen, 1991.
12 Shen J, Bishop D P, Heine E and Hollfelder B, 'Some factors affecting the control of proteolytic enzyme reactions on wool', *J. Text. Inst.*, 1999, **90**(1), No. 3, 404–11.
13 Makinson R, *Shrinkproofing of Wool*, New York, Basel, M Dekker Inc., 1979.
14 Haefely H R, 'Enzymatische Behandlung von Wolle', *Textilveredlung*, 1989, **24**(7/8), 271.
15 Anon., 'Shrink-resist processes for wool Part II Commercial methods', *Wool. Sci. Rev.*, 1960, **18**, 18.
16 Otten H G and Blankenburg G, 'Ueber die Verfahren zur Antifilzausruestung von Wolle', *Z. Ges. Textilind*, 1962, **64**, 503–9.
17 Kurashiki Spinning Co. Ltd, Japan, 'Production of descaled animal fiber', *Jpn. Kokai Tokkyo Koho*, 6pp, JP 58144105 A2 830827.
18 Inoue Y, 'Modification of animal fiber', *Jpn. Kokai Tokkyo Koho*, 5pp, JP 61266676 A2 861126.
19 Connell D L, Palethorpe H A, Szpala A and Thompson A P, 'Method for the treatment of wool', Eur. Pat. Appl., 16pp, EP 358386 A2 900314.
20 Nakanishi T and Iwasaki K, 'Shrink proofing method for animal hair fiber product', *Jpn. Kokai Tokkyo*, 4pp, JP 04327274 A2 921116.
21 Ciampi L, Haefeli H R and Knauseder F, 'Enzymatic treatment of wool', PCT Int. Appl., 16pp, WO 8903909 A1 890505.
22 Fornelli S and Souren I, 'Process for the treatment of textile material', Eur. Pat. Appl., 6pp, EP 530150 A1 930303.
23 Schindler P, *Textil-Revue*, 1991, **45**, 12.
24 Aulbach M J, 'Prozeß- und Produktoptimierung in der Wollausruestung Teil II: Die Filzfrei-Ausruestungsanlage fuer Wollkammzuege', *Melliand. Textilber.*, 1993, **11**, 1156.
25 Fornelli S, 'Enzymatic treatment of protein fibres – state-of-the-art biotechnology', *Internat. Dyer*, 1993, **10**, 29.
26 Fornelli S, 'Eine Art von IQ fuer Enzyme – Enzymatisches Behandeln von Proteinfasern', *Melliand. Textilber.*, 1994, **2**, 120.
27 Riva A, Cegarra J and Prieto R, 'The role of an enzyme in reducing wool shrinkage', *JSDC*, 1993, **109**(5/6), 210.
28 Heine E, 'Neue biotechnologische Produkte fuer die Wollveredlung', *DWI Reports*, 1999, **122**, 179–85.
29 Jovancic P, Jocic D and Dumic J, 'The efficiency of an enzyme treatment in reducing wool shrinkage', *J. Text. Inst.*, 1998, **89**(1), 390–401.
30 Breier R, 'Lanazym – Rein enzymatische Antifilzausruestung von Wolle von der Idee zur erfolgreichen Umsetzung in die Praxis', *DWI Reports*, 2000, **123**, 49–62.
31 King R D and Brockway B E, 'Treatment of wool materials', Eur. Pat. Appl., 4pp, EP 276547 A1 880803.
32 Ogawa M, Ito N and Seguro K, 'Modification of zootic hair fiber', *Jpn. Kokai Tokkyo Koho*, 4pp, JP 03213574 A2 910918 Heisei.

33 Kondo T, Sakai C and Karakawa T, 'The process for modifying animal fibers', Eur. Pat. Appl., 16pp, EP 134267 A1 850320.
34 Anon, *Text Horizons*, 1992, **4/5**, 7.
35 Saito T and Kawase M, 'Method for modifying animal hair fiber structure', *Jpn. Kokai Tokkyo Koho*, 4pp, JP 04174778 A2 920622 Heisei.
36 Benesch A, Ger. Offen., 10pp, DE 2123607 711125.
37 Unitika Ltd, Japan, 'Modification of wool fiber', *Jpn. Kokai Tokkyo Koho*, 3pp, JP 57071474 A2 820504 Showa.
38 Wiedemann A, 'Stable, aqueous cellulase and protease compositions', UK Pat. Appl., 14pp, GB 2258655 A1 930217.
39 Bishop D P, Shen J, Heine E and Hollfelder B, 'The use of proteolytic enzymes to reduce wool-fibre stiffness and prickle', *J. Text. Inst.*, 1998, **89**, 546–53.
40 Goddinger D, Heine E, Mueller B, Schaefer K, Thomas H and Hoecker H, Proc. Eur. Workshop 'Tech. for Env. Protection', 31.1.–3.2., 1995, Bilbao, Spain, p. 72.
41 Cegarra J, Gacen J, Cayuela D and Bernades A, *IWTO Meeting*, March, 1994, New Delhi, Report No. 3.
42 Fornelli S, 'Bio-Lana und Magic-Enzyme', *Textilveredlung*, 1992, **27**, 308.
43 Levene R, 'Enzyme-enhanced bleaching of wool', *JSDC*, 1997, **113**(7–8), 206–10.
44 Riva A, Alsina JM and Prieto R, 'Enzymes as auxiliary agents in wool dyeing', *JSDC*, 1999, **115**(4), 125–9.
45 Heine E, Hollfelder B and Hoecker H, 'Enzymatische Modifizierung von Wolle', *DWI Rep.*, 1995, **114**, 101–7.
46 Sawicka-Zukowska R and Zakrzewski A, 'Method of removing solid plant impurities from fibres and textiles in particular from butry wool', Pol, 3pp, PL 147498 B1 890831.
47 Sedelnik N, Latkowska L and Sawicka-Zukowska R, 'Verfahren zum Reinigen eines Rohgewebes von vegetabilischen und zellulose-haltigen Verunreinigungen', *Ger. Offen.*, 10pp, DE 3543501 A1 860619.
48 Zhu Pin Rong and Zhou Jing Hua, 'Burr-removal by cellulase', *8th Int. Conf. Wool. Text. Res.*, 1990, **3**, 195.
49 Brahimi-Horn M C, Guglielmino M L, Gaal A M and Sparrow L G, 'Potential uses of enzymes in the early processing of wool', *8th Int. Conf. Wool. Text. Res.*, 1990, **3**, 205.
50 Liebeskind M, Hoecker H, Wandrey C and Jaeger A G, 'Strategies for improved lignin peroxidase production in agitated pellet cultures of *Phanerochaete chrysosporium* and the use of a novel inducer', *FEMS Microbiol. Lett.*, 1990, **71**, 325–30.
51 Sedelnik N, 'Application of the mechanical and of the biological methods to remove vegetable impurities from wool', *Fibers and Textiles in Eastern Europe*, 1998, **6**(1), 39–41.
52 Heine E et al., 'Enzymkatalysierter Abbau von Vegetabilien in Wolle', *DWI Reports*, 2000, **123**, 475–9.
53 Gulrajani, M L, 'Efficacy of proteases on degumming of dupion silk', *Ind. J. Fibre Text. Res.*, 1998.
54 Krasowski A, Mueller B, Foehles J and Hoecker H, 'Entbastung von Seide im Ultraschallfeld', *Melliand. Textilber.*, 1999, **6**, 543.
55 Wiedemann A, 'Stabiles Enzympraeparat', Ger. Offen., 4pp, DE 4226162 A1 930218.

56 Saito T and Kawase M, 'Coloring of cellulosic fiber structure containing protein fiber', *Jpn. Kokai Tokkyo Koho*, 4pp, JP 04257378 A2 920911.
57 Riva A, Cegarra J and Prieto R, 'Einfluß von Enzymbehandlungen auf das Faerben von Wolle', *Melliand. Textilber.*, 1991, **11**, 934.
58 Cegarra J, Riva A, Gacen J and Naik A, *Tinctoria*, 1992, **89**(4), 64.
59 (a) Heine E, Hollfelder B and Hoecker H, 'Enzymatische Modifizierung von Wolle', *DWI Reports*, 1995, **114**, 101.
   (b) Heine E, Berndt H and Hoecker H, 'Einsatz von Enzymen in der Wollverarbeitung und Wollveredlung', *Schrift DWI*, 1990, **105**, 57.
   (c) Heine E, Hollfelder B and Hoecker H, 'Alternative wool treatment by means of enzymes and combined processes', *9th Int. Conf. Wool. Text. Res.*, Biella Italy, 28th June–5th July, 1995, **3**, 247–54.
60 Jayme G and Balser K L, 'Beitrag zur Elektronenmikrskopie und zum Feinbau der Baumwollfaser', *Melliand. Textilber.*, 1970, **51**, 3.
61 Hess K, Mahl H and Guetter E, 'Elektronenmikroskopische Darstellung groger Laengsperioden in Cellulosefasern und ihr Vergleich mit den Perioden anderer Faserarten', *Kolloid-Z Z Polym.*, 1957, **155**, 1.
62 Lee Y H and Fan L T, *Adv. Biochem. Eng.*, 1980, **17**, 101.
63 Ryu D D R and Mandels M, *Enzyme Microbiol. Technol.*, 1980, **2**, 91.
64 Nicolai M, Nechwatal A and Mieck K-P, 'Biofinish-Prozesse in der Textilveredlung – Moeglichkeiten und Grenzen', *Textilveredlung*, 1999, **5/6**, 19–22.
65 Kumar A, Purtell C and Lepola M, 'Enzymic treatment of man-made cellulosic fabrics', *Text. Chem. Col.*, 1994, **26**(10), 25–8.
66 Breier R, 'Finishing of Lyocell fibers', *Chemiefasern/Textilindustrie*, 1994, **44/96**.Jg.11/12, 812–15.
67 Browning H R, Ger. Offen., 20pp, DE 2148278 720330.
68 Kurashiki Spinning Co. Ltd., 'Modification of cellulosic fiber', *Jpn. Kokai Tokkyo Koho*, 3pp, JP 58054082 A2 830330 Showa.
69 Yamagishi M, *Gijutsu Kako (Osaka)*, 1988, **23**(3), 6.
70 Uragami Y and Tanaka I, 'Method for improving cotton-containing fabric', *Jpn. Kokai Tokkyo Koho*, 3pp, JP 02169775 A2 900625 Heisei.
71 Clarkson K A, Weiss G L and Larenas E A, 'Methods for treating cotton-containing fabrics with cellulase', PCT Int. Appl., 108pp, WO 9206183 A1 920416.
72 Seko M, Takama M and Saito K, 'Method for modifying cellulosic fiber', *Jpn. Kokai Tokkyo Koho*, 6pp, JP 04333673 A2 921120 Heisei.
73 Pedersen G L, Screws Jr G A and Cedroni D M, 'Biopolishing of cellulosic textiles', *Can. Text. J.*, 1992, **109**(10), 31.
74 Anon, *Textile Horizons*, 1994, **14**(2), TH42.
75 (a) Pedersen G L, Screws Jr G A and Cedroni D M, *L'industrie Textile*, 1993, **6**, 53.
   (b) 'Biopolishing von Cellulosetextilien', *Melliand. Textilber.*, 1993, **74**(12), 1277.
   (c) *Textile Asia*, 1993, **24**(12), 50.
76 Bazin J, Sasserod S and Schmidt M, 'Cellulasebehandlung von Cellulosewaren mit Hilfe von Enzymen', *Text. Prax. Int.*, 1992, **47**(10), 972.
77 Videbaek Th and Andersen L D, 'A process for defuzzing and depilling cellulosic fabrics', PCT Int. Appl., 17pp, WO 9320278 A1 931014.
78 Cox Th C, 'Strength loss resistant methods for improving the softening of cotton

toweling and related fabrics', PCT Int.Appl., 28pp, WO 9313261 A1 930708.
79 Mori R, Haga T and Takagishi T, 'Bending and shear properties of cotton fabrics subjected to cellulase treatment', *Text. Res. J.*, 1999, **69**(10), 742–6.
80 Clarke D, 'Enzyme treatment for removing pills from garment dyed goods', *Internat. Dyer*, 1993, **7**, 20.
81 Almeida L and Cavaco-Paulo A, 'Weichmachen von Baumwolle durch enzymatische Hydrolyse', *Melliand. Textilber.*, 1993, **5**, 404.
82 Koo H, Ueda M, Wakika T, Yoshimura Y and Igarashi T, 'Cellulase treatment of cotton fabrics', *Text. Res. J.*, 1994, **64**(2), 70.
83 Schmidt M, Hempel W H and Wurster J, 'Treatment of cellulosic fibre materials with cellulase', *Dtsch Faerber-Kalender*, 1994, **98**, 13.
84 Breier R, 'Fashionable handle/surface dissonance by cellulase enzymes', *Melliand. Textilber.*, 1999, **5**, 411–18.
85 Radhakrishnaiah P, Menge X, Huang G, Buschle-Diller G and Walsh W K, 'Mechanical agitation of cotton fabrics during enzyme treatment and its effect on tactile properties', *Textile Res. J.*, 1999, **69**(10), 708–13.
86 Cavaco-Paulo A, Almeida L and Bishop D, 'Effects of agitation and endoglucanase pretreatment on the hydrolysis of cotton fabrics by a total cellulase', *Text. Res. J.*, 1996, **66**(5), 287–94.
87 Cavaco-Paulo A, Almeida L and Bishop D, 'Hydrolysis of cotton cellulose by engineered cellulases from *Trichoderma reesei*', *Text. Res. J.*, 1998, **68**(4), 273–80.
88 Heikinheimo L, Cavaco-Paulo A, Nousiainen P, Siika-aho M and Buchert J, 'Treatment of cotton fabrics with purified *Trichoderma reesei* cellulases', *JSDC*, 1998, **114**(7/8), 216–20.
89 Kumar A, Yoon M-Y and Purtell C, 'Optimizing the use of cellulase enzymes in finishing cellulosic fabrics', *Text. Chem. Col.*, 1997, **29**(4), 37–42.
90 Buschle-Diller G and Traore M K, 'Influence of direct and reactive dyes on the enzymatic hydrolysis of cotton', *Text. Res. J.*, 1998, **68**(3), 185–92.
91 NOVO brochure: 'Enzymes at work'.
92 Tyndall R M, 'Improving the softness and surface appearance of cotton fabrics and garments by treatment with cellulase enzymes', *Text. Chem. Col.*, 1992, **24**(6), 23.
93 Ehret S, 'Blue jeans: a comprehensive denim finishing programme', *Internat. Dyer*, 1994, **7**, 27.
94 Zeyer C, Rucker J W, Joyce T W and Heitmann J A, 'Enzymatic deinking of cellulose fabric', *Text. Chem. Col.*, 1994, **26**(3), 26.
95 Bajaj P and Agarwal R, 'Innovations in denim production', *Am. Dyestuff Reporter*, 1999, **5**, 26–38.
96 Anon, *Wirkerei- u Strickerei-Technik*, 1994, **44**(2), 138.
97 Gebhart P, Etschmann M and Sell D, 'Entfernung von Bleichmittelresten mit Enzymen bringt Vorteile', *Melliand. Textilber.*, 2000, **1–2**, 56–8.
98 Roesch G, 'Aus der Praxis der Vorbehandlung von Baumwollgeweben', *Text. Prax. Int.*, 1988, **43**(1), 61.
99 Schulz G, 'Entstehung halogenorganischer Verbindungen (AOX) beim Bleichen von Baumwolle', *Text. Prax. Int.*, 1990, **1**, 40.
100 Bach E and Schollmeyer E, 'Kinetische Untersuchungen zum enzymatischen Abbau von Baumwollpektin', *Textilveredlung*, 1992, **27**(1), 2.
101 Bach E and Schollmeyer E, 'Vergeich des alkalischen Abkochprozesses mit der

enzymatischen Entfernung der Begleitsubstanzen der Baumwolle', *Text. Prax. Int.*, 1993, **3**, 220.
102 Roessner U, 'Enzymatischer Abbau von Baumwollbegleitsubstanzen', *Melliand. Textilber.*, 1993, **2**, 144.
103 Li Y and Hardin I R, *Text. Chem. Col.*, 1998, **30**(9), 23–9.
104 Sawada K, Tokino S, Ueda M and Wang X Y, 'Bioscouring of cotton with pectinase enzyme', *JSDC*, 1998, **114**(11), 333–6.
105 Hartzell M M and Hsieh Y-L, 'Enzymatic scouring to improve cotton fabric wettability', *Text. Res. J.*, 1998, **68**(4), 233–41.
106 Lange N K, 'Biopreparation in action', *Int. Dyer*, 2000, **2**, 18–21.
107 Etters J N, Condon B D, Husain P A and Lange N K, 'Alkaline pectinase: key to cost-effective environmentally friendly preparation', *Am. Dyestuff Rep.*, 1999, **6**, 19–23.
108 Buschle-Diller G, El Mogahzy Y, Inglesby M K and Zeronian S H, 'Effects of scouring enzymes, organic solvents, and caustic soda on the properties of hydrogen peroxide bleached cotton yarn', *Text. Res. J.*, 1998, **68**(12), 920–9.
109 Dezert M H, Viallier P and Wattiez D, 'Continuous control of an enzymatic pretreatment on linen fabric before dyeing', *JSDC*, 1998, **114**(10), 283–6.
110 Akin D E, Rigsby L, Henriksson G and Eriksson K-E L, 'Structural effects on flax stems of three potential retting fungi', *Text. Res. J.*, 1998, **68**(7), 515–19.
111 Akin D E, Rigsby L and Perkins W, 'Quality properties of flax fibers retted with enzymes', *Text. Res. J.*, 1999, **69**(10), 747–53.
112 Buschle-Diller G, Fanter C and Loth F, 'Structural changes in hemp fibers as a result of enzymatic hydrolysis with mixed enzyme systems', *Text. Res. J.*, 1999, **69**(4), 244–51.
113 Kumar A and Harnden A, 'Cellulase enzymes in wet processing of lyocell and its blends', *Text. Chem. Col. & Am. Dyest. Rep.*, 1999, **1**(1), Sept., 37–41.
114 Mochizuki M, Hiramo M, Kanmuri Y, Juko K and Tokiwa Y, 'Hydrolysis of polycaprolactone fibers by lipase: effects of draw ratio on enzymatic degradation', *J. Appl. Polymer Sci.*, 1995, **55**(2), 289–96.
115 Hsieh Y-L and Cram L A, 'Enzymatic hydrolysis to improve wetting and absorbency of polyester fabrics', *Text. Res. J.*, 1998, **68**(5), 311–19.
116 Ko F K, *Text. Asia*, 1997, **28**(4), 38–40, 43.
117 Tirrell D A, 'Putting a new spin on spider silk', *Science*, 1996, **271**(1), 39–40.
118 Schulz S, 'Die Chemie von Spinnengift und Spinnseide', *Angew. Chem.*, 1997, **109**, 324–37.
119 Heinemann K, Guehrs K-H and Weisshart K, *Chem. Fib. Int.*, 2000, **50**(2), 44–8.
120 Prince J T, McGrath K P, DiGirolamo C M and Kaplan D L, 'Construction, cloning, and expression of synthetic genes encoding spider dragline silk', *Biochemistry*, 1995, **34**, 10 879.
121 Anon, *Techn. Text. Int.*, 1999, **8**(7), 23–6.
122 Congalton D, 2. Conf. Int. Text. Conf. SENAI/CETIQT Brazil, July 21–23, 1999, 1–7, 25.

16
Tailor-made intelligent polymers for biomedical applications

ANDREAS LENDLEIN

16.1 Introduction

The introduction of resorbable, synthetic suture materials represented important progress in surgery in the early 1970s. Those sutures consisted of poly(α-hydroxyacid)s like polyglycolide and the copolyesters of L,L-dilactide and diglycolide.[1,2] Originally commercialized by American Cyanamid, Ethion Inc. and Davis & Geck, these aliphatic polyesters have been successfully applied until today and have become a well-accepted standard. Due to the time-consuming and cost-intensive process for the approval of novel biomedical devices by federal administration, only a few further biodegradable polymers have reached the market since then. Polyanhydrides are an example of a group of materials which has been introduced to clinics during the 1980s.[3] Based on a polyanhydride matrix, implantable drug delivery systems like Gliadel™ (Guilford Pharmaceutical Co., Baltimore) and Septicin™ (Abbott Laboratories, Illinois) have been developed. Gliadel™ is applied for the treatment of brain cancer (glioblastoma multiforme). Gliadel™ pellets are used to fill cavities caused by surgical treatment of brain tumours and in addition to combat remaining tumour cells. With Septicin™ implants, it is possible to cure chronic bone infections.

To be degradable, a biomaterial needs bonds which are cleavable under physiological conditions. In the case of the aliphatic polyesters mentioned above, as well as the polyanhydrides, these are hydrolysable bonds. There are two mechanisms for hydrolytic degradation – bulk degradation and surface erosion.[1] Both mechanisms differ in the ratio between the rate of diffusion of water in the polymer matrix and the rate of hydrolysis of the cleavable bond. If the rate of diffusion is higher than the rate of hydrolysis, a water uptake of a few per cent, typically 1–3 wt.%, can be observed. The hydrolysable bonds within the bulk will be degraded almost homogeneously. This mechanism is called bulk degradation. For hydrophobic polymers, the rate of diffusion of water into the polymer matrix can be significantly lower than the rate of

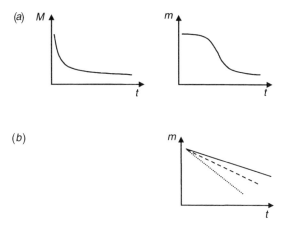

16.1 General schematic representation of the degradation behaviour of degradable polymers: (a) bulk degradation, (b) surface erosion.

cleavage of the hydrolysable bonds. Here the degradation is taking place only within a thin surface layer of the implant. In the case of surface erosion the degradation rate therefore depends on the surface area of a device. While polyanhydrides show surface erosion, polyhydroxyacids undergo bulk degradation. A general scheme of the degradation behaviour for both mechanisms is given in Fig. 16.1.

The number of potential applications for biodegradable implant materials is increasing constantly. One reason can be seen in the growing confidence of clinicians in the concept of degradable biomaterials based on the successful application of the established polymers. Another motivation can be found in completely new therapeutic methods, which have been developed taking advantage of the concept of biodegradable polymers such as tissue engineering.[4,5] Tissue engineering is an interdisciplinary approach aiming at the generation of new functional living tissue, which can be transplanted into a patient in terms of reconstructive surgery. This new tissue should be fabricated using living cells associated to a degradable porous scaffold. The scaffold should determine the three-dimensional shape of the resulting tissue, and should be degraded while the cells are growing and replacing the artificial structures.

The requirements for an implant material are determined by the respective application. The key properties of degradable biomaterials are their mechanical properties, degradation rate and behaviour, their functionality and their biocompatibility. For each application, a specific set of the properties mentioned is needed. With the growing number of potential applications, the number of required materials with specific combinations of properties is also increasing. As the variability of properties is limited for those biomaterials established, a new generation of biodegradable implant materials is demanded.

In the following section, promising candidates for the next generation of degradable biomaterials will be introduced. This class of biomaterials allows variation of macroscopic properties within a wide range by only small changes in the chemical structure. As an additional functionality, these new materials show shape memory properties.

16.2 Fundamental aspects of shape memory materials

Shape memory materials are able to memorize a second, permanent shape besides their actual, temporary shape. After application of an external stimulus, e.g. an increase in temperature, such a material can be transferred into its memorized, permanent shape. The process of programming and restoring a shape can be repeated several times. This behaviour is called the thermally induced 'one-way' shape memory effect.

The shape memory effect has been reported for different materials, such as metallic alloys,[6–10] ceramics,[11,12] and glasses[11], polymers[18–30] and gels[13–17].

Shape memory alloys (e.g. CuZnAl-, FeNiAl-, TiNi-alloys) are already being used in biomedicine as cardiovascular stents, guidewires and orthodontic wires. The shape memory effect of these materials is based on a martensitic phase transformation.

Several types of shape memory gels are described in the literature.[13–17] Two different concepts are explained below. In the first system the shape memory effect is 'one-way' and originates from the chemical structure of the polymer network. The other system is an example of a reversible 'two-way' shape memory effect. However, this effect is being achieved by the design of the gel specimen as a bilayer system.

The first system can be prepared by a radical copolymerization of stearyl acrylate and acrylic acid with N,N-methylenebisacrylamide as a crosslinker.[16] Due to the intermolecular aggregation of the stearyl acrylate side chains, a crystalline lamellar structure can be observed in the dry as well as the swollen state in DMSO at room temperature. The swelling ratio of a gel film grows with increasing temperature up to 47 °C corresponding to the melting point of the stearyl side chains. This crystallizable side chain is the physical cross-link which can be used to fix a temporary shape. The permanent shape is being determined by the covalent cross-links of the polymer network. In this way, a thermally induced one-way shape memory effect can be programmed.

A reversible shape memory effect can be achieved using modulated gel technology.[14] These gels consist of two components, typically in the form of layers. The first component is not sensitive to an external stimulus (substrate element), while the second part is responsive to a selected stimulus (control element). The design of the gel specimen is optimized in such a way that a small change of the control element causes a large movement of the substrate element. An example of such a system is a partially interpenetrating system. The non-responsive part consists of a polyacrylamide gel. The control element is an interpenetrating network of the same polyacrylamide gel with a crosslinked poly(N-isopropylacrylamide) (NIPA). It is a specific property of the ionic NIPA gel (containing a small amount of sodium acrylate) to drastically shrink at temperatures higher than 37 °C. Since this change in volume of the control component is reversible, the shape memory effect is also reversible.

The shape memory effect of polymers, e.g. heat-shrinkable films or tubes,[18] is not a specific bulk property, but results from the polymers' structure and morphology. The effect is persistent in many polymers, which might differ significantly in their chemical composition. However, only a few shape memory polymer systems have been described in the literature. One example is segmented polyurethanes.[19–30] The thermal transition, which triggers the shape memory effect, can be a glass transition[21–26] as well as a melting point.[20–27] Segmented polyurethanes have found some applications, e.g. as chokes in cars. However, they are not suitable as degradable biomaterials for two reasons. On the one hand, the urethane bonds of their hard segments are hardly hydrolysable. On the other hand, the degradation products would be highly toxic low molecular weight aromatic compounds.

16.3 Concept of biodegradable shape memory polymers

Biodegradable, stimuli-sensitive polymers have great potential in minimal invasive surgery. Degradable implants can be brought into the body through a small incision in a compressed or stretched temporary shape. Upon heating up to body temperature, they switch back to their memorized shape. Repeat surgery for the removal of the implant is not required, since the materials will degrade after a predetermined implantation time period.

Structural concepts for tissue-compatible and biodegradable polymers, thermoplastic elastomers,[31] and thermosets[32] with shape memory capabilities will be introduced. Their thermal and mechanical properties and degradation behaviour will be explained. An important precondition for the shape memory effect of polymers is elasticity. An elastic polymeric material consists of flexible segments, so-called network chains, which are connected via netpoints or junctions. The permanent shape of such a polymer is determined by the netpoints. The network chains take a coil-like conformation in unloaded condition. If the polymer is stretched, the network chains become extended

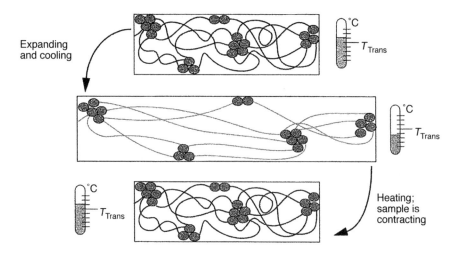

16.2 General schematic representation of the shape memory effect.

and oriented. In the case of an ideal entropy elastic material, the original shape is recovered after the stress is released.

An elastic polymer gains a shape memory functionality if the deformation of the material can be stabilized in the temperature range which is assigned for the specific application. This can be realized by using the network chains as a kind of molecular switch. For this purpose it should be possible to vary the segment flexibility as a function of temperature. Ideally, this process should be reversible. One way to obtain a switch functionality is given by the introduction of a thermal transition of the network chains at a temperature T_{trans}. Above T_{trans}, the segments are flexible, while below the transition temperature, the flexibility of the network chains can be limited to a certain extent. Below a glass transition temperature T_g, the flexibility of the network is frozen. If the thermal transition is a melting point, the network may become partially crystalline at temperatures below the melting temperature T_m. The so-formed crystalline domains prevent the segment chains from spontaneously recovering a coil-like conformation. The process of programming a temporary shape and the recovery of a permanent shape is shown in Fig. 16.2. Above T_{trans}, the segments are flexible and the polymer can be deformed elastically. The programmed shape is fixed by cooling the material to a temperature below T_{trans}. Upon heating above T_{trans}, the permanent shape can be recovered.

For biomedical applications, a thermal transition of the segment chains in the range between room and body temperature is of great interest. Suitable segments for degradable shape memory polymers can be found by regarding

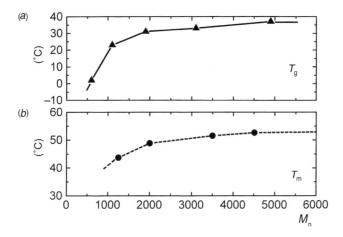

16.3 Dependence of thermal transition on molecular weight: (a) oligo[glycolide-co-(L-Kwsir-lactide)]diol having a glycolate content of 15 mol%, (b) oligo(ε-caprolactone)diol.

the thermal properties of well-established degradable biomaterials. From this assortment, two promising candidates can be extracted: poly(ε-caprolactone), which has a T_m of 61 °C, and the amorphous copolyesters of diglycolide and dilactides, showing glass transition temperatures T_g in the range from 35 °C to 55 °C. A fine-tuning of the respective thermal transition can be managed by variation of the molecular weight and the comonomer ratio (see Fig. 16.3).

Appropriate macrodiols are produced via ring opening polymerization of cyclic diesters or lactones initialized by low molecular weight diols (see Fig. 16.4).[33] The sequence structure of cooligomers can be influenced by application of a transesterification catalyst. The molecular weight of the oligomers can be controlled by the monomer/initiator ratio. Typically, the molecular masses Mn being obtained are between 500 g mol^{-1} and 10 000 g mol^{-1}. The net points can either be of physical or chemical nature. In the case of physical crosslinks, e.g. crystallizable segments with $T_m \gg T_{trans}$, the resulting polymer represents a thermoplastic elastomer. These materials can be melt processed, e.g. by extrusion or mould injection. Here, the permanent shape can be changed several times. In contrast, the permanent shape of a covalently crosslinked polymer network cannot be changed after the crosslinking process.

Important characteristics to be adjusted are the mechanical properties of the polymers in their permanent and temporary shape, the thermal transition temperature T_{trans}, the rate and the mechanism of the degradation process, and the shape memory properties.

16.4 Synthesis of macrodiols via ring-opening polymerization of lactones or cyclic diesters.[33] Courtesy of Wiley-VCH.

16.4 Degradable thermoplastic elastomers having shape memory properties

16.4.1 Synthesis

In order to synthesize biodegradable multiblock copolymers, the oligoesterdiols and cooligoester diols are linked by bifunctional junction units, e.g. diisocyanates, diacidchlorids or phosgene (see Fig. 16.5). High molecular weight polymers in the range of $M_w = 100\,000\,\text{g}\,\text{mol}^{-1}$ need to be obtained in order to get the desired mechanical properties. The resulting thermoplastic copolyesterurethanes are tough and show high elongations at break ε_R. These linear multiblock copolymers are phase segregated and consist of crystallizable hard segments (T_m) and amorphous switching segments ($T_{trans} = T_g$), e.g. poly[(L-lactide)-co-glycolide] with a glycolate content of 15 mol %. The permanent shape of these materials is obtained by melting the polymer followed by cooling to a temperature $T_m > T > T_{trans}$. The shape memory polymer can now be brought

Tailor-made intelligent polymers for biomedical applications

16.5 Synthesis of multiblock copolymers via polyaddition reaction. Courtesy of Wiley-VCH.

into its temporary shape, which is being fixed by cooling below T_{trans}. The permanent shape can be recovered by heating the material above T_{trans}.

16.4.2 Thermomechanical properties of thermoplastic elastomers

The shape memory effect can be determined quantitatively by cyclic thermo-mechanical tests. These measurements are performed in a tensile tester equipped with a thermo-chamber. At a temperature above T_{trans}, a bone-shaped sample is fixed between two clamps and stretched. If the maximum elongation has been reached, the sample is cooled down to a temperature below T_{trans}. The clamps then return to their initial distance. The sample reacts with bending. After reheating to a temperature above T_{trans} but below T_m of the hard segment, the next cycle can be started. Figure 16.6 shows an example for the result of such a cyclic thermomechanical test.

16.4.3 Degradability

As shown in Fig. 16.7, accelerated hydrolytic degradation experiments with different copolyester-urethanes in buffer solution of pH 7 at 70 °C showed that these materials are hydrolytically degradable. The degradation rate varies within a wide range. In contrast to the degradation behaviour of several polyhydroxyacids, mass loss of the investigated shape memory polymers starts early and shows linear behaviour during the whole degradation period.

16.4.4 Toxicity testing

In a first set of experiments, the multiblock-copolymers proved to be non-toxic. The CAM (chorioallantoic membrane) test is a sensitive test for cell toxicity. It is performed by placing a sterilized sample of the polymer on the

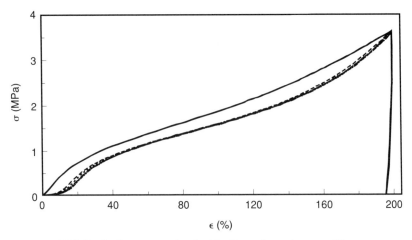

16.6 Cyclic thermomechanical testing of a polymer with an oligo[glycolide-*co*-(L-lactide)] as switching segment with $T_g = 35\,°C$; hard segment content: 22 w/w %; (Cycles: $n = 1$ ———; $n = 2$ ------; $n = 3$ ······; $n = 4$ ·····; $n = 5$ ·········).

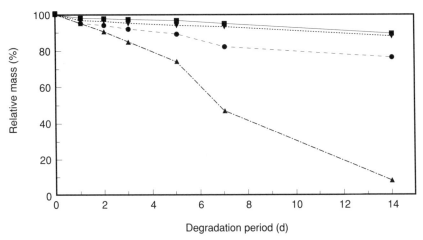

16.7 Hydrolytic degradation of shape memory polymers at 70 °C in buffered solution of pH 7: loss in relative sample mass (■ PDC27; ▼ PDC31; ● PDC40; ▲ PDL30).

chorioallantoic membrane of a fertilized chicken egg for two days. After incubation, the growth of blood vessels around the polymer sample is observed. In case of a non-toxic polymer, the blood vessels remain unchanged and their development is not restricted. In case of incompatibility, the sample causes changes in the number and shape of blood vessels, and the formation of a thrombus might occur (see Fig. 16.8).

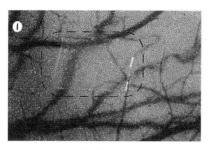 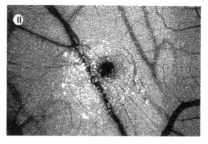

16.8 CAM test after 48 h incubation time: (a) multiblock copolymer, edge length of the sample 1–2 mm, (b) control experiment: incompatible sample causes thrombus (dark spot).[1] Courtesy of Wiley-VCH.

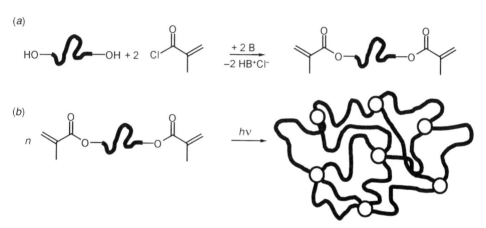

16.9 Synthesis of polymer networks with shape memory properties: (a) synthesis of the dimethyacrylate macromonomers, (b) cross-linking of the macromonomers.

16.5 Degradable polymer networks having shape memory properties

Based on the same switching segments as mentioned for the thermoplastic elastomers, a group of shape memory polymer networks can be prepared. Instead of crystallizable hard segments, covalent cross-links are introduced. For this purpose, the macrodiols can be turned into macrodimethacrylates, which can be cross-linked by photocuring. An example for the synthesis of biodegradable shape memory polymer networks is shown in Fig. 16.9. A potential educt is poly(ε-caprolactone) dimethacrylate with molecular weights between 1000 and 10 000. By copolymerization with n-butylacrylate, AB-networks can be obtained.[32] The permanent shape of these polyester networks is fixed via photocuring. The thermo-mechanical properties of the network can be tuned by the choice of the molecular weight of the respective macrodimethacrylates. The temporary shape can be formed by deformation of the sample under temporary heating above T_m.

16.6 Conclusion and outlook

Biodegradable shape memory polymers are candidates for the next promising generation of implant materials. The fact that these materials belong to a polymer system allows the adjustment of certain properties in a wide range, e.g. mechanical properties and degradation behaviour. Today, such materials can be synthesized in a kilogram scale.

In contrast to metallic shape memory alloys like NiTi-alloys, the polymers presented here combine the features of degradability and high elasticities, with elongations at break up to 1500%. Furthermore, shape memory polymers can be programmed much faster, allowing the individual adaptation of an implant to the patient's needs during surgery. Compared to hydrogels, these materials exhibit much higher mechanical strength.

From the point of view of economy and costs in healthcare systems, biodegradable shape memory polymers have two major advantages. Implants based on these materials can be brought into the body by minimally invasive methods, e.g. belly button surgery, allowing more careful treatment of patients; in addition, repeat surgery for the removal of the implant can be circumvented. The high potential of shape memory polymers for biomedical applications will therefore have a decisive influence on the way in which medical devices are designed in the future.

Acknowledgements

The author would like to thank Dr Steffen Kelch for his support in creating the manuscript for this chapter, as well as the author's research team, which has contributed to the experimental results being described. For financial support, the author is grateful to BMBF (BioFuture project no. 0311867) and Penguin-Foundation, as well as Fonds der Chemischen Industrie.

References

1. Lendlein A, 'Polymere als Implantatmaterialien', *Chemie in unserer Zeit*, 1999, **33**, 279–95.
2. Shalaby S W and Johnson R A, 'Synthetic absorbable polyesters'. In S W Shalaby (ed.), *Biomedical Polymers*, Hanser Publishers, München, 1994.
3. Laurencin C T, Sobrasua I E M and Langer R S, *Poly(anhydrides)*. In J O Hollinger

(ed.), *Biomedical Applications of Synthetic Biodegradable Polymers*, CRC Press, Boca Raton, 1995.
4 Bronzino J D, *The Biomedical Engineering Handbook*, CRC Press, Boca Raton, 1995.
5 Silver F H and Doillon Ch, *Biocompatibility: Interactions of Biological and Implantable Materials*, VCH-Wiley, New York, 1989.
6 Perkins J, *Shape Memory Effects in Alloys*, Plenum Press, New York, 1975.
7 Lipscomb P and Nokes L D M, *The Application of Shape Memory Alloys in Medicine*, MEP, Suffolk, 1996.
8 Linge L and Dahm S, 'Practical aspects of using 'super-elastic' archwires for edgewise technique', *Fortschr. Kieferorthop.*, 1994, **55**, 324–9.
9 Quandt E, Halene C, Holleck H, Feit K, Kohl M and Schlossmacher P, 'Sputter deposition of TiNi and TiNiPd films displaying the two-way shape memory effect', *The 8th International Conference on Solid-State Sensors, Actuators and Eurosensors IX*. Stockholm, 1995.
10 Cederström J and Van Humbeeck J, 'Relationship between shape memory material properties and applications', *Journal Phys. IV, Coll. C2, Suppl. J. Phys. III*, 1995, **5**, 335–41.
11 Itok A, Miwa Y and Iguchi N, 'Shape memory phenomena of glass-ceramics and sintered ceramics', *J. Japan Inst. Metals*, 1990, **54**, 117–24.
12 Swain M V, 'Shape memory behaviour in partially stabilized zirconia ceramics', *Nature*, 1986, **322**, 234–6.
13 Osada Y and Matsuda A, 'Shape memory in hydrogels', *Nature*, 1995, **376**, 219.
14 Hu Z, Zhang X and Li Y, 'Synthesis and application of modulated polymer gels', *Science*, 1995, **269**, 525–7.
15 Kagami Y, Gong J P and Osada Y, 'Shape memory behaviors of crosslinked copolymers containing stearyl acrylate', *Macromol. Rapid Commun.*, 1996, **17**, 539–43.
16 He X, Oishi Y, Takahara A and Kajiyama T, 'Higher order structure and thermo-responsive properties of polymeric gel with crystalline side chains', *Polymer J.*, 1996, **28**(5), 452–7.
17 Sawhney A, Pathak C P and Hubbell J A, 'Bioerodible hydrogels based on photopolymerized poly(ethylene glycol)-*co*-poly(α-hydroxy acid) diacrylate macromers', *Macromolecules*, 1993, **26**, 581–7.
18 McLoughlin R (RayChem, Ltd.), Method of making heat shrinkable articles, US Pat 4 425 174, 1984.
19 Sakurai K, Shirakawa Y, Kashiwagi T and Takahashi T, 'Crystal transformation of styrene-butadiene block copolymer', *Polymer*, 1992, **35**(19), 4288–9.
20 Kim B K, Lee S Y and Xu M, 'Polyurethanes having shape memory effect', *Polymer*, 1996, **37**, 5781–93.
21 Takahashi T, Hayashi N and Hayashi S, 'Structure and properties of shape-memory polyurethane block copolymers', *J. Appl. Polym. Sci.*, 1996, **60**, 1061–9.
22 Hayashi S (Mitsubihi Heavy Industries, Ltd.), Shape memory polyurethane elastomer molded article, US Pat 5 145 935, 1992.
23 Shirai Y and Hayashi S, Development of polymeric shape memory material, *Mitsubishi Tech. Bull.*, 1988, **184**.
24 Tobushi H, Hayashi S and Kojima S, 'Mechanical properties of shape memory polymer of polyurethane series', *JSME Int. J.*, Series I, 1992, **35**(3).
25 Ito K, Abe K, Li H L, Ujihira Y, Ishikawa N and Hayashi S, 'Variation of free

volume size and content of shape memory polymer – polyurethane – upon temperature', *J. Radioanalyt. Nucl. Chem.*, 1996, **211**(1), 53–60.
26 Tobushi H, Hayashi S, Ikai A and Hara H, 'Thermomechanical properties of shape memory polymers of polyurethane series and their applications', *J. Phys. IV, Coll. C12, Suppl. J. Phys. III*, 1996, **6**, 377–84.
27 Li F, Hou J, Zhu W et al., 'Crystallinity and morphology of segmented polyurethanes with different soft-segment length', *J. Appl. Polymer Sci.*, 1996, **62**, 631–8.
28 Hayashi S, Kondo S and Giordano C, 'Properties and applications of polyurethane-series shape memory polymer', *Antec 94*, 1994, 1998–2001.
29 Jeong H M, Lee J B, Lee S Y and Kim B K, 'Shape memory polyurethane containing mesogenic moiety', *J. Mater. Sci.*, 2000, **35**, 279–83.
30 Jeong H M, Lee S Y and Kim B K, 'Shape memory polyurethane containing amorphous reversible phase', *J. Mater. Sci.*, 2000, **35**, 1579–83.
31 Lendlein A, Grablowitz H and Langer R, in preparation.
32 Lendlein A, Schmidt A and Langer R 'AB-polymer networks based on oligo (ε-caprolactone) segments showing shape-memory properties', *Proc. Natl. Acad. Sci. USA*, 2001 **98**(3), 842–7.
33 Lendlein A, Neuenschwander P, Suter U W, 'Hydroxy-telechelic copolyesters with well defined sequence structure through ring-opening polymerization', *Macromol. Chem. Phys.*, 2000, 201, 1067–76.

17
Textile scaffolds in tissue engineering

SEERAM RAMAKRISHNA

17.1 Introduction

Use of textile structures in the medical field is not recent because sutures, which for centuries have been used for the closure of wounds or incisions, are fundamentally textile structures. The emergence of textiles apart from sutures, for various biomedical applications, became of real significance in the early 1950s. Nowadays they are commonly used in various biomedical applications,[1] and are generally referred to as 'medical textiles'. Based on the application, they may be grouped into three broad categories (Table 17.1), namely healthcare and hygiene textiles, extracorporeal devices and surgical textiles.[2] In terms of volume of usage, surgical textiles are much smaller than healthcare and hygiene textiles. However, scientifically, surgical applications are far more challenging. In these applications, the textiles are expected to fulfil a number of requirements, including surface biocompatibility (chemical structure, topography, etc.), mechanical compatibility (elastic modulus, strength, stiffness, etc.), non-toxicity, durability in in vivo (human body environment) conditions and sterilizability. Due to recent advancements in textile engineering and biomedical research, the use of textiles in surgery is growing. They are routinely used to direct, supplement or replace the functions of living tissues of the human body. Soft tissue replacements or implants such as vascular grafts, skin grafts, hernia patches and artificial ligaments are made of textile structures.[3–8] Moreover, polymers reinforced with textiles, called polymer composite materials, are also considered in hard tissue replacements or implants such as dental posts, bone grafts, bone plates, joint replacements, spine rods, intervertebral discs and spine cages.[9,10] Table 17.2 is a partial list of some of the most common implant applications of textiles. Some implant applications are shown schematically in Fig. 17.1. As can be seen from Table 17.2, the implantable textiles are made from a variety of synthetic biomaterials, which are essentially non-living (avital) type. Although the synthetic biomaterials are fairly successful, the profound differences between them and the living tissues of the human body

Table 17.1 Classification of medical textiles

Category	Sub-category	Applications
Healthcare and hygiene textiles		Bedding, protective clothing, surgical gowns, clothes, wipes, etc.
Extracorporeal devices		Artificial kidney, artificial liver, artificial lung, bioreactors, etc.
Surgical textiles	Non-implantable textiles	Wound dressings, plaster casts, bandages, external fracture fixation systems, etc.
	Implantable textiles	Sutures, vascular grafts, ligament and tendon prostheses, bone plates, heart valves, hernia patches, joint replacements, artificial skin, etc.

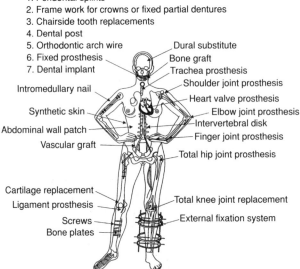

17.1 Schematic illustration of various implants.

lead to problems such as infection, loosening, failure and finally rejection of implants. On the other hand, transplantation (transfer of a tissue from one body to another, or from one location in a body to another) is not always practical due to a shortage of donor tissue, and the risk of rejection and disease transfer. Hence, there is a need to develop biological substitutes (living or vital materials) to avoid these problems. The newly developed field of 'tissue engineering' combines mammalian cells and certain synthetic biodegradable

Table 17.2 Implant applications of textile structures

Application	Materials	Textile structures {monofilament (m), yarn (y), weave (w), briad (b), knit (k), non-woven (n)}
Abdominal wall	Polyester	w
Blood vessel (vascular graft)	Polyester, polytetrafluoroethylene (PTFE), polyurethane	w, k
Bone plant	Carbon, PGA	y, w, b, k
Cartilage	Low density polyethylene, polyester, PTFE, carbon	w, b
Dental bridge	Ultrahighmolecular weight polyethylene (UHMWPE), carbon, glass, aramid	w, n
Dental post	Carbon, glass	y, b
Dural substitute	Polyester, PTFE, polyurethane, collagen	n, w, k
Heart valve (sewing ring)	Polyester	k, w
Intervertebral disc	Polyester, PTFE	w, k
Intramedullary rod	Carbon, glass	y, b
Joint	Polyester, carbon, UHMWPE	w, k
Ligament	Polyester, carbon, glass, aramid	b, w, k
Orthodontic arch wire	Glass	y, b
Skin	Chitin	n, w, k
Spine rod	Carbon	y, b
Suture	Polyester, PTFE, polyamide, polypropylene, polyethylene, collagen, polylactic acid (PLA), polyglycolic acid (PGA)	m, y, b
Tendon	Polyester, PTFE, polyamide, polyethylene, silk	b, w, y

materials (materials that eventually disappear after being introduced into a living tissue or organism) to produce living (vital) synthetic tissue substitutes or replacement tissues.[11-13] It is envisaged that such tissue substitutes will merge seamlessly with the surrounding host tissues, eliminating problems associated with contemporary biomaterials and transplantation. Recognizing the potential of tissue engineering, researchers worldwide are harnessing techniques to produce tissue engineered skin,[14] cartilage,[15] nerve,[16] heart valve[17] and blood vessels.[18,19] It is also envisaged that, using tissue engineering techniques, it will eventually be possible to construct entire replacement organs such as the liver[20] and bladder.[21]

Table 17.3 Various scaffolds used in tissue engineering

Tissue engineered biological substitute	Scaffold materials	Scaffold structures {yarn (y), weave (w), braid (b), knit (k), non-woven (n)}
Bladder[21]	PGA	Textile (n)
Blood vessel[18,19]	Polyester (Dacron); polyurethane; ePTFE; PGA; PLA; PGLA (Vicryl)*	Textile (n, w, b, k)
	Collagen	Textile (k)
Bone[61,62,64]	PGA; PLGA* + hydroxyapatite fibres	Textile (n); Foam
	PLLA	Foam
Cartilage[15,65–68]	PGA; PLLA; PGLA*	Textile (n)
Dental[69]	DL-PLA; PGLA (Vicryl)*	Foam (porous membrane); textile (n)
Heart valve[17]	PGA	Textile (n, w)
Tendon[70]	PGA	Textile (n, y)
Ligament[67,71]	Collagen	Textile (y)
	PGA, PLAGA*	Textile (b, n)
Liver[20,72]	PGA; PLA; PGLA; polyorthoesters; polyanhydride	Foam; textile (n)
	PLGA*	3D Printed
Nerve[16,73,74]	Collagen–glycosaminoglycan	Foam
	PGA	Textile (n)
Skin[14,73]	PGA; PGLA (Vicryl)*; Nylon	Textile (w)
	Collagen–glycosaminoglycan	Foam

*PLAGA, PGLA, PLGA are co-polymers of polyglycolic acid (PGA) and polylactic acid (PLA).

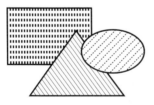

17.2 Knitted fabric scaffold seeded with osteoblastic cells.

The basic concept of tissue engineering is to regenerate or grow new tissues and organs by culturing isolated cells from the tissue or organ of interest on porous biodegradable scaffolds or templates (Fig. 17.2). The scaffold acts as an extracellular matrix for cell adhesion and growth and/or regeneration. An important challenge is to pursuade the cells transplanted onto scaffolds to multiply and produce correct tissue matrices, which can take up and secrete protein, generate force and resistance, constrain permeability and exhibit other life processes. It has been recognized that the engineering of the scaffold is an important aspect, as it provides the optimal requirements for the survival, proliferation and differentiation of cells, and for the formation of tissue in in vitro or in vivo conditions depending on the intended application. Hence, there has been a multitude of research work carried out in the last decade to design and develop various types of optimum scaffolds for tissue engineering (Table 17.3). They may be broadly categorized into three groups based on the processing methods: (1) foams/sponges,[22–27] (2) three-dimensional (3D) printed substrates/templates[28] and (3) textile structures.[29–39] Textile structures form an important class of porous scaffolds used in tissue engineering.[40–42] This chapter reviews various textile scaffolds from the viewpoint of tissue engineering requirements and possible future developments.

17.2 Ideal scaffold system

Making biological substitutes using the tissue engineering approach fundamentally encompasses several phases, namely: selection of scaffold material; fabrication of scaffold; preparation of scaffold; cell harvest from animal or human patient; cell seeding onto the scaffold; cell proliferation and differentiation; growth of mature tissue; surgical transplantation; and implant adaptation and assimilation. The following describes features of an ideal scaffold system.[43,44] Specific requirements vary from one tissue to another.

- The material used for the scaffold should be biocompatible, not inducing an unfavourable tissue response in the host. The material should be ultra-pure, and easily and reliably reproducible into a variety of sizes and structures.
- In most applications, the support of a scaffold is needed only for a limited time. These temporary scaffolds cannot be removed easily because of tissue grown into the porous structure. Therefore, scaffolds have to be manufactured out of a biodegradable material in which the degradation rate has to be adjusted to match the rate of tissue formation. The scaffold should maintain its volume, structure and mechanical stability long enough to allow adequate formation of tissue inside the scaffold. However, none of the degradation products released should provoke inflammation or toxicity.
- Scaffolds must provide a reproducible microscopic and macroscopic structure with a high surface-area to volume ratio in order to allow a significant amount of cell–surface interaction. The scaffold processing method should not affect the biocompatibility or the desired degradation

behaviour of the material. It should also allow the manufacture of scaffolds with controlled interconnected pore structure, pore size distribution and pore geometry, since these are important factors in tissue growth or regeneration.

- The average pore size and the macroscopic dimensions of a scaffold are important factors which are associated with cell proliferation and nutrition supply, from tissue culture media in vitro and through newly formed blood vessels in vivo, to cells and tissue. The pore size of such scaffolds should be sufficient to allow cells to grow in multiple layers in order to form a three-dimensional tissue. The optimal pore size may be highly variable, depending also on the intended application of the scaffold. For instance, it has been hypothesized that in orthotopic sites, pore sizes below 400 μm lead to bone formation and pore sizes above 400 μm lead to fibrous tissue ingrowth.[45–48] In addition to pore size, porosity, which more reflects the interconnectivity of the scaffold, is also important. High porosity maximizes the volume of tissue ingrowth and minimizes the amount of scaffold material used. It also facilitates transport of nutrients and cellular waste products. Another parameter is the pore morphology, which may be meaningful in favouring the ingrowth of certain cell types.[49]
- Scaffold surface chemistry should be suitable for cell attachment and cell proliferation.
- In certain tissue engineering applications, external electro-mechanical stimulations are often used to promote cell proliferation and tissue development. The scaffold should be able to retain its shape and structure under these electro-mechanical conditions.
- Further, the flexibility of such a scaffold should be close to that of its surrounding tissue so, once the vascularization starts, no extreme change in the mechanical properties between the host tissue and the scaffold can be experienced by the ingrowing tissue. Such forces could be harmful, not only for the vascularization process, but also because they could induce the formation of a different tissue from the desired one.

17.3 Scaffold materials

As most cells are substrate dependent, the scaffold structure as well as the material has control over the cell adhesion and function. The various scaffold materials used in tissue engineering can be grouped into natural and synthetic materials. Collagen, chitin, starch, etc. are a few examples of natural materials (see Table 17.4). Natural materials are isolated from human, animal or plant tissues, which typically result in high costs and large batch-to-batch variations. In addition, these materials exhibit a very limited range of properties and are

Table 17.4 Biodegradable polymers used in tissue engineering

Category	Sub-groups and typical polymers
Natural polymers	Cellulose, starch, chitin, collagen and fibrin[12,43,73,75-74]
Synthetic polymers	Polyesters[24,26,30-32,53,66,75,82,85-98]
	Poly(glycolic acid) (PGA) and copolymers
	Poly(lactic acid) (PLA) and copolymers
	Poly(alkylene succinates)
	Poly(hydroxybutyrate) (PHB)
	Poly(butylene diglycolate)
	Polyanhydrides[99]
	Polyorthoesters[100-102]
	Polyiminocarbonates[103,104]
	Polyphosphoesters[105]
	Polyphosphazenes[106-112]

often difficult to process. Synthetic materials are further classified into degradable[23,50] and non-degradable[51] types. The non-degradable materials, such as polyethylene, polyethyleneterephthalate (PET) and polytetrafluoroethylene (PTFE) may carry a risk of permanent tissue reaction.[33] On the other hand, synthetic biodegradable polymers, such as polyesters, polyanhydrides and polyorthoesters (see Table 17.4) offer control over structure and properties. They can be processed into various shapes and microstructures, such as desired surface area, porosity, pore size and pore size distribution. They can be tailored with degradation times ranging from several days up to years. Their surface properties can be altered to adapt to the biological requirements for cell adhesion, growth and function. Therefore, synthetic biodegradable polymers have been widely investigated in tissue engineering research.[52-54] From the literature (Tables 17.3 and 17.4) it is evident that biodegradable polyester-based materials dominate the tissue engineering applications compared to other biodegradable polymers. This is mainly due to the fact that polyesters of poly(α-hydroxy acids) are used successfully in various implant applications and have already been approved by the US Food and Drug Administration (FDA). Another factor could be the familiar processing and characteristics of these materials to many tissue engineering researchers. However, it is to be noted that the mechanical properties and degradation profiles of these polyesters are insufficient for certain applications. Moreover, certain copolymers may release toxic products during degradation. As tissue engineering applications continue to grow, it is important to find and develop alternative biodegradable polymers that meet the specific requirements of various tissues.

17.4 Textile scaffolds

The need for scaffolds in tissue engineering is undisputed as cells cannot survive on their own and are substrate dependent. However, there is no universal scaffold that meets all the requirements of various tissues, as the optimum tissue engineering conditions vary from tissue to tissue. In other words, the targeted tissue dictates the optimum scaffold design. For example, for hard tissues, such as bone, scaffolds need to have high stiffness in order to maintain the space they are designated to provide the tissue with enough space for growth. If scaffolds are used as a temporary load-bearing device, they should be strong enough to maintain that load for the required time without showing any symptoms of failure. Used in combination with soft tissues, the flexibility and the stiffness of the scaffold have to be within the same order of magnitude as the surrounding tissues in order to prevent the scaffold from either breaking or collapsing and from stress shielding the adjacent tissues. The choice of scaffold for a tissue therefore depends on its characteristics. In addition to the mechanical properties, the optimum design of a scaffold for a specific tissue application requires consideration of microstructural, chemical and biological aspects. It is often difficult to isolate these aspects as they are interdependent and sometimes their effects are unknown. The following sections critically look into some of these aspects of different textile scaffolds.

17.4.1 Microstructural aspects

The microstructural aspects of scaffolds includes pore size, porosity, pore size distribution, pore connectivity and reproducibility of pores. These aspects are vital, as they provide the optimal spatial and nutritional conditions for the cells, and determine the successful integration of the natural tissue and the scaffold. For example, Hubbell and Langer[37] showed in their experiments that in animals, the size and alignment of pores in the scaffold greatly influence the amount and rate of vascular and connective tissue growth. Fibrovascular tissues require a pore size greater than 500 μm for rapid vascularization, whereas the optimal porosity for bone bonding materials is considered to be between 70 and 200 μm.[55] In another study,[56] osteoblasts cultured in calcium phosphate ceramic prefer a pore size of 200 μm, and it has been proposed that this pore size possesses a curvature that optimizes the compression and tension of the cell's mechanoreceptors. However, there is concern that optimal pore size in ceramics may not generalize for all scaffold materials.[57] For example, when poly(lactic acid) was implanted in calvarial rat defects, pore sizes of 300–350 μm supported bone ingrowth while smaller sizes did not. Yet, in another study, osteoblasts showed no significant difference in proliferation or function when seeded on poly(lactic glycolic acid) foams with pore sizes of either 150–300 µm, 300–500 µm or 500–710 µm. The importance of determining optimum values for the specific cells or tissue cannot be underestimated. In another study related to skin tissue engineering, Nerem et al.[18] showed that the endothelial morphology depends on the pattern of the scaffold. The interconnectivity of pores determines the transport of nutrients and waste and thus influences the success of tissue engineering. The reproducibility of scaffolds is also very important as it determines the dimensional stability of the scaffold as well as the consistency of the tissue formation. Table 17.5 compares the various microstructural aspects of foams and textile structures. Owing to the processing techniques employed, in general, each batch of foam will have one particular of porosity. It is possible to tailor the porosity to a certain extent. However, within the same foam, organizing or grading the porosity in a particular fashion may be difficult to achieve with the current processing techniques. On the other hand, textile structures can be tailored to give the required porosity in terms of size, quantity and distribution pattern. For example, in a typical textile scaffold, three levels of porosity can be achieved. The arrangement of fibres in the yarn determines the accessible space for cells. The inter-fibre space (or groove between two adjacent fibres) may be considered as the first level of porosity. In our study[58] it was found that the fibroblasts preferentially organize themselves along the length of the fibres, grouping along the groove created by two adjacent fibres. Figure 17.3 shows SEM pictures of polyethylene terephthalate (PET) fibre yarn before and after seeding with fibroblasts. What is more interesting is that fibroblasts are capable of bridging fibres which are as far as 40 µm apart (Fig. 17.4). The inter-fibre gap or first level of porosity in a textile scaffold can be controlled by changing the number of fibres in the yarn and also the yarn packing density. Further variations in porosity can be achieved by using twisted, untwisted, textured, untextured, continuous or spun yarns.

The gap or open space between the yarns (it is open space inside the loop in the case of knits) forms the second level of porosity. In the case of knitted scaffolds, the porosity can be varied selectively by changing the stitch density and the stitch pattern. In the case of braided scaffolds, porosity can be varied by controlling the bias angle of the interlacing yarns. In the case of woven scaffolds, it is possible to change the porosity by controlling the inter-yarn gaps through a beating action. In our preliminary study[58] involving the seeding of woven, braided and knitted scaffolds with hepatocytes, it was observed that cells attach preferentially at the inter-yarn gaps or pores in the case of woven and braided scaffolds, whereas they clump together on the ridges of curved yarns in the case of knitted scaffolds (Fig. 17.5). It may be noted that woven and braided scaffolds share similar surface topographies formed by the interlacing yarns. Knitted scaffolds, however, comprise curved yarns, which had a significant effect on the behaviour of hepatocytes. The same

Table 17.5 Microstructural aspects of scaffolds

Fabrication	Foam/sponge	Textile structures			
		Non-woven	Weave	Braid	Knit
Pore size (μm)	0.5–500	10–1000	0.5–1000	0.5–1000	50–1000
Porosity (%)	0–90	40–95	30–90	30–90	40–95
Pore distribution	Random to uniform	Random	Uniform	Uniform	Uniform
Reproducibility of porosity	Poor to good	Poor	Excellent	Excellent	Good to excellent
Pore connectivity	Good	Good	Excellent	Excellent	Excellent
Processability	Good	Good	Excellent	Excellent	Good
Other comments	Current techniques are associated with processing undesirable residues such as solvents, salt particles	Equipment cost is high. Control over porosity is always questionable	Shapes are limited	Limited to tubular or uniform cross-sectional shapes	Limited by the low bending properties of current biodegradable fibres

Textile scaffolds in tissue engineering

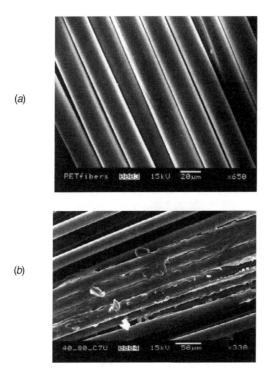

17.3 PET fibre yarn (a) before and (b) after seeding with fibroblasts (Cos7 kidney cells from SV40-transformed African green monkey).

experiment was repeated with fibroblasts to investigate the effect of cell type. Unlike the hepatocytes, the fibroblasts were found to attach to the ridges of yarns irrespective of the scaffold type. The different behaviour of fibroblasts and hepatocytes may be due to their different cell sizes and shapes. It may be noted that the diameter of fibroblasts ranges from 10 μm to 20 μm, and they flatten out after attachment. The hepatocytes are larger with diameters in the range 15 μm to 30 μm, and they retain their spherical structure even after attachment to the scaffold.

Furthermore, a third kind of porosity can be introduced by subjecting the textile structures to secondary operations such as crimping, folding, rolling, stacking, etc. In other words, the flexibility of microstructural parameters is tremendous in the case of textile scaffolds.

Bowers et al.[59] investigated the effect of surface roughness. It has been reported that a higher percentage of osteoblast-like cells cultured on commercially pure Ti attached to rougher surfaces than to smooth surfaces. Another study using the same material showed higher osteocalcin content and ALPase activity on smooth, polished surfaces than on rough surfaces.[60] In our

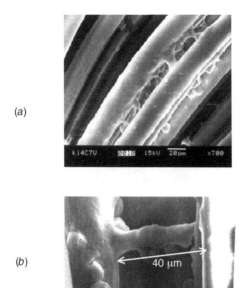

17.4 Fibroblasts bridging adjacent PET fibres.

study[58] involving woven, braid and knitted scaffolds, it was observed that the fibroblasts prefer to attach to the ridges of scaffolds rather than the valleys. This may be due to the cell's attempt to minimize distortion to its cytoskeleton in response to the topography of the scaffold. Further systematic study is needed to fully understand the influence of scaffold topology on tissue engineering.

17.4.2 Mechanical aspects

Similar to the microstructural aspects, the mechanical aspects of scaffolds, such as structural stability, stiffness and strength, have considerable influence on the cellular activity. For example, in tissues like bone, cell shape is influenced by mechanical forces.[11] Cell shape modification takes place as a result of external forces including gravity, and also of internal physical forces. Cell shape modification also depends on the nature (constant or cyclic), type (uniaxial, biaxial, multiaxial, etc.) and magnitude of the mechanical stimulation. The mechanical stimulation also affects the release of soluble signalling factors and the deposition of extracellular matrix constituents. Researchers are

Textile scaffolds in tissue engineering

(a)

(b)

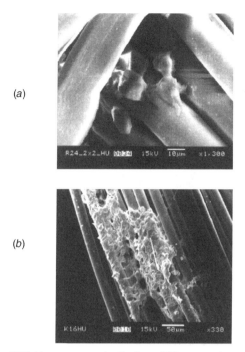

17.5 Hepatocytes (cells from Wistar rats) attached preferentially at (a) interyarn gaps in a woven scaffold and (b) ridges of curved yarn in a knitted scaffold.

making use of these observations in the case of bone tissue engineering. They are applying external mechanical stimulation to promote tissue formation. Therefore, in bone tissue engineering, the scaffolds are designed to withstand severe physiological loads.[55] In blood vessel applications, the scaffold needs to be strong enough to withstand physiologically relevant pulsatile pressures and at the same time match the compliance or elasticity values of a native blood vessel. The mechanical aspects of various scaffolds are compared in Table 17.6. Among the scaffolds, the woven fabrics are normally rigid and inflexible due to the tight interlacing of yarns. The next stiff and strong scaffold is the braid. Knits, non-woven and foams display the lower end of the mechanical properties. Of all the scaffolds, knits display considerable deformability and good compliance owing to their looped yarn arrangement. Hence, they are suitable for bladder and blood vessel tissue engineering applications. Researchers are using woven scaffolds for tissue engineering of bone and acetabular cups. It is to be noted that the mechanical behaviours of scaffolds can be varied significantly by controlling the various microstructural aspects stated earlier. In other words, both the microstructural and mechanical aspects are

Table 17.6 Mechanical aspects of scaffolds

Fabrication	Foam/sponge	Textile structures			
		Non-woven	Weave	Braid	Knit
Stiffness	Low	Low	High	High	Medium
Strength	Low	Low	High	High	Low
Structural stability	Good	Poor to good	Excellent	Excellent	Poor to good
Drapeability	Poor	Good	Poor	Poor	Excellent
Other comments	Isotropic behaviour	Isotropic behaviour	Anisotropic, with good properties parallel to fibres and poor properties normal to fibres	Anisotropic, with good properties in axial direction and poor properties in transverse direction	The behaviour can be tailored from isotropic to anisotropic

interrelated and it is less meaningful to understand them individually. Further work is necessary to understand how the scaffolds behave in in vitro or in vivo environments, and how they contribute to the growth of tissue.

17.4.3 Other aspects

There is increasing evidence that scaffold surface chemistry influences cellular activity.[61] Boyan et al.[113] showed that osteoblast response varies with the material on which cells are cultured, and attributed this to differences in the surface chemistry, charge density and net polarity of the charge. Some variations have been attributed to the proteins present in the medium that adsorb onto the surface to different degrees or with different structural arrangements.[62] In one study, osteoblasts were cultured on glass modified with the RGD peptide or non-adhesive, scrambled sequence and in the presence or absence of BMP-7.[63] The culture with a combination of RGD substrate and BMP-7 showed a substantial increase in mineralization in 21 days over all other combinations of treatments. Because of its role in both attachment and differentiation, RGD incorporation may contribute greatly to scaffold osteoinductivity and bone regeneration. Technologies for the incorporation of peptides on to the scaffold surfaces are being further perfected.

In our laboratory, a systematic study[58] was made involving unmodified polyethylene terephthalate (PET) textile scaffolds and YIGSR (Tyr–Ile–Gly–Ser–Arg) peptide conjugated PET textile scaffolds. Three types of scaffolds, namely woven, braided and knitted fabrics, were seeded with fibroblasts (Cos7 kidney cells from SV40-transformed African green monkey) and hepatocytes (cells from Wistar rats) separately. All three types of scaffolds indicated a 35% to 46% increase in the number of fibroblasts attached when conjugated with peptide bonds compared to the unmodified scaffolds. However, no appreciable change in the hepatocyte attachments was found with the peptide surface modification of scaffolds. This study clearly indicates that the cellular activity also depends on the source of cells (bone, liver, blood vessels, etc.), number of cell types (pure, co-cultured or mixed cell type cultures), species (e.g. rat, rabbit, chicken, human), sex and age (i.e. embryonic, neonatal or adult). Furthermore, it is generally believed that, depending on the characteristics of the cell culture and culture period used, different reactions may be expected. The current literature clearly indicates that a combination of various factors, such as scaffold material, structure, physical, chemical, mechanical, and biological properties, cell types, in vitro or in vivo conditions, etc., determines the success of tissue engineering.

17.5 Conclusions

Scaffolds play a central role in tissue engineering. Textile structures are particularly attractive to tissue engineering because of their ability to tailor a broad spectrum of scaffolds with a wide range of properties. Preliminary studies clearly demonstrate the suitability of textile scaffolds for tissue engineering purposes. There is no universal scaffold that meets the requirements of the various tissues of the human body. Further systematic study is necessary to design an optimal scaffold for each tissue application.

Acknowledgements

The author wishes to thank Professor Kam W. Leong, Johns Hopkins University, USA, and Dr J. Mayer, Swiss Federal Institute of Technology, Switzerland for their valuable suggestions and discussions. The author also appreciates the support of Dr Henry Yu and contributions of Miss Wee Su-Leng Serene, National University of Singapore, in preparing this chapter.

References

1 Planck H, General aspects in the use of medical textiles for implantation. In *Medical Textiles for Implantation*, ed. H Planck, M Dauner and M Renardy, Springer-Verlag, Heidelberg, Germany, 1990, 1–16.
2 Adanur S, *Wellington Sears Handbook of Industrial Textiles*. Technomic Publishing Co. Inc., USA, 1995.
3 Arnoczky S P, Torzilli PA et al., 'Biologic fixation of ligament prostheses and augmentations: An evaluation of bone in-growth in dog', *Am. J. Sports Med.*, 1988, **16**(2), 106–12.
4 Berry J L, Berg W S et al., 'Evaluation of dacron covered and plain bovine xenografts as replacements for the anterior cruciate ligament', *Clin. Orthop.*, 1988, **236**, 270–8.
5 Marios Y, Roy R et al, 'Histopathological and immunological investigations of synthetic fibers and structures used in three prosthetic anterior cruciate ligaments: in vivo study in the rat', *Biomaterials*, 1993, **14**(4), 255–62.
6 Pietrucha K, 'New collagen implant as dural substitute', *Biomaterials*, 1991, **12**(3), 320–3.
7 Molina J E, Edwards J E, Bianco R W, Clack R W, Lang G and Molina J R, 'Composite and plain tubular synthetic graft conduits in right ventricle-pulmonary artery position: fate in growing limbs', *J. Thorac. Cardiovasc. Surg.*, 1995, **110**(2), 427–35.
8 Menger M D, Hammersen F et al., 'In vivo assessment of neovascularization and incorporation of prosthetic vascular biografts', *Thorac. Cardiovasc. Surg.*, 1992, **40**(1), 19–25.
9 Ramakrishna S, Ramaswamy S, Teoh S H, Hastings G W and Tan C T,

'Application of textiles and textile composites concepts for biomaterials development', *Int. Conf. New Textiles for Composites, TEXCOMP 3 Conference Series*, Aachen, Germany, 1996, 27/1–27/27.
10 Ramakrishna S, Mayer J, Wintermantel E and Leong K W, 'Biomedical applications of polymer composite materials: a review', *Composites Science and Technology* (in press).
11 Langer R and Vacanti J, 'Tissue engineering', *Science*, 1993, **260**, 920–6.
12 Peppas N A and Langer R, 'New challenges in biomaterials', *Science*, 1994, **263**, 1715–20.
13 Langer R and Vacanti J, 'Artificial organs', *Sci. Amer.*, 1995, September, 100–3.
14 Naughton G K, Bartel R and Mansbridge J, 'Synthetic biodegradable polymer scaffolds'. In *Synthetic Biodegradable Polymer Scaffolds*, ed. A Ataala, D J Mooney, J P Vacanti and R Langer, Birkhauser, Boston, USA, 1997, 121–47.
15 Cao Y, Ibarra C and Vacanti J P, 'Tissue engineering of cartilage and bone'. In *Synthetic Biodegradable Polymer Scaffolds*, ed. A Ataala, D J Mooney, J P Vacanti and R Langer, Birkhauser, Boston, USA, 1997, 199–214.
16 Dunn R L, 'Clinical applications and update on the poly(α-hydroxy acids)'. In *Biomedical Applications of Synthetic Biodegradable Polymers*, ed. J O Hollinger, CRC Press Inc., USA, 1995, 17–31.
17 Shinoka T and Mayer J E, 'New frontiers in tissue engineering: tissue engineered heart valves'. In *Synthetic Biodegradable Polymer Scaffolds*, ed. A Ataala, D J Mooney, J P Vacanti and R Langer, Birkhauser, Boston, USA, 1997, 188–98.
18 Nerem R M, Braddon L G, Seliktar D and Ziegler T, 'Tissue engineering and the vascular system'. In *Synthetic Biodegradable Polymer Scaffolds*, ed. A Ataala, D J Mooney, J P Vacanti and R Langer, Birkhauser, Boston, USA, 1997, 165–85.
19 Weinberg C B, O'Neil K D, Carr R M, Cavallaro J F, Ekstein B A, Kemp P D and Rosenberg M, 'Matric engineering: remodeling of dense fibrillar collagen vascular grafts in vivo'. In *Tissue Engineering, Current Perspectives*, ed. E Bell, Birkhauser, Boston, USA, 1993, 190–8.
20 Lee H and Vacanti J P, 'Tissue engineering of liver'. In *Synthetic Biodegradable Polymer Scaffolds*, ed. A Ataala, D J Mooney, J P Vacanti and R Langer, Birkhauser, Boston, USA, 1997, 235–51.
21 Atala A, 'Tissue engineering in the genitourinary system'. In *Synthetic Biodegradable Polymer Scaffolds*, ed. A Ataala, D J Mooney, J P Vacanti and R Langer, Birkhauser Boston, USA, 1997, 149–64.
22 Dagalakis N, Flink J, Stasikelis P, Burke J F and Yannas I V, 'Design of artificial skin. Part III. Control of pore structure', *J. Biomed. Mater. Res.*, 1980, **14**, 107–31.
23 Mikos A G, Thorsen A J, Czerwonka L A, Bao Y and Langer R, 'Preparation and characterization of poly(L-lactic acid) foams', *Polymer*, 1994, **35**, 1068–77.
24 Thomson, R C, Yaszemski M J, Powers J M and Mikos A G, 'Fabrication of biodegradable polymer scaffolds to engineer trabecular bone', *J. Biomater. Sci., Polymer Edition*, 1995, **7**, 23–38.
25 Lo H, Ponticiello M S and Leong K W, 'Fabrication of controlled release biodegradable foams by phase separation', *Tissue Eng.*, 1995, **1**, 15–28.
26 Mooney D J, Baldwin D F, Suh N P, Vacanti J P and Langer R, 'Novel approach to fabricate porous sponges of poly(D,L-lactic-*co*-glycolic acid) without the use of organic solvents', *Biomaterials*, 1996, **17**, 1417–22.

27 Widmer M S, Evans G R D, Brandt K, Savel T, Patrik C W and Mikos A G, 'Porous biodegradable polymer scaffolds for nerve regeneration', *Proc. 1997 Summer Bioengineering Conference*, Volume 35, ed. K B Chandran, R Vanderby Jr and M S Hefzy, The American Society for Mechanical Engineers, New York, 1997, 353–4.
28 Wu B M, Borland S W, Giordano R A, Cima L G, Sachs E M and Cima M J, 'Solid free-form fabrication of drug delivery devices', *J. Controlled Release*, 1996, **40**, 77–87.
29 Katz A R and Turner R, 'Evaluation of tensile and absorption properties of polyglycolic acid sutures', *Surg. Gynecol. Obstet.*, 1970, **131**, 701–16.
30 Frazza E J and Schmitt E E, 'A new absorbable suture', *J. Biomed. Mater. Res. Symp.*, 1971, **1**, 43–58.
31 Reed A M and Gilding D K, 'Biodegradable polymers for use in surgery – poly(glycolic)/poly(lactic acid) homo and co-polymers: 1', *Polymer*, 1981, **22**, 505–9.
32 Reed A M and Gilding D K, 'Biodegradable polymers for use in surgery – poly(glycolic)/poly(lactic acid) homo and copolymers: 2. In vitro degradation', *Polymer*, 1981, **22**, 494–8.
33 Cima L G, Vacanti J P, Vacanti C, Ingber D, Mooney D and Langer R, 'Tissue engineering by cell transplantation using degradable polymer substrates', *J. Biomed. Eng.*, 1991, **113**, 143–51.
34 Mikos A G, Bao Y, Cima L G, Ingber D E, Vacanti J P and Langer R, 'Preparation of poly(glycolic acid) bonded fiber structures for cell attachment and transplantation', *J. Biomed. Mater. Res.*, 1993, **27**, 183–9.
35 Shalaby S W and Johnson R A, 'Synthetic absorbable polyesters'. In *Biomedical Polymers: Designed-to-Degrade Systems*, ed. S W Shalaby, Carl Hanser Verlag, New York, USA, 1994.
36 Freed L E, Vunjak-Novakovic G, Biron R J et al., 'Biodegradable polymer scaffolds for tissue engineering', *Bio Technology*, 1994, **12**, 689–93.
37 Hubbell J A and Langer R, 'Tissue engineering', *C&EN*, 1995, March 13, 42–54.
38 Piskin E, 'Biomaterials in different forms for tissue engineering: an overview'. In *Porous Materials for Tissue Engineering*, ed. D M Liu and V Dixit, Trans Tech Publications, Switzerland, Materials Science Forum, 1997, **250**, 1–14.
39 Dauner M, 'Textile scaffolds for biohybrid organs', *Proc. Techtextil Symposium: Health and Protective Textiles*, France, 1998, **2**, 67–72.
40 Mayer J, Karamuk E, Bruinink A, Wintermantel E and Ramakrishna S, 'Structural and mechanical aspects of textile scaffold systems for tissue engineering', *9th Int. Conf. Biomedical Engineering*, Singapore, 1997, 617–20.
41 Mayer J, Karamuk E, Bruinink A, Wintermantel E and Ramakrishna S, 'Textile scaffolding for tissue engineering: influence of structural deformation in the microscopic range', *Proc. Topical Conference Int. Conf. Biomaterials, Carriers for Drug Delivery, and Scaffolds for Tissue Engineering*, Los Angeles, USA, 1997, 96–8.
42 Leong K W and Ramakrishna S, 'Scaffold engineering', *Ann. Rev. Biomed. Eng.* (in press).
43 Widmer M S and Mikos A G, 'Fabrication of biodegradable polymer scaffolds for tissue engineering'. In *Frontiers in Tissue Engineering*, ed. C W Patrick, A G Mikos and L V Mcintire, Pergamon Press, USA, 1998, 107–20.
44 Chaignaud B E, Langer R and Vacanti J P, 'The history of tissue engineering using synthetic biodegradable polymer scaffolds and cells'. In *Synthetic Biodegradable*

Polymer Scaffolds, ed. A Ataala, D J Mooney, J P Vacanti and R Langer, Birkhauser, Boston, USA, 1997, 1–14.
45 Ducheyne P, 'Success of prosthetic devices fixed by ingrowth or surface interaction', *Acta Orthop. Belg.*, 1985, **51**, 144–61.
46 Schliephake H, Neukam F W and Klosa D, 'Influence of pore dimensions on bone ingrowth into hydroxyapatite blocks used as bone graft substitutes: a histometric study', *Int. J. Oral. Maxillofac. Surg.*, 1991, **20**, 53–8.
47 Eggli P S, Muller W and Schenk R K, 'Porous hydroxyapatite and tricalcium phosphate cylinders with two different pore size ranges implanted in cancellous bone of rabbits: a comparative histomorphometric and histologic study of bony ingrowth and implant substitution', *Clin. Orthop. Rel. Res.*, 1988, **232**, 127–38.
48 Collier J P, Mayor M B, Chae J C, Surprenant V A, Surprenant H P and Dauphinais L A, 'Macroscopic and microscopic evidence of prosthetic fixation with porous-coated materials', *Clin. Orthop.*, 1988, **235**, 173–80.
49 Kadiyala S, Lo H and Leong K W, 'Formation of highly porous polymeric foams with controlled release capability'. In *Tissue Engineering Methods and Protocols*, ed. J R Morgan and M L Yarmush, Humana Press, USA, 1999, 57–65.
50 Ma P X and Langer R, 'Degradation, structure and properties of fibrous nonwoven poly(glycolic acid) scaffolds for tissue engineering'. In *Proc. of MRS Symposium Polymers in Medicine and Pharmacy*, ed. A G Mikos, K W Leong, M L Radomsky, J A Tamada and M J Yaszemki, Materials Research Society, USA, 1995, **394**, 99–104.
51 Pongor P, Betts J, Muckle D and Bentley G, 'Woven carbon surface replacement in the knee: independent clinical review', *Biomaterials*, 1992, **13**, 1070–6.
52 Ma P X and Langer R, 'Fabrication of biodegradable polymer foams for cell transplantation and tissue engineering'. In *Tissue Engineering Methods and Protocols*, ed. J R Morgan and M L Yarmush, Humana Press, USA, Methods in Molecular Medicine, 1999, **18**, 47–56.
53 Wong W H and Mooney D J, 'Synthesis and properties of biodegradable polymers used as synthetic matrices for tissue engineering'. In *Synthetic Biodegradable Polymer Scaffolds*, ed. A Atala, D J Mooney, J P Vacanti and R Langer, Birkhauser, Boston, USA, 1997, 51–82.
54 Leong K W, 'Chemical and mechanical considerations of biodegradable polymers for orthopaedic applications'. In *Biodegradable Implants in Fracture Fixation*, ed. K S Leung, L K Hung and P C Leung, World Scientific Publishing Co. Pte. Ltd., Singapore, 1994, 45–56.
55 Wintermantel E, Bruinink A, Eckert C L, Ruffieux K, Petitmermet M and Mayer J, 'Tissue engineering supported with structured biocompatible materials: goals and achievements'. In *Materials in Medicine*, ed. M O Speidel and P J Uggowitzer, vdf Hochschulverlag AG an der ETH Zurich, 1998, 1–136.
56 Dennis J E, Haynesworth S E, Young R G and Caplan A I, 'Osteogenesis in marrow derived mesenchymal cell porous ceramic composites transplanted subcutaneously: effect of fibronectin and lamin on cell retention and rate of osteogenic expression', *Cell Transpl.*, 1992, **1**, 23–32.
57 Robinson B, Hollinger J O, Szachowicz E and Brekke J, 'Calvarial bone repair with porous D,L-polylactide', *Otolaryngol. Head Neck Surg.*, 1995, **112**, 707–13.
58 Wee S S, 'Development of fibrous scaffolds for liver tissue engineering', *B. Eng. Dissertation*, National University of Singapore, Singapore, 2000.

59 Bowers K T, Keller J C, Randolph B A, Wick D G and Michaels C M, 'Optimization of surface micromorphology for enhanced osteoblast responses in vitro', *Int. J. Oral Maxillofac. Imp.*, 1992, **7**, 302–10.
60 Stanford C M, Keller J C and Solursh M, 'Bone cell expression on titanium surfaces altered by sterilization treatments', *J. Dent. Res.*, 1994, **73**, 1061–71.
61 Bostrom R D and Mikos A G, 'Tissue engineering of bone'. In *Synthetic Biodegradable Polymer Scaffolds*, ed. A Ataala, D J Mooney, J P Vacanti and R Langer, Birkhauser, Boston, USA, 1997, 215–34.
62 Kadiyala S, Lo H and Leong K W, 'Biodegradable polymers as synthetic bone grafts'. In *Bone Formation and Repair*, ed. C T Brighton, G Freidlaender and J M Lane, American Academy of Orthopedic Surgeons, USA, 1994, 317–24.
63 Norde W, 'The behavior of proteins at interfaces, with special attention to the role of the structure stability of the protein molecule', *Clin. Mater.*, 1992, **11**, 85–91.
64 Dee K C, Rueger D C, Anderson T T and Bizios R, 'Conditions which promote mineralization at the bone-implant interface: A model in vitro study', *Biomaterials*, 1996, **17**, 209–15.
65 Vacanti C A, Kim W S and Mooney D, 'Tissue engineered composites of bone and cartilage using synthetic polymers seeded with two cell types', *Orthopaed. Trans.*, 1993, **18**, 276.
66 Kim W S, Vacanti J P, Cima L et al., 'Cartilage engineered in predetermined shapes employing cell transplantation on synthetic biodegradable polymers', *Plast. Reconstr. Surg.*, 1994, **94**, 233–7.
67 Freed L E and Vunjak-Novakovic G, 'Tissue engineering of cartilage'. In *The Biomedical Engineering Handbook*, ed. J D Bronzino, CRC Press, Boca Raton, 1995, 1788–803.
68 Laurencin C T, Ambrosio A M A, Borden M D and Cooper Jr J A, 'Tissue engineering: orthopedic applications', *Ann. Rev. Biomed. Eng.*, 1999, **1**, 19–46.
69 Laurencin C T, Ko F K, Borden M D, Cooper Jr J A, Li W J and Attawia M A, 'Fiber based tissue engineered scaffolds for musculoskeletal applications: *in vitro* cellular response'. In *Biomedical Materials – Drug Delivery, Implants and Tissue Engineering*, ed. T Neenan, M Marcolongo and R F Valentini, Materials Research Society, Warrendale, USA, 1999, **550**, 127–35.
70 Gottlow J, 'Guided tissue regeneration using bioresorbable and non-resorbable devices: initial healing and long term results', *J. Periodontol.*, 1993, **64**, 1157.
71 Cao Y, Vacanti J P, Ma P X et al., 'Tissue engineering of tendon', *Proc. Mat. Res. Soc. Symp.*, 1995, **394**, 83–9.
72 Bellincampi L D, Closkey R F, Prasad R, Zawadsky J P and Dunn M G, 'Viability of fibroblast-seeded ligament analogs after autogenous implantation', *J. Orthop. Res.*, 1998, **16**, 414–20.
73 Kim S S, Utsunomiya H, Koski J A et al., 'Survival and function of hepatocytes on a novel three-dimensional synthetic biodegradable polymer scaffold with an intrinsic network of channels', *Ann. Surg.*, 1998, **228**(1), 8–13.
74 Chamberlain L J and Yannas I V, 'Preparation of collagen–glycosaminoglycan copolymers for tissue engineering'. In *Tissue Engineering Methods and Protocols*, ed. J R Morgan and M L Yarmush, Humana Press, USA, Methods in Molecular Medicine, 1999, **18**, 3–17.
75 Tountas C P, Bergman R A, Lewis T W, Stone H E, Pyrek J D and Mendenhall H V,

'A comparison of peripheral nerve repair using an absorbable tubalization device and conventional sutures in primates', *J. Appl. Biomat.*, 1993, **4**, 261.
76 Maquet V and Jerome R, 'Design of macroporous biodegradable polymer scaffolds for cell transplanatation'. In *Porous Materials for Tissue Engineering*, ed. D M Liu and V Dixit, Trans Tech Publications, Switzerland, Materials Science Forum, 1997, **250**, 15–42.
77 Atala A, Cima L G, Kim W et al., 'Injectable alginate seeded with chondrocytes as a potential treatment for vesicoureteral reflux', *J. Urol.*, 1993, **150**, 745–7.
78 Atala A, Kim W, Paige K T, Vacanti C A and Retik A B, 'Endoscopic treatment of vesicoureteral reflux with a chrondrocyte-alginate suspension', *J. Urol.*, 1994, **152**, 641–3.
79 Kung I M, Wang F F, Chang Y C and Wang Y J, 'Surface modifications of alginate/poly(L-lysine) microcapsular membrane with poly(ethylene glycol) and poly(vinyl alcohol)', *Biomaterials*, 1995, **16**, 649–55.
80 Polk A, Amsden B, De Yao K, Peng T and Goosen M F, 'Controlled release of albumin from chitosan-alginate microcapsules', *J. Pharm. Sci.*, 1994, **83**, 178–85.
81 Pachence J M, 'Collagen-based devices for soft tissue repair', *J. Biomed. Mater. Res.*, 1996, **33**, 35–40.
82 Bell E, Rosenberg M, Kemp P et al., 'Recipes for reconstructing skin', *J. Biomech. Eng.*, 1991, **113**, 113–19.
83 Mooney D J and Rowley J A, 'Tissue engineering: integrating cells and materials to create functional tissue replacements'. In *Controlled Drug Delivery: Challenges and Strategies*, ed. K Park, American Chemical Society, USA, 1997, 333–46.
84 Krewson C E, Chung S W, Dai W and Saltzman W M, 'Cell aggregation and neurite growth in gels of extracellular matrix molecules', *Biotechnol. Bioeng.*, 1994, **43**, 555–62.
85 Yannas I V, 'Applications of ECM analogs in surgery', *J. Cell. Biochem.*, 1994, **56**, 188–91.
86 Hansbrough J F, Cooper M L, Cohen R et al., 'Evaluation of a biodegradable matrix containing cultured human fibroblasts as a dermal replacement beneath meshed skin graft on athymic mice', *Surgery*, 1992, **111**, 438–46.
87 Mooney D J, Organ G M, Vacanti J P and Langer R, 'Design and fabrication of biodegradable polymer devices to engineer tubular tissues', *Cell Transplantation*, 1994, **3**, 203–10.
88 Johnson L B, Aiken J, Mooney D et al., 'The mesentery as a laminated vascular bed for hepatocyte transplantation', *Cell Transplantation*, 1994, **3**, 273–81.
89 Mooney D J, Park S, Kaufmann P, Sano K, McNamara K, Vacanti J P and Langer R, 'Biodegradable sponges for hepatocyte transplantation', *J. Biomed. Mater. Res.*, 1995, **29**, 959–65.
90 Zhang X et al., 'Biodegradable polymers for orthopedic applications', *Rev. Macromol. Chem. Phys.*, 1993, **C33**, 81.
91 Gilding D K and Reed A M, 'Biodegradable polymers for use in surgery – polyglycolic/poly(lactic acid) homo- and copolymers', *Polymer*, 1979, **20**, 1459–64.
92 Engelberg I and Kohn J, 'Physico-mechanical properties of degradable polymers used in medical applications: a comparative study', *Biomaterials*, 1991, **12**, 292–304.
93 Cooper M L, Hansbrough J F, Speilvogel R L, Cohen R, Bartel R L and Naughton G, 'In vivo optimization of a living dermal substitute employing cultured human

fibroblasts on a biodegradable polyglycolic acid or polyglactic mesh', *Biomaterials*, 1991, **12**, 243–8.
94 Kim I M and Vacanti J P, 'Tissue engineering'. In *The Biomedical Engineering Handbook*, ed. J D Bronzino, CRC Press, Boca Raton, USA, 1995.
95 Ma P X, Schloo B, Mooney D and Langer R, 'Development of biomechanical properties and morphogenesis of in vitro tissue engineered cartilage', *J. Biomed. Mater. Res.*, 1995, **29**, 1587–95.
96 Puelacher W C, Mooney D, Langer R, Upton J, Vacanti J P and Vacanti C A, 'Design of nasoseptal cartilage replacements synthesized from biodegradable polymers and chrondrocytes', *Biomaterials*, 1994, **15**, 774–8.
97 Merrell J C, Russell R C and Zook E G, 'Polyglycolic acid tubing as a conduit for nerve regeneration', *Ann. Plast. Surg.*, 1986, **17**, 49–58.
98 Pham H N, Padilla J A, Nguyen K D and Rosen J M, 'Comparison of nerve repair techniques: sutures vs. avitene-polyglycolic acid tube', *J. Reconstr. Microsurg.*, 1991, **1**, 31–6.
99 Vert M, Christel P., Chalot F and Leray J, 'Bioresorbable plastic materials for bone surgery'. In *Macromolecular Biomaterials*, ed. G W Hastings and P Ducheyne, CRC Press, Boca Raton, USA, 1984.
100 Leong K W et al., 'Polyanhydrides'. In *Encycl. Polym. Sci. Eng.*, ed. J I Kroschwitz, Wiley, New York, USA, 1989.
101 Heller J, 'Controlled drug release from poly(orthoester) – a surface eroding polymer', *J. Controlled Release*, 1985, **2**, 167–77.
102 Daniels A U et al., 'Mechnical properties of biodegradable polymers and composites proposed for internal fixation of bone', *J. Appl. Biomater.*, 1990, **1**, 57.
103 Heller J and Daniels A U, 'Poly(orthoesters)'. In *Biomedical Polymers: Designed to Degrade Systems*, ed. S W Shalaby, Carl Hanser Verlag, Munich, Germany, 1994, 35.
104 Kohn J, 'Pseudopoly(aminoacids)'. In *Biodegradable Polymers as Drug Delivery Systems*, ed. M Chasin and R Langer, New York, USA, 1990, 195.
105 Pulapura S, Li C and Kohn J, 'Structure–property relationships for the design of polyiminocarbonates', *Biomaterials*, 1990, **11**, 666.
106 Kadiyala S et al., 'Poly(phosphoesters) as bioabsorbable osteosynthetic materials'. In *Tissue Inducing Biomaterials*, ed. L Cima and E Ron, MRS Series, 1992 252, 311.
107 Allcock H R, 'Phosphazene high polymers'. In *Comprehensive Polymer Science*, ed. G Allen, 4, Pergamon Press, New York, USA, 1989.
108 Allcock H R, Gebura M, Kwon S and Neenan T X, 'Amphiphilic polyphosphazenes as membrane materials: Influence of side group on radiation cross-linking', *Biomaterials*, 1988, **9**, 500–8.
109 Allcock H R, Kwon S, Riding G H, Fitzpatrick R J and Bennett J L, 'Hydrophilic polyphosphazenes as hydrogels: radiation cross-linking and hydrogel characteristics of poly[bis(methyethoxythoxy)-phosphazene', *Biomaterials*, 1988, **9**, 509–13.
110 Lora S, Carenza M, Palma G, Caliceti P, Battaglia P and Lora A, 'Biocompatible polyphosphazene by radiation-induced graft copolymerization and heparinization', *Biomaterials*, 1991, **12**, 275–80.
111 Laurencin C T, Norman M E, Elgendy H M et al. 'Use of polyphosphazenes for skeletal tissue regeneration', *J. Biomed. Mater. Res.*, 1993, **27**, 963–73.
112 Razavi R, Khan Z, Haenerle C B and Beam D, 'Clinical applications of polyphosphazene-based resilient denture liner', *J. Prosthodontics*, 1993, **2**, 224–7.

113 Scopelianos A G, 'Polyphosphzenes new biomaterials'. In *Biomedical Polymers: Designed-to-Degrade Systems*, ed. S W Shalaby, Carl Hanser Verlag, New York, USA, 1994.
114 Boyan B D, Hummert T W, Dean D D and Schwartz Z, 'Role of material surfaces in regulating bone and cartilage cell response', *Biomaterials*, 1996, **17**, 137–46.

Index

actuators, 3, 4, 8
adaptive and responsive textile structures, 226

bio-compatibility, 285
bio-materials, 296–297
bio-processes and products, 5
bio-processing, 254

combat casualty care, 236
cotton, 263–270
coupled mode theory, 152
crosslinked polyols, 85

degradability, 284
degradable thermosets, 287
dimensional memories, 87
distributive measurement systems, 195
down, 78
dynamic moisture transfer, 75

electrically active polymer materials, 8
　actuation, 8
　artificial muscle, 15
　large deformation, 28
　nonionic polymer gel, 8
embroidery, 218, 220
enzymatic textile processing, 254, 255–256
excimer laser, 127

Fabry-Perot interferometric sensors, 177
far infra-red radiation ceramics, 60
fibre Bragg grating (FBG), 124, 151, 174
　densification, 129
　fabrication, 125–127
　modulus and hardness, 131, 140
　strength, 130, 133
　trapped-electron centres, 127
　Weibull distribution, 131
fibre optic sensors, 124, 177, 235

germania-doped silica optical fibre, 125
global positioning system, 249
graft membrane, 111

GSM, 249, 250

healthcare and telemedicine, 237
heat-storage and thermoregulated textiles and clothing
　development history, 36
　fibre spinning, 44
　hydrated inorganic salts, 38
　latent heat-storage materials, 38
　linear chain hydrocarbon, 40
　PEG and PTMG, 39
　PET-PEG block copolymer, 40
　phase change fibres, 41
　phase change materials, 65, 248
　polyhydric alcohol, 39
　properties, 47
　sensible heat-storage materials, 38
hollow fibre, 63
hollow fibre membranes, 211, 214
　gas separation, 200, 202
　materials, 207
　resistance model, 204
hydrogel, 93

insulation values, 79
integrated processes, 5
integrated products, 5
IPN, 95, 97

Mach-Zehnder interferometric sensors, 177
measurement effectiveness, 187–191
measurement reliability, 191–195
medical textiles, 221, 291
micro-capsules, 45
milkweed floss, 78
multi-axial strain measurements by embedded FBGs, 184–186

nano-indentor, 131

optical responses of fibre Bragg grating sensors (FBGs) under
　axial tension, 156–158

316   Index

optical responses of fibre Bragg grating sensors (FBGs) under (*cont.*)
   bending, 165–166
   lateral compression, 161–165
   torsion, 158–161

permeation, 116
personalized mobile information processing systems, 238
plasma, 111
polarization, 152
poly(acrylic acid), 95,97
polymeric optical fibre, 151
poly(vinyl alcohol), 95, 97

quality function deployment, 232

radiation, 110
reflection/transmission spectra, 153
reflective index modulation, 127

sensors, 3, 4, 177, 178
signal processing, 3, 5
smart materials and structures, 2
stimuli-responsive, 93
shape memory materials, 280–284

silk, 270–271
simultaneous measurement of temperature and strain, 181
single-model optical fibre, 151
smart textile composites, 174
snowmobile suit, 246
sports and athletics, 237
stimuli-responsive membrane, 109

temperature and strain coupling, 181
thermal insulation, 68
thermal memories, 86
tissue engineering, 222, 291
   scaffolds, 293, 294
      textile scaffolds, 298–305
   tissue regeneration, 305

vital signs monitoring systems, 235, 238

waterproof breathable coatings, 70
wearable electronics, 246
wet shrinkage, 88
wool, 256–263

zirconium carbide, 61